THE CONTOURS OF AMERICA'S COLD WAR

THE CONTOURS OF
AMERICA'S COLD WAR

Matthew Farish

 University of Minnesota Press
Minneapolis
London

The University of Minnesota Press gratefully acknowledges the generous assistance provided for the publication of this book from the University of Toronto Faculty Association.

For publication information on previously published material in this book, see pages 329–30.

Published by the University of Minnesota Press
111 Third Avenue South, Suite 290
Minneapolis, MN 55401-2520
http://www.upress.umn.edu

Library of Congress Cataloging-in-Publication Data

Farish, Matthew.
    The contours of America's Cold War / Matthew Farish.
        p.   cm.
    Includes bibliographical references and index.
    ISBN 978-0-8166-4842-9 (hc : alk. paper) — ISBN 978-0-8166-4843-6 (pb : alk. paper)
        1. United States — Civilization — 1945- .   2. Cold War — Social aspects — United States.   3. Atomic bomb — Social aspects — United States — History — 20th century.   4. Nuclear warfare — Social aspects — United States — History — 20th century.   5. Popular culture — United States — History — 20th century.   6. War and society — United States — History — 20th century.   I. Title.
    E169.12.F37   2010
    973.918 — dc22   2010038340

For Mom, Dad, Jill, and Paul

CONTENTS

INTRODUCTION

A History of Cold War Spaces

> Geographical knowledge — in its resistance to change and in its
> effort to make sense of the world — has operated conservatively.
> But it has also functioned conservatively in its service to and
> defense of national power.
>
> —SUSAN SCHULTEN, *The Geographical Imagination*
> *in America, 1880–1950*

In June 1959, on the eve of a trip to the Soviet Union (one that would feature the famous Kitchen Debate), American Vice President Richard Nixon
and his family joined Rear Admiral Charles C. Kirkpatrick of the U.S. Navy
on the first voyage of the world's largest atomic submarine fleet. The tour
took them past a graveyard of sunken ships and led one reporter for the
Christian Science Monitor to exclaim that the spectacle was "sheer fun,
as though the real purpose of technological achievement, after all, was
human happiness." Although the submarine project had been blessed by
the federal government and aided by General Dynamics, a manufacturer
of nuclear reactors, the purveyor of these amusements was Walt Disney,
and the site was the Tomorrowland section of Disneyland.[1]

The submarines were not Disney's first atomic product. "Tomorrowland" was also a segment of Disney's popular television program, and in
January 1957, the show had aired a feature titled *Our Friend the Atom,* a
combination of live action and animation sponsored by General Dynamics. It was a promotional vehicle for Dwight Eisenhower's Atoms for Peace
campaign, launched by the president in December 1953 with a stirring
speech at the United Nations.[2] On television, in the companion book, and
in a Tomorrowland exhibit at Disneyland, *Our Friend the Atom* explained
atomic energy in the terms of household technologies and the performance
of tasks, domesticating a destructive and arcane science.[3] As the president

came under increased pressure to limit atomic testing, demilitarizing the atom became a necessary political gesture.

According to the journalist Eric Schlosser, Disneyland's designer was

> America's most popular exponent of Cold War science. For audiences living in fear of nuclear annihilation, Walt Disney became a source of reassurance, making the latest technical advances seem marvelous and exciting. His faith in the goodness of American technology was suc-cinctly expressed by the title of a film that the Disney studio produced for Westinghouse Electric: *The Dawn of Better Living.*[4]

To bolster his credentials, Disney hired Wernher von Braun and Heinz Haber, German scientists who had been recruited to military positions in the United States after the Second World War, to consult on space-oriented television features and to design a Tomorrowland ride called Rocket to the Moon.[5] By the late 1950s, the gregarious Von Braun, whose rocket work at Peenemünde had depended on thousands of slave laborers, includ-ing Allied prisoners, was America's "Mr. Space," his face paired with an image of a rocket launch on the cover of *Time* magazine.[6] The less famous Haber was an alumnus of the Luftwaffe Institute for Aviation Medicine, where Dachau inmates were used in experiments simulating high-altitude flight. He eventually became the chief science consultant to Disney Pro-ductions, hosting the film *Our Friend the Atom* and authoring the book of the same title.[7] The latter's dust jacket condensed Haber's biography con-siderably. He was simply "born in Germany... In 1947 he came to the United States as a research scientist with the Air Force School of Aviation Medicine."[8]

Walt Disney's more critical biographers have documented his role as an FBI informant, his affection for J. Edgar Hoover (a "foster father"), his opposition to labor activism in the film industry, his affiliations with organized crime, and his anti-Semitism.[9] He also held long-standing and occasionally contentious relationships with branches of the armed forces. During the Second World War, after accepting a contract from the Naval Bureau of Aeronautics for twenty animated training films, Disney grew annoyed with interference from Washington. Once freed from a formal military presence in his studio, he chose a project that resembled his propa-ganda films, but one that he could control completely: an animated version of Alexander De Seversky's 1942 best seller *Victory through Air Power.* At the film's conclusion, a "mighty air fleet" leaves Alaska to obliterate Japan "in a roaring blaze of Technicolor," as Thomas M. Pryor put it in a glowing *New York Times* review.[10] Although the adaptation was not particularly successful in the United States, it found devotees across the Atlantic, where Winston Churchill reputedly used the film to convince Franklin Roosevelt that long-range bombing was an appropriate strategic measure.[11]

In his classic study *The Disney Version*, Richard Schickel claims that Walt Disney's fondness for air power stemmed from its sanitized efficiency.[12] This trait was certainly also present in Disney's most significant physical creation, the Magic Kingdom, which opened four years before Nixon's 1959 visit. It was not a typical amusement park, where social rules and roles might be subverted amid crowds. Instead, Disney strove for "just the opposite—not freedom but control and order."[13] Millions of visitors delighted in the placid nostalgia of Main Street U.S.A. and the more colorful Western violence of Frontierland. They were inspired to challenge and navigate the hostile environments of Adventureland's jungle and Tomorrowland's moonscape.[14] Despite Disney's directive to "leave today behind and enter a world of yesterday, tomorrow and Fantasy," the America of 1955—of the Cold War—nonetheless haunted this vivid and yet carefully managed space not just in the combination of cultural tradition and mission but also in the purpose of the site itself.[15] Its mythic narrative of history, culminating in the benevolent technological utopia of Tomorrowland's Monsanto House of the Future, was a lesson in national modernization with global implications.[16] Disneyland, its creator predicted, was "based upon and dedicated to the ideals, the dreams and the hard facts that created America. And it will be uniquely equipped to dramatize these dreams and facts and send them forth as a sort of courage and inspiration to all the world."[17]

Like Disney's other ventures, Disneyland embodied and projected certain impressions of America and Americans, a role that was never more obvious than when Soviet Premier Nikita Khrushchev announced his desire to visit the Magic Kingdom while reciprocating Nixon's Russian visit in September 1959. Khrushchev's hopes were dashed "because of security concerns," and in a characteristically heated exchange with Twentieth Century-Fox president Spyros Skouras, an angered Soviet premier asked, "What is it, do you have rocket-launching pads there?"[18] No mention was made of the submarines.

The Cold War as Historical Geography

In this brief Disney history lie the outlines of this book. Walt Disney's various creations and affiliations demonstrate the historical inseparability of culture from politics and science, and they confirm the pervasive and persistent militarization that characterized the United States during the 1940s and 1950s. But they also inspire a geographic approach to the period — whether in reference to the cartographic drama of *Victory through Air Power* or the various spaces of Disneyland. These, after all, are two superlative examples of *imaginative geographies,* influential presentations of the world and its contours that are made possible by particular forms of knowledge about that world. However propagandistic or fantastic they might seem, both the film and the theme park are only fictional in the sense that they have been fabricated. With sufficient circulation, or attendance, they could alter attitudes or policies.[19]

This book is a history of the imaginative geographies that came to define the American Cold War. It is a genealogical study that aims to demystify four spatial scales: the globe, the region, the continent, and the city. These terms are not only familiar, part of our popular lexicons, but also much older than the Cold War. And yet it was in the middle of the twentieth century, in the United States, when they were fundamentally redefined, *collectively, in strategic terms.* Although they were not and have never been singular or unchallenged, as Cold War categories, these banal words were extraordinarily potent — and, reconfigured, they remain so today. This authority derives from their unprecedented alignment alongside and in the name of a fifth realm: the nation. By the end of the 1950s, where my account concludes, all had become terrain for the American Cold War, a joint solidification that limited other, less hegemonic geographies. At no other juncture in history had global, regional, continental, and urban spaces been wrapped into a single "regime of truth" that delineated the parameters of reliable knowledge in the antagonistic and divisive terms of national strategy and in turn defined strategic knowledge literally along specific geographic lines.[20] This precedent is important because it was consequential. "To dictate definition," David Livingstone writes, "is to wield cultural power."[21]

Flip through periodicals from the early Cold War — magazines like *Time, Life,* and *Newsweek* — and you will repeatedly encounter maps, of various scales, that are rife with contours. Perhaps the most notable ex-amples are maps featuring impossibly perfect rings that suggest degrees of atomic destruction, viewed directly from above in the form of a target. The heart of the target, usually urban, is the point of detonation. From there, devastation spreads, and whether it moves uniformly or is rendered in more nuanced ways depends on the model used to simulate the event. As a confrontational condition, the Cold War itself was understood in such terms, with America—both an imagined community and a list of specific locations, many of them prominent cities—at the geographic center.[22] A national space was the source of the Cold War's contours but also the site of absolute danger. Beyond lay a series of frontiers—the continent, the region, the globe, and even outer space—which were strategic foci. Each frontier suggested unique but interdependent threats, demanding both aggressive action and retreat behind lines of defense, such that the United States was nominally dedicated to fighting *and* preventing the Cold War at every scale. Contours, like scales, are manufactured distinctions, but they are also relational, and their influence depends on these relations.[23] America itself is certainly a relational or comparative concept, transform-ing as a result of competing assertions made in its name.[24]

In the abstraction of a contour map, the space between two lines is homogeneous, but this is of course a severe simplification.[25] A history of categories is also inescapably a study of the places that came to be associ-ated most prominently with those categories, from the Slavic region of Soviet area studies and the Arctic realm at the edge of the North American continent to the (apparently) uniquely threatened "atomic cities" of New York and Washington, D.C. These were sites where general geographic imaginaries met specific landscapes. Despite their authority and longevity, the interlocking categories of the American Cold War were not fixed or timeless; they are thus well suited to a *genealogical* inquiry. A genealogical attitude, writes Richard Ashley, permits social inquiry "to find its focus in the posing of 'how' questions, not 'what' questions."[26] In this sense, it is not precisely the spaces of the Cold War that matter but the *spacings*—the way in which categories were created and "placed together."[27]

To borrow a phrase from a leading scholar of the Cold War, "we now know" much about the diplomatic, scientific, and cultural histories of the era.[28] The case is not closed, of course—nor has the risk of uncritical tri-umphalism diminished.[29] But what happens if we follow an alternative approach and turn these histories into geographies? To supplement Ashley's focus, *where* was America's Cold War—specifically but also categorically? To proceed thusly and to ask how categories were articulated and given meaningful form requires the combination of the foreign, strategic realm and the domestic, social realm so commonly divided by political reasoning. A historical geography of America's Cold War should not ignore the space titled America, especially when its borders hum with both authority and contingency. Much mapmaking is a "practice of persuasion" by which those who draw boundaries generate influence for the limits they inscribe as well as for themselves.[30]

From its first usage as a phrase to its contemporary manifestations, the Cold War has been consistently treated as a historical epoch, an era identified, as all epochs are, by a beginning and, now, an end, no matter how disputed these moments remain. But it has always also been a spatial concept. This is in one sense obvious: a very basic form of geography lies at the heart of classic Manichean descriptors such as good versus evil, capitalist versus communist, West versus East, and First World versus Second.[31] Above all, the Cold War was understood as a contest between two formidable states that were easily identifiable on the globes and maps of the day, a dualism that suppressed Cartesian anxieties.[32] Those in possession of such cartographic representations could, as with any other conflict, *see* its geographical dimensions and reaffirm its premise of difference: we are here, they are over there; this is friendly territory, and this is hostile. Such geographical thinking could also be made active: what happens if they situate missiles here, topple the government in this country, or attack us from this direction?

Each of the great Cold War contrasts were powerful and politically useful ways of dividing the world and often provided the basis for more specific spatial strategies, such as the "points" targeted by the well-known American policy of containment. But such binaries also hide more complicated forms of geography; it is for this reason that Cold War geopolitics has actually been described as "antigeographical."[33] Similarly, and not coincidentally, historians of geographical thought have repeatedly noted the thin forms of spatial thinking that characterized the discipline in the United States during the early Cold War.[34] But if geography, as both an idea and a subject of study, was "lost" in America during the 1950s—and perhaps over the duration of the longer American Century, according to Neil Smith — this does not mean that it cannot be found.[35]

Spatial categories were central to the conduct of the American Cold War — influencing military deployments, diplomacy, espionage, finance, and so on — but they were also crucial to the imaginative geographies that preceded and permitted this activity.[36] Only if the Soviet Union was understood as a wholly distinct space, containing a wholly distinct society, for example, could it be treated in the stark terms of enmity. This distinctiveness, like political boundaries drawn on a map, had to be represented and reproduced — a largely intellectual act.[37] But while the Cold War is often attributed to architects, a better term, and one that encourages a shift from distant design to direct maintenance, and from individuals to institutions, is *engineers*.[38]

My interest in the Cold War began as a graduate student when I sought to position the midcentury history of geography in relation to other social sciences. As conventionally told, this was a period of great upheaval for American geographers: an older approach based on areal differentiation was giving way to a science of space rooted in geometrical patterns and positivist laws.[39] What, I wondered, did all of this have to do with the Cold War? Other scholars were productively linking geographical thought and practice to both the history of empire and significant military episodes such as the First World War, and academics beyond geography were drawing provocative connections between their own disciplines and Cold War demands. A good deal of this scholarship is put to use here. But substantially less work existed on equivalent networks of affiliation for geography's "quantitative revolution." Historians of geography seemed content to recount a largely internalist battle between a tradition of regional description and an upstart science of spatial laws, with the latter prevailing in many quarters by the early 1960s.[40]

I soon realized, however, that my interest was misguided, or at least too constricted. A hint was provided by Edward Soja's attempt to return space to a position of prominence in social theory. Describing a midcentury geography of "passivity and measurement," he notes that such narrow, "applied" knowledge was still of great use to "intelligence, planning, and administration . . . cementing a special relationship with the state that probably arose first in an earlier age of imperial exploration." But this type of geography is not Soja's concern, and it is not as distinct from "action and meaning," or from other fields of inquiry, as he indicates.[41] Writing on another discipline with its own rich and disturbing Cold War genealogy, Philip Mirowski remarks that "you cannot understand modern economics simply by confining your attention to self-identified economists; it is always far more fruitful to keep your eyes on the military and the operations researchers."[42] Geography, in the basic etymological sense of *earth-writing*, cannot be limited to specific personalities or academic locations. This does not mean abandoning professional geographers but rather situating them alongside others who also contributed to the construction of Cold War spaces—a tactic I employ directly in chapters 2 and 3.

What further strengthens this broader approach is the relatively recent attention directed toward representational and performative presumptions in the social sciences and humanities. This work addresses the practices through which "understanding is articulated and maintained in specific cultural contexts and translated and extended into new contexts."[43] Such extensions and translations of knowledge depend heavily on claims to authority—claims to know the world—and then actions, or interventions, based on those claims.[44]

And yet in much recent writing, the Cold War is fast assuming fixity. It is congealing in time and space as sets of dates and periods, on the one hand, and as lists of blocs, alliances, and rivalries, on the other. This coagulation has been distinctly advanced by the study of a *post*–Cold War era and its dimensions. A new world of proliferating insecurities appears to require intellectual innovations to account for its shifting and deterritorialized alignments. There is undoubtedly a need to move beyond the overarching metanarrative of a singular Cold War, explained solely through a series of mobile terms such as *containment* and *domino*. But the mutual dependence on theory and a contemporary perspective is a hazardous habit. It risks implying that the Cold War has been resigned to the rigidity of a historical period, characterized by a wholly different order of things, including a straightforward superpower competition that is made to seem positively, unquestionably realist in comparison to more recent vertigo.

This narrow treatment of the Cold War resembles what the philosopher Michel Foucault called a "project of total history... one that seeks to reconstitute the overall form of a civilization," or in this case, a conflict between two civilizations. The impulse to describe the American Cold War in this way, to draw "all phenomena around a single center," is tempting given my interest in linking a national center to various strategic scales.[45] My objective is not to show the creation of coherence from a core of truth, but the opposite: to "question how meaning and order are imposed" by considering how a series of linked spaces provided a simple geographic configuration for the Cold War.[46] This schema was presented by its advocates as a sensible and sovereign vision of the world and its continuing history.[47]

The rigid framing of the Cold War as total history has been both aided and challenged by the rush of detail that has flooded historical literature as a result of recent declassification, including material from former Soviet archives. This result is not surprising, because the Cold War itself was understood by its protagonists as both a "struggle for the world" and a "struggle for the word."[48] But while new archival discoveries are to be welcomed, it is also important, as Chris Philo puts it, "to shed fresh light on these historical happenings by injecting them with a measure of geographical sensitivity."[49] A study of the Cold War's multiple spaces demonstrates that there were also multiple Cold Wars, many not easily rendered in the language of victory.[50]

Toward a Spatial History of Strategy

In an interview first published in 1977, Foucault claimed that a history of spaces "would at the same time be the history of *powers* (both these terms in the plural) — from the great strategies of geopolitics to the little tactics of the habitat."[51] Much of Foucault's own work was *spatial history* — a phrase that he never used formally. His genealogical studies of limits, boundaries, classifications, and enclosures explored the themes of illness, crime, and deviance but also gestured to strategic practices, particularly those that defined and managed territories. In a contemporaneous dialogue with the editors of the French journal *Hérodote,* Foucault concluded that

> the formation of discourses and the genealogy of knowledge need to be analysed, not in terms of types of consciousness, modes of perception and forms of ideology, but in terms of tactics and strategies of power. Tactics and strategies deployed through implantations, distributions, demarcations, control of territories and organizations of domains which could well make up a sort of geopolitics.[52]

What this dense observation suggests is the imperative to write a history of spaces that is equally a spatialization of history.[53]

One of Foucault's *Hérodote* interlocutors was the geographer Yves Lacoste, who famously argued, in the same year as the Foucault interview, that geography "serves, first and foremost, to wage war." It had always been essential to acts of organized violence, and military history is littered with examples of armies and planners mixing terrain and tactics to great or disastrous effect. But Lacoste was also suggesting something else: that despite its emergence as a valid intellectual field, geography was still primarily a military idea and tool, a strategic form of knowledge about the world.[54] Both he and Foucault were proposing an expansive definition of geography that was tied only partially to the academy. In a less heralded passage from the same conversation, Foucault mentioned a research project he was planning to undertake that would directly address "the army as a matrix of organization and knowledge."[55] While he never completely fulfilled this goal, his later writings on governmentality and his discussions and lectures on security and war (some only recently published in English) were clearly concerned with the broader implications of the same subject.

As Lacoste (in all likelihood) put it to Foucault in the *Hérodote* interview, a "circulation of notions can be observed between geographical and strategic discourses." Foucault, for his part, replied that his "spatial obsessions" allowed him to "capture the process by which knowledge functions as a form of power and disseminates the effects of power."[56] To consider Lacoste's circulation, therefore, is to open up possibilities for critically evaluating what the political geographer Gearóid Ó Tuathail describes as "knowledge born out of the practical management problems of government, problems addressing the administration, surveillance and control of populations, territories and colonies."[57] Put simply, militaries continue to be central to the production and popularization of geographical knowledge — not only to formal cartographic representations but to more elusive spatial principles, such as the perception that the world is a composite of hostile environments. Despite the substantial, ongoing production of military geographies and despite the recent reexamination of geography's history from a more critical perspective, it is fair to suggest that geographers have yet to adequately account for the abiding associations between our field and the contemplation and conduct of war.

In his 1975–76 lectures at the Collège de France, Foucault inverted Clausewitz to assert that politics was "the continuation of war by other means," suggested that discourse was a struggle or battle, and proposed that war produces a form of truth that operates as a weapon.[58] His references were drawn overwhelmingly from the early modern period, and little reference was made to his own century or, for that matter, to the United States. It was there, in 1953, that Edward Barrett, who worked in the Office of War Information during World War II and then served as assistant secretary of state for public affairs under Harry Truman, published *Truth Is Our Weapon,* a popular study of persuasion, propaganda, and psychological warfare.[59]

The publication date of Barrett's book was not arbitrary. My choice of rough historical markers, from the American entry into World War II to Eisenhower's January 1961 farewell address, is a deliberate attempt to address a time of almost unabashed militarization in the country, when signs, subjects, and structures of violence proliferated.[60] Eisenhower was hardly a staunch opponent—except perhaps on budgetary grounds—of the "military–industrial complex" formally identified in his speech, but as a general, president of Columbia University, and then president of the United States he was certainly qualified to note its novel pervasiveness in every aspect and arena of American life.[61] Following hard on an extraordinarily destructive global war, the Cold War was at once a paradigmatic case of near-constant political confrontation and an arena for the continued control and influencing of human subjects, in the United States and beyond, using techniques derived from the military.

In *Truth Is Our Weapon,* Barrett was determined to demonstrate the relevance of ideas in a battle for human minds, and in a much different register, this is also my goal. My history of geographical thought is specifically concerned with the *human sciences,* a field of knowledge, according to Foucault, "that takes as its object man as an empirical entity."[62] In a promotional document for the Human Relations Area Files (HRAF), one of the period's most ambitious attempts to forge a "science of man," the project is explicitly described as an attempt to match, for the study of human affairs, the spectacular successes of science in harnessing, conquering, and coming to terms with nature. Pushing the boundaries of science necessitated a turn inward, to "man himself" and his own complex cultural and behavioral systems.[63] But as in the case of scientific research, the far reaches of this social inquiry were not merely metaphorical; there was a distinct, material correlation between intellectual and geopolitical frontiers. In other words, the Cold War human sciences created and supported knowledge — in this case spatial knowledge — that was both authoritative and instrumental. Strategy, by the middle of the twentieth century, was still very much concerned with territory, but also with populations, and depended on forms of classification such as the HRAF, which usefully combined these factors.

The human sciences are pivotal to liberal thought because they construct autonomous units of understanding, in the form of social *bodies*, which provide the targets for subtle administration and regulation — what Foucault called governmentality. Governmental objects are most powerfully constituted by means of quantification or related methods that allow for stability and synthesis. The modern scale of government is undoubtedly the national state, but the domain of government is not simply equivalent to the territorial limits of the state.[64] Various scales, spaces, and societies have long been named and fashioned as American, while others — not least the Soviet Union of the Cold War — have been given oppositional designations, amenable to the distant scrutiny characteristic of area studies but also, perhaps, to certain techniques, derived from the human sciences, that were both universal by nature and national in origin. This contradictory status is less confusing when we consider the prevalence of what became known as psychological warfare in both domestic and foreign environments — indeed, at each scale on which the Cold War was contested.

As social scientists grappled with the contradictions of midcentury modernity, and in some cases the demonization of academic inquiry, many found safety in the formidable theories of their scientific relatives, and many moved into laboratories modeled on those of the same intellectual stalwarts. The result was that at every scale, "the social," and its terrain, was increasingly *simulated*, meaning that the human sciences were also very much concerned with machines. But simulation additionally meant that the spaces of the Cold War were more than the product of intangible musings and more than outlines on a map; they were constructed and in the process acquired solidity but also simplicity. It was within the orderly microworlds of a laboratory or, less formally, a closed study group that the fragility of strategic geographic knowledge was most evident, but these sites were also where military models could be most clearly and cleanly expressed.

The uncertainties of fragility and force associated with militarization are equally characteristic of the term *frontier,* a word whose American variant certainly implies regenerative and often violent activity.[65] Some enthusiastic Cold Warriors were fond of shifting frontiers from a geopolitical past to a scientific future: in a 1947 address, MIT's president Karl Compton described "the end of an epoch in the history of mankind," one of "geographic expansion and exploitation of natural resources." Now, Compton went on, "man's new resources must come largely through his scientific and engineering skill."[66] Such gestures limited frontiers to a depoliticized realm of authorless and value-free research affected only by financial support, promoting a culture of Cold War scientism in which ideas born in laboratories were understood, during an era at the "end of ideology," to extend across a great swath of American social life as well.[67] The risks associated with modern science, most notably atomic energy, were compartmentalized from the public, rendered placid in books like *Our Friend the Atom,* or wielded carefully by the prophets of urban disaster, who were depicted as expert managers of these risks.[68] But risks are not easily contained by laboratory walls or national borders. And this is precisely the point: as Foucault argued in a classic essay, frontiers can also be used to subvert "the outside-inside alternative" so characteristic of modernity, placing the security of terms such as *science* or *the state* in crisis.[69] If the political dualism of the Cold War was a spatial equivalent of self–other distinctions, then uncovering the intellectual work required to constantly shore up inevitably permeable strategic boundaries opens possibilities for less conventional forms of "ethico-political" thinking.[70]

In a brilliant survey of American history since the 1930s, Michael Sherry defines militarization as "the process by which war and national security became consuming anxieties and provided the memories, models, and metaphors that shaped broad areas of national life."[71] One important consequence is that divisions (however fabricated) between military and civilian spaces and institutions are blurred or even erased. While Sherry notes that militarization was not a uniquely American experience, the United States—during the Second World War, of course, but also in the decades that followed—is a choice candidate for the designation, *especially* given America's status as a "liberal" space. Liberal political theory, haunted by the contradictions of violence and reason, is premised on the attempt to remove disorder from within the boundaries of the state, a project that is doomed to fail.[72] Cold War America is a striking example of that failure. While perhaps not a strict garrison state, to use a term that has received some attention from historians, it may have been one by another name.[73]

A history of militarization's geographies must look beyond military leaders and institutions, beyond the hundreds of bases and installations established across the United States and around the globe since 1941, and even beyond the myriad towns and cities dependent on the military–industrial complex. It must consider what C. Wright Mills recognized as a "military definition of the situation"—that Americans, in the words of Catherine Lutz, "all inhabit an army camp, mobilized to lend support to the permanent state of war readiness that has been with us since World War II."[74] In the decade following the Second World War, the Department of Defense became the largest single patron of American scientific research, but it also funded the human sciences lavishly and took a direct and persistent interest in cultural and social themes. Meanwhile, ostensibly nonmilitary initiatives such as civil defense positioned the nuclear family as an institution that could shoulder a burden of survival in partnership with the armed forces. Disaster scholars argued that every city and every home was a potential target while advocating responses that would contain panic and chaos to certain areas in the event of an atomic attack. The idea of America as an insecure space at the heart of a Cold War map is thus highly appropriate.

The American Cold War coalesced in the numerous interdisciplinary panels, advisory boards, and study groups created to address the novelties and challenges of an atomic world. These congregations were composed of physicists and policymakers, soldiers and social scientists, who all sought to comprehend the shape of this new globe. Their success was dependent on a collective organizational structure, and while rarely flush with professional geographers, these groups definitely debated, produced, and reinforced geographic ideas. Such intellectual labor was profoundly meaningful. As David Engerman wisely notes, attending to the production and dissemination of "American knowledge" reveals its role as a *form* of "global power."[75] Perceptual layers of security and enmity were given strength not only on the frozen ground of the high Arctic or across an archipelago of military installations but also in Harvard seminar rooms, RAND Corporation laboratories, and in the pages of periodicals and letters. In these venues, as in Disneyland, it was impossible to distinguish science from culture and politics. But the spatial varieties of Cold War knowledge have not been prioritized by historians.

Air Force Chief of Staff Hoyt Vandenberg neatly, if unintentionally, captured the ubiquitous spirit of Cold War militarization in a 1953 speech to the Advertising Council of New York. Referring to the recruitment of volunteers for the Ground Observer Corps, the army of civilian "sky-watchers" on alert for signs of Soviet planes, Vandenberg admitted that an appropriate marketing campaign

> meant asking them not merely to be realistic, but to be imaginative; to accept the fact that the nature of war has changed and that because of this change our Nation, at the very moment it has reached a position and influence unparalleled in our history, has become vulnerable as never before. In consequence of this vulnerability, our citizens must now assume responsibilities that are new and strange.[76]

Such reasoning supported a pervasive condition that was not demobilization or mobilization but an uncertain, prolonged search for an impossible national security. The contours of the Cold War were comforting but also profoundly precarious.

The development of a state's discrete identity is dependent on and inseparable from the representation *and location* of dangerous others. As Edward Said wrote in *Orientalism*, it "is enough for 'us' to set up these boundaries in our own minds; 'they' become 'they' accordingly, and both their territory and their mentality are designated as different from 'ours.'"[77] By considering the spatial forms of American Cold War knowledge, we can critically assess representations of "other" spaces and additionally call into question definitions of America and American:

> Those events and actors that come to be "foreign" through the imposition of a certain interpretation are not considered as "foreign" simply because they are situated in opposition to a pregiven social entity (the state). The construction of the "foreign" is made possible by practices that also constitute the "domestic." In other words, foreign policy is "a specific sort of *boundary-producing political performance*."[78]

As a "struggle related to the production and reproduction of identity," the Cold War was premised on the endless performance of multiple boundaries but also on the granting of priority and legitimacy to certain boundaries.[79]

Those entities placed outside by foreign policy are frustratingly "always slipping back across the porous and invisible borders to disturb and subvert from the inside."[80] While there is no escaping what Michel de Certeau called *tactics*—the opponents who refuse to disappear from an agonistic horizon—this book attends to a second part of his deliberately militarist formulation. Although I use the term *strategy* in its Cold War context, de Certeau's broader definition is also relevant:

> I call a *strategy* the calculation (or manipulation) of power relationships that becomes possible as soon as a subject with will and power (a business, an army, a city, a scientific institution) can be isolated. It postulates a *place* that can be delimited as its *own* and serve as the base from which relations with an *exteriority* composed of targets or threats.[81]

However powerful, then, the strategic spaces considered in the rest of this book were ultimately failed attempts at geographic division and distinction.

In the subsequent chapters, I draw substantially on archival and period sources from a range of repositories. Archives are contradictory, incomplete sites, where interpretations are forged and not provided. In the case of the Cold War, security restrictions remain an obstacle. Almost all of the material cited here is unclassified or has been previously declassified.[82] In addition, certain individuals—although perhaps not always the expected ones—float through the chapters but are hardly icons for valorization. And yet, it cannot be avoided that this is, on balance, a study of masculine elites and thus is tilted toward a cluster of dominant modes of knowing that are by no means homogenous but equally are hardly diverse. Many of these men shared a faith in the ability of intellectual inquiry to reveal truth, but also to shape it; this made it possible to speak of a seemingly paradoxical *Cold War social science.*

For all the inevitable political diversity of the United States, it is striking to consider just how many viewpoints were amenable to, or appropriated by, the geographies of militarization, even with the end of the Second World War. Such coalescence may well be confirmed by many of the films, magazine articles, and popular nonfictional texts that I weave together with archival documents, but this methodology also lessens the disciplinary role of the archive as a distinct repository of historical authority. While it is pointless to identify a single point of origin for something so overdetermined as, say, the idea of American cities as strategic spaces, it is still clear that such imaginative geographies were produced and maintained across the intellectual networks of what Stuart Leslie has called the military–industrial–academic complex.[83] Amidst mass audiences, who no doubt interpreted them variably, these geographies were decontextualized—often owing to an invigorated demand for secrecy—and made dependent on overarching and simplistic narratives such as ideological confrontation.

Chapter 1, "Global Views: Geopolitics, Science, and Culture," considers several dimensions of American globalism—the idea that America's strategic interests were universal. Although these ambitions were certainly not novel in the 1940s and 1950s, they became fully realized when the list of environments central to American strategy included the globe itself. But claims to universal awareness are always situated, and chapter 1 shows how an American global geography was constructed and made ordinary. Whether in the form of popular cartographic techniques, manifestos of aerial might, or revised theories of international relations and diplomacy, globalism could be buttressed by claims of geopolitical reality, scientific necessity, and cultural supremacy all at once.

Still, both the Second World War and the Cold War clearly demanded finer scales of spatial understanding. Chapter 2, "Regional Intelligence: The Militarization of Geographical Knowledge," examines a globe divided into regions America confronted as it went to war in 1941. In both this and the next chapter, I argue that the disciplinary history of geography can be productively set next to a much broader intellectual record of universal aspirations and regional necessity in the human sciences. But chapter 2 is also a detailed examination of the precepts driving the ambitious collection of a particularly geopolitical type of areal *intelligence.* At the Smithsonian Institution's Ethnogeographic Board, the Army Specialized Training Program, the Office of Strategic Services, and other sites of area-based research and education, the world was divided into regions not just for the sake of military convenience but in the service of a grand if fragmented project that sought an authoritative geography of hostility and difference.

In chapter 3, "Illuminating the Terrain: Social Science Finds Its Targets," I carry this thread into the early years of the Cold War, when area studies not only flowered but also shifted to focus on the troublesome regions coming to define a confrontation with international communism. Under the sign of strategic intelligence, the result was an uneasy but influential mixture of abstract rigorism and political practicality, and it involved a remarkable cast of actors from universities, foundations, and military and intelligence agencies. I position programs at Harvard and Columbia, as well as several nonacademic locations, alongside a broader corpus of cutting-edge work in social science, demonstrating the value and pervasiveness of regional structures. The result is a profile of Cold War social science that reveals a profound engagement with strategic pursuits and their regional contexts, an interest perhaps best epitomized by the overt advocacy of modernization theory and the covert practices of psychological warfare.

Not surprisingly, many of the same tensions and presumptions of area studies were present in concurrent attempts to secure a continental space. Chapter 4, "The Cybernetic Continent: North America as Defense Laboratory," combines the laboratory studies of MIT and the RAND Corporation with the regional investigation of the North American Arctic in a historical geography of Cold War continental defense. Once northern landscapes were enframed by radar technology and other scientific initiatives, with the assistance of a largely acquiescent Canadian partner, a continental space could be conceived, and perhaps defended, in the manner of a cybernetic system.

Should a response to attack from the north fail, a final set of spaces would, it was widely believed, be specifically targeted: American cities. While certain metropolitan regions were thought to be particularly threatened, they were complemented by a "City X" of civil defense plans and simulations. In chapter 5, "Anxious Urbanism: Strategies for the Atomic City," I argue that the portentous history of post–World War II urbanism requires attention to the concerns exacerbated if not caused by the recitation of Cold War dangers, chiefly nuclear weapons. In the novel interdisciplinary field of disaster scholarship, cities were turned into laboratories for the study of behavioral abnormality, precisely as they were actually being emptied and redesigned to account for other perceived forms of difference.

In the Conclusion, I turn briefly to a last landscape understood as a site of both national opportunity and terrible hazards. By the late 1950s, outer space had supplanted the Arctic as the most fantastic and hostile environment in the minds of Americans. But this "new ocean" is best understood as an extension and synthesis of already existing geographies.[84] Beginning with the film *Forbidden Planet* (1956) and ending with the outer-space plans of President Eisenhower's scientific advisors, the Conclusion traces the exploratory urges, military necessity, cultural prestige, and scientific possibility that consistently characterized representations of outer space in the 1950s. These were all factors, I argue, that were similarly present at other scales and in other Cold War spaces.

1 GLOBAL VIEWS
Geopolitics, Science, and Culture

> Practical men hold global views, though they may not verbalize
> them, and in the light of their views they make important deci-
> sions. The whole direction of a nation's effort may be determined
> by the global thought filters of its leaders. A global view is more
> than just a filing system for information. Necessarily it becomes a
> system of evaluation. As such it may also be a system of distortion.
> — STEPHEN B. JONES, "Views of the Political World"

Long before the term *globalization* acquired popular currency in the 1960s, the earth was represented as a unitary sphere. To conceive of the globe as a single object requires a perspective removed from the planet's surface — a view named by Denis Cosgrove as the "Apollonian eye." This disinterested and rational outlook is paradoxical. Its universality is proclaimed from and for a particular location such that the object becomes an *objective.* "The imperial imperative," Cosgrove argues, has consistently been "figured through the image of the globe."[1] Likewise, to see the world as an ordered whole, from a position of supposed detachment, is the basis for what John Agnew calls the "modern geopolitical imagination," which in turn makes possible finer — although still startlingly crude — forms of geographic differentiation, such as those between known and unknown lands or democratic and totalitarian spheres. These spatial partitions match with temporal signatures: parts of the globe are backward or advanced, modernizing or developed, or, in a division made more acute by twentieth-century flight and aerial photography, target and home.[2]

It was not until the Second World War and the early Cold War, however, that the entire planet became an *American* strategic environment — what the modernization theorist and psychological warrior Daniel Lerner aptly labeled "a global arena of national action."[3] To be able to think in such terms, linking the nation to the globe, required a certain geographic sensibility: a modern manifestation of the much older tendency to visualize the world as a whole, supplemented by a newer, if equally important, belief that the world's remote corners were in some way relevant to the security of the homeland. American strategists of the 1940s and 1950s presented a more fully realized version of the world famously described by the British geographer and imperialist Halford Mackinder some fifty years earlier, a geopolitical system featuring social explosions reverberating "from the far side of the globe."[4] Elsewhere, Mackinder described these shocks as part of a self-regulatory "closed circuit . . . complete and balanced in all its parts."[5] This closure was, by the time of the Second World War, more pronounced, as Mackinder himself acknowledged in a prominent 1943 article on the "round world."[6]

If the frame of Cold War does not do complete justice to the history of the world in the second half of the twentieth century, despite "crucial intersections between the Cold War, decolonization, and global social awakenings," it is certainly the case that the Cold War was regularly presented by its most prominent protagonists as planetary in scope.[7] This chapter brings together several overlapping global views — ways of seeing that took the world as a singular unit — which provided the strategic template for America's Cold War. The first view, found in popular cartography, was obviously the most directly visual of the group, although the others — airpower doctrine, geopolitical and international relations theories, science policy, and cultural diplomacy — all relied on cartographic expressions for persuasive authority. My aim is not to exhaustively historicize each of these views or to artificially distinguish them but instead to show how a set of diverse but contiguous modes of thought contributed to and solidified the idea of an *American globe*. The finer strategic scales considered in subsequent chapters — scales that frequently provided important and necessary support for the most comprehensive of Cold War concepts, such as containment — were granted solidity underneath the umbrella of globalism. In this sense, although global views seem to harbor a universal quality, these perspectives were never beyond geography, even as American strategic horizons were undoubtedly expanding.

In its inheritance of "the European *mission civilatrice*" after the Second World War, the United States continued to rewrite the language of empire in more explicitly commercial terms.[8] But the global outlook that was the very foundation of Cold War strategy also had military, scientific, and cultural roots, as Cosgrove suggests with his discussion of the Apollo space program and its iconic photographs of Earth. Similarly, just as historians have recently shown the extent to which the Cold War was "total," requiring "contributions from all sectors of society," examining American global views of the 1940s and 1950s means blurring the distinctions between formal, practical, and popular forms of geography.[9]

The inefficacy of such separate classification is apparent if we consider a 1954 speech in which President Dwight Eisenhower described the geography of the nascent Cold War. It was, he argued, a landscape of reduced physical challenges (owing to new technologies) and growing social divisions: "The world, once divided by oceans and mountain ranges, is now split by hostile concepts of man's character and nature," two "world camps" lying "farther apart in motivation and conduct than the poles in space."[10] Such familiar language not only affirmed the binary spatial division on which the condition of Cold War was premised but also, by endorsing the worldly aspirations of each "camp," the planetary nature of this condition. Eisenhower's rhetoric was not simply hyperbole but a fine example of Cold War geographical thought, forcefully tying the United States to a global arena and simultaneously identifying the Soviet Union as an obstacle to a perfect match with that space. His words indicate that the Cold War's geography was profoundly imaginative and dependent on *perception*—a trait that could not be erased by even the most hardheaded of strategists, just as they could not confine strategic thought to the realms of government, universities, or think tanks. Finally, because the earth was divided by conduct, or behavior, one's opponents were not easily enclosed. Their pervasive presence, or potential, meant that the world itself must necessarily become an object of surveillance and a scene of suspicion.

The antithetical character of Cold War rivalry voiced by Eisenhower was also supplemented by an alternative theme in American strategy, a useful *integrative* intermediary between isolationism and imperialism that nonetheless implied incorporation.[11] If containment, in theory and in practice, provides an obvious example of Cold War geopolitics, then integration relied more directly on cultural and scientific (or social scientific) ideals and was occasionally presented in the dreamy language of global humanism. But integration was also an assimilative project rooted in a particularly iconic version of America whereby foreign areas and populations, as detailed in the next two chapters, could be domesticated and modernized, marked as successes on a checklist leading to a harmony that was more calculative than cosmopolitan.

In the same way that Enlightenment presumptions of human difference were balanced by universal philosophies, a celebration of Cold War cultural diversity could easily be wrapped into a broader cartography of identification that did not quite extend to the Soviet Union. Eisenhower's 1953 inaugural address, for instance, while premised on basic distinctions between good and evil, or freedom and slavery, was also rife with the rhetoric of international unity, shared faith, and "equal regard" for "all continents and peoples."[12] These tropes provided rationalizations for American strategy in the name of political self-determination or economic liberalization and hastened the concealment of other strategic methods inside the covert categories of espionage and psychological warfare. Referring to a more recent time, Amy Kaplan notes that the idea of a "homeland may contract borders around a fixed space of nation and nativity, but it simultaneously also expands the capacity of the United States to move unilaterally across the borders of other nations."[13] And as a result of the Cold War's ubiquity, the United States and the Soviet Union were both repeatedly required to attend to anticolonial struggles in the various locations collectively labeled the third world. Although this book does not directly address these interventions, it is not because I believe they can be fitted neatly to a metanarrative of Cold War bipolarity. I am instead interested in the power of a global strategic geography that made such reductions seem logical or inevitable.

"A Globe in Practice"

In July 1956, *Scientific American* reported that the U.S. Army Map Service had just calculated the "longest line ever surveyed," stretching from Finland to the southern tip of Africa. The task was grounded in a long tradition of scientific adventure: south of Egypt the surveyors "were hindered by grass fires and aroused buffaloes." But such perils, and their specific locations, were ultimately erased from the record of certainty, as "data" were "reduced to summary form with the aid of a large computer." This process of measurement and abstraction enabled the military cartographers to revise the estimated radius of the earth and transfer sites of fieldwork—and, by extension, the globe itself—to the storage banks of a computer. Obtaining greater standardization and universality was a scientific accomplishment, but one with practical consequences. Not only would maps become more accurate, but the new measurements would also be used to plot "the course of the earth satellites to be launched during the International Geophysical Year." Unmentioned in the article were the more ominous military applications of sharpened coordinates for intelligence satellites and intercontinental ballistic missiles (ICBMs).[14]

The satellites mentioned by *Scientific American* were, in 1956, still the stuff of congruent science fiction scenarios and classified reports, and when a space vehicle was launched fifteen months later, it was, alarmingly, not an American product. Yet the exploits of the Army Map Service suggested that one did not have to leave the globe to grasp it as a whole; indeed, the radius of the world was calculated to be shorter than previously thought by some 128 meters. For readers of the magazine, this claim was probably not surprising, because it was bolstered by a far more pervasive — if inevitably vague — impression of time–space compression during the 1940s and 1950s. Perceptions of increased proximity were, of course, more acute for some and seemed particularly relevant in a United States held up as the leader of liberal globalization after the Second World War. But in order to completely comprehend a supposedly shrinking world, the geographic unit of the globe had to be understood as *alterable,* as a malleable abstraction, through a process of detachment.

Despite Halford Mackinder's popularity, many of the dominant geopolitical models of the early twentieth century fell out of favor during the Second World War, with the full (if frequently belated) realization that Germany's National Socialists subscribed to a particularly perverse form of reasoning based on the combination of territorial expansion and racial purity. Despite the taint of this *Geopolitik,* the dual characteristics of the modern geopolitical imagination — a totalizing perspective and the division of the globe into positive and negative realms — did not suffer for proponents during the early Cold War. Only the term *geopolitics* was disavowed, and arguments rooted in environmental justifications gave way to the less controversial variables of cultural, political, and technological difference.

In the case of technology, it is not surprising that geopolitical arguments rooted in physical determinants such as soil and climate were less popular in the age of air power. During the 1930s and 1940s, the "view

from above" was still very much linked to a specific archetype of the airman, with Amelia Earhart being the major exception to this masculine
caricature. The airman's popular position was shifting from one of blissful
freedom and adventure—a mythical status that had been already complicated by the minor but spectacular role of aerial combat in the First World
War, and by colonial operations in the 1920s—to a moving point on strategic maps. No longer always a dashing solo voyageur, the airman became
tied to the populations and structures that he might destroy; to systems of
production, maintenance, and team-based missions; and to national causes
rather than eccentric ambitions. The apotheosis of this changed individual
was undoubtedly Paul Tibbets, pilot of the *Enola Gay,* the B-29 Super-
fortress that dropped an atomic bomb on the Japanese city of Hiroshima.[15]
Rather "than an as-yet-undefined freedom for everybody," *Time* reported
in 1943, American airmen "want more freedom for the U.S." Crucially,
this "qualified" and specifically national freedom included "ownership of
bases" beyond American borders.[16] One such example was the island of
Tinian, a former Spanish and German possession taken from the Japanese
by the United States in 1944 and effectively converted into a massive air-
base where the *Enola Gay* was launched.

The militarization of the airman was awkwardly matched by his role
as a harbinger of liberal democracy, particularly as a result of expanding
commercial air transport networks. Prefiguring *Scientific American,* a 1943
cartography guide produced by one aviation company proclaimed that
"[we] are turning more and more to the globe—and to maps which are
projected in such a way that they show true *distances.*"[17] And yet distance,
in the "air age," was not static but a function of mobility, or speed, in the
same sense that the telegraph had transformed perceptions of time, space,
and information in the nineteenth century. Realizing and representing distance accurately was therefore not just an abstract mathematical principle:
it was both a privilege and a responsibility. By rising above the provincial,
debased earth, the aviator might, as Librarian of Congress Archibald Mac-
Leish wrote, in the 1944 collection *Compass of the World: A Symposium on
Political Geography,* make the globe "truly round: a globe in practice, not
in theory."[18]

Inspired by the global modernism of the 1930s, MacLeish and his fellow contributors to *Compass of the World* were determined to set out a
new vision for the postwar world, and their essays collectively stood as an

Pilots aboard a U.S. Navy aircraft carrier receive last-minute instructions before attacking industrial and military installations in Tokyo, February 17, 1945. National Archives and Records Administration (photo 208-N-38374).

idealistic dissent against the state-centric realism that was fast becoming dominant in the study of international relations and in the pronouncements of politicians.[19] But *Compass of the World* was nonetheless premised on American primacy. The book's tenets were visualized by the dramatic cartography of Richard Edes Harrison, a prolific contributor to periodicals such as *Fortune* and *Life*. His maps often presented views from a fixed point beyond the earth rendered at unusual angles — a persuasive "illusion of depth and perspective," as one representative atlas put it, which Harrison achieved by first photographing a large globe.[20] The uncluttered appearance of these maps departed from familiar conventions so much that they were more akin to advertising than to anything previously treated as cartography. Harrison actually "preferred to identify himself as an artist rather than a mapmaker."[21] But his compositions were still understood *as maps* by those who consumed them. And these were geographic representations with a special power: they turned readers into pilots and were reprinted in military periodicals and used to train actual airmen, helping them to "visualize regions that had not been photographed from the air." This popular and practical cartography of the Second World War outlasted the conflict itself, as newspapers and magazines continued to print similar maps that were resoundingly geopolitical, clearly displaying biases and the advantages of perspective, into the 1950s and beyond.[22] Wilbur Zelinsky may not have been exaggerating when he argued that "perhaps more than any other individual, Harrison helped to alter quite literally the ways in which Americans perceive our planet."[23]

Compass of the World brought geopolitics together with scientific and cultural authority to fashion a prospectus for the projection of American power over the horizons of the globe itself. The world, in other words, became a strategic environment, with America at its center. The suppleness of this vision was such that it could be taken up in the cause of imperialism and militarism or, alternatively, under the banner of a humanistic internationalism based on integration and brotherhood. Harrison's maps could reinforce national boundaries, or they could reveal the falsity of these divisions.[24] But as world war turned into Cold War, the first of these options began to dominate the second, even in related but distinct debates over postcolonial equality and the spread of liberal capitalism. The most ambitious expressions of cosmopolitan identity were rarely severed from an American affiliation signaling dominance and responsibility.

In *The Geography of the Peace,* his short 1944 manifesto on the likely shape of the postwar spatial order, Yale's Nicholas Spykman argued that the "basis of world planning for peace must be world geography," but equally, the predicament of global war meant that military planning must also "consider the whole world as a unit and must think of all fronts in their relations with each other." Those adept at the interpretation of maps could serve the state well under both conditions. The key was that the "entire earth's surface" be the unit of study. But all global maps, he acknowledged, included biases, and their creators and users should justify cartographic choices. For his part, Spykman expressed affection for cylindrical adaptations of the Miller projection, named after O. M. Miller of the American Geographical Society. These views portrayed the United States as central and besieged by other landmasses. While additional arguments were required to justify the impression of being surrounded, Spykman indicated that these would not be difficult to detail. But his endorsement of the Miller

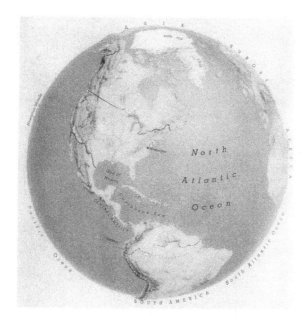

"Outward from the U.S." From Richard Edes Harrison, *Look at the World: The Fortune Atlas for World Strategy* (New York: Knopf, 1944), 13. Originally published in *Fortune* magazine. Copyright 1940, 1944 Time Inc. All rights reserved. Courtesy of Fortune | Money Group.

Projection not only identified the specific dangers that motivated mid-century American strategy, it also lent credence to a more general vision of the globe itself as a battlefield. Once it was acknowledged that a perspective of encirclement could be produced from anywhere in the world, that "every point is surrounded by all other points," an approach to international relations prioritizing ubiquitous threats to a particular national space was bound to follow.[25] Because the United States was not only an emerging superpower but also new to the awareness of *proximate* threats, it was easy for Spykman to emphatically situate America as the heart — and the target — of a dangerous world.

Maps of a World at War

Although he had a "wide audience at *Fortune*" by 1940, Richard Edes Harrison was particularly prolific and prominent during the years of American participation in the Second World War.[26] His *Look at the World: The Fortune Atlas for World Strategy* was released in the same year (1944) as *Compass of the World* and became not only a companion volume but also

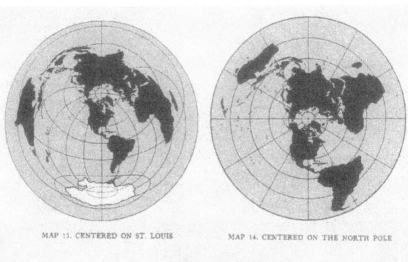

MAP 13. CENTERED ON ST. LOUIS   MAP 14. CENTERED ON THE NORTH POLE

MAP 15. CENTERED NEAR THE PANAMA CANAL

America encircled. From Nicholas J. Spykman, *The Geography of the Peace* (New York: Harper, Brace, and Co., 1944), 20. Courtesy of Houghton Mifflin Harcourt Publishing Company.

a publishing sensation. *Life* called Harrison's atlas a "geographical mile-stone . . . peculiarly appropriate to the present day."[27] The war was also a boon for many other American purveyors of geographical knowledge. Clark University's Wallace Atwood noted in 1944 that interest "in a broad world-wide study of geography with a human point of view is sweeping over the country like a great tidal wave."[28] In a February 1942 radio address, Presi-dent Franklin Roosevelt—himself lavishly provisioned by the National Geographic Society—famously instructed Americans to "take out and spread before you a map of the whole earth."[29] Geographical awareness, at least that which could be derived from either a globe or one of the many war atlases sold to the public, became a valuable and much-discussed commodity, particularly in a society constantly chastised by "experts" for spatial illiteracy.[30] As Harrison and the geostrategist Robert Strausz-Hupé put it in the same year, American "psychological isolationism . . . can be in large measure traced to our failures in mapmaking and the teaching of geography—the prerequisites of education in international relations. The world is round. By the skillful presentation of its 'roundness' strategic realities are made clear."[31]

Also in 1942, Roosevelt was presented with a Christmas gift from Army Chief of Staff George C. Marshall: a 50-inch, 750-pound globe. The giant sphere, Marshall wrote Roosevelt on December 12, would allow the presi-dent to "accurately chart the progress of the global struggle of 1943 to free the world of terror and bondage." Roosevelt replied thankfully on Decem-ber 30, noting that he had placed the globe "in my office directly behind my chair. I can swing around and figure distances to my great satisfaction." A similar globe was given to British Prime Minister Winston Churchill. According to Arthur Robinson, the idea for the globe came from William J. Donovan, director of the Office of Strategic Services (OSS), who enlisted Robinson, the head of the OSS Map Division, to design the object. Look-ing back on the task, Robinson mused that "sitting close to the large globe" would have given the two leaders "a view rather like that of an astronaut today. That certainly would have helped them contemplate the immense strategic and logistical problems of a truly global conflict."[32] As a physical aid to this omniscient view of "strategic unity," Marshall's globe was sig-nificant precisely because of its size.[33] It was ever so slightly closer to the dimensions of the earth itself, and the resulting authority granted Roos-evelt and Churchill was one entrusted only to leaders of their stature.

The U.S. Army presented a 50-inch, 750-pound globe to President Franklin D. Roosevelt, December 1942. Courtesy of the Franklin D. Roosevelt Presidential Library and Museum.

For his part, Harrison "was able to translate the conflict's new realities into graphic images for the public," shattering what he called the "static condition bordering on senility" afflicting many professional geographers and cartographers.[34] That war had generated tremendous interest in geography was not surprising. It was the unusual images the public was consuming, and the sources of these images, that were significant and novel. Older modes of representation, it seemed, could not accurately represent altered geographical realities, particularly for the readers of newspapers and magazines. But Harrison's radical visual sensibilities did not necessitate a cartographic decentering of America. Rather, his projections revealed that the "entire conflict pivots around the U.S.," and that the world was divided into two distinct camps: "those who are for us . . . and those who are against us."[35] Among the new map types revealing the destiny and dangers facing the United States was one situated above the North Pole, showing the tight alignment of the Northern Hemisphere's landmasses. This view, as I show in chapter 4, became central to Cold War strategy.

These and other perspectives shattered isolationist impulses and heightened domestic impressions of vulnerability by demonstrating the potential for siege from all directions, attacks that could extend deep into the American heartland. Equally, the cartographic output of the National Geographic Society had been recognized as "an important war weapon," according to society president Gilbert Grosvenor.[36]

Not surprisingly, Harrison's bold journalistic methods prompted criticism from geographers and cartographers whom he challenged and even dismissed. Predictably, some, such as the University of Chicago's Charles Colby, confronted him for failing to conform to standardized cartographic conventions or for the sacrifice of mathematical precision. Harrison responded, appropriately, that all mapmaking was subjective.[37] But in so doing, he was also implicitly acknowledging the veracity of another accusation: that his work was propagandistic. By drawing the United States closer to enemy spaces and forces converging on North America, Harrison's maps rendered the nation automatically internationalist, forced to contend with new forms of war and aviation technology. He and others considering the cartography of global combat were not afraid to admit that a map could be, as the commentator Hans Weigert put it directly, "a psychological weapon."[38] Secure in the knowledge that their representations were helping to shape the contours of a world at war, these writers and artists chose what they believed to be the appropriate perspectives. Despite Harrison's admission of partiality, he believed his angles—the North Polar Azimuthal Equidistant Projection, for instance—to be closer to the *necessary truth* required for successful strategic planning.[39]

Harrison's maps, and others like them, were a component of what the New York Public Library cartographer Walter Ristow and many others called "air-age geography," a new form of integrated spatial study that Ristow contrasted against older variants of physical, economic, and political geography. Although the air age had been heralded before the Second World War, for Ristow it was a post–Pearl Harbor phenomenon. But it was indisputably global: war and long-range aircraft had rendered "all regions intimately related." Hemispheric divisions were outdated, and he argued that geographers should "emphasize global problems" in teaching and research.[40] This approach possessed distinct moral and political elements. Widening the horizons of map, mind, and text would, Wallace Atwood proposed, "help prevent the recurrence of isolationism."[41] There were

Allen W. Dulles, director of the Central Intelligence Agency, in his Washington, D.C., office on July 28, 1954. Behind him is a polar projection map, "which he often consults." Courtesy of the Associated Press.

academic justifications too. Researchers began to look to the "geography of the air" as a scientific space whose study had strategic implications.[42]

The literature on air-age geography marks a transition from a deterministic accent on the natural environment to an emphasis on the power of technology, particularly the airplane. In one sense, such vertical movement was slight; humans were still dwelling and warring on the surface of the earth. But this qualification did not stop a litany of commentators, many writing for school-aged audiences, from embracing a boundless "third dimension," with significant ideological connotations for the understanding of geopolitics in an age without an "Eastern or Western hemisphere."[43] Macmillan's Air-Age Education Series included some of the most inflated examples of this rhetoric. The foreword to each book proclaimed that "wide seas, dangerous reefs, precipitous mountains, frozen wastes, and jungle depths, all barriers to earthbound generations, have become features of the landscape below the global sweep of the airplane travelers in the ocean of air."[44] The guide for teachers applying this new curriculum listed national isolationism, ignorance, egocentricity, poor diplomacy, and lack of regional and urban planning as some of the effects of "our non-geographical thinking."[45]

While the outspoken Richard Edes Harrison was quick to denounce some of the excesses and betrayals of the "'Air Age' group," in the process revealing his own penchant for methodological policing and promotion, many publishers ignored his caution, especially early in the Second World War, when books, articles, and atlases replete with the rhetoric of globalism piled high in American shops and schools.[46] As Harrison claimed, most of the best geographers could not contribute "to this output except in an advisory way, the reason of course being that this undermanned profession is busy fighting a war *with* maps and has little time for a war *about* them."[47]

Despite Harrison's reservations, geographical facts were inseparable from geopolitical theories. The concept of an integrated, shrinking, and traversable planet was filtered through the conditions of world war and the Cold War, conflicts whose dimensions meant that a war atlas was not complete unless it was global, regardless of whether it was three-dimensional. These atlases, some indebted to Harrison's new methods, were, like his maps, still partially dependent on an earthbound form of geopolitics:

The grand design of this war was largely a product of strategic geography. Strategic geography means, roughly, the influence of the earth's physical make-up on the operations of war. The techniques of warfare had changed, but the scene of battle was still the earth itself.[48]

An equivalent statement had been made by Nicholas Spykman in 1938:

So we find that, although the entire policy of a state does not derive from its geography, it cannot escape that geography. . . . With these facts foreign policy must reckon. It can deal with them skillfully or ineptly; it can modify them; but it cannot ignore them. For geography does not argue. It simply is.[49]

Spykman's definition of geography, akin to a stage on which the drama of "security policy" is played, was, like much scholarship of the prewar period, married to environmental constraints.[50] In this respect, the hindrances of space and the authority accompanying the occupation of certain regions could be radically reduced only by technologies such as the railway or the airplane. These innovations *overcame* geography, but not completely; a political layer was inscribed atop a physical template.

Impressions of air-age geography were not the marginal postulations of cloistered intellectuals. From Germany to Great Britain and the United States, in the decades leading up to the Second World War geopolitics was—like academic geography more generally—consistently presented as useful for statecraft. Geopolitical theories could be found in Nazi doctrines as well as in the primers distributed to American soldiers. The Army Specialized Training Program's textbook *Geographical Foundations of National Power* (1944), for instance, described "the world that matters" as including only "the parts where the combination of rich natural resources and advanced industrial techniques has produced high standards of living and concentrations of political power." Measuring these elements over time revealed a state's relative "power potential." Power, then, still sprung from the earth, and even as geographers moved away from determinism, elements of an older tradition still lingered, particularly in derivative, practical, or popular publications.[51]

From its official beginnings at the turn of the twentieth century, geopolitical writing was obsessed with states. This habit persisted after the Second World War, partly as a result of, and not despite, the formation of new organizations built to manage international order. Realist variants of international relations theory added to the pervasiveness of this *territorial trap* by presenting unitary, timeless states as the basis of a global system and by perpetuating the equivalence of sovereignty, security, and state boundaries. The state remained the "container of society" and carried the characteristics of a human individual.[52] This geopolitical anthropomorphism was nicely compatible with similarly simple theories of national character, which, as I show in the next chapter, were elastic enough that they could be expanded beyond the state to encapsulate a larger regional realm.

The struggle to shake off determinist shackles, meanwhile, led to the search for a more scientific political geography, which was set against not only the horrific excesses of German *Geopolitik* but also the crude theories of George Renner and others who seemed to win favor with popular periodicals. This was a battle for reputation, because the various sides of the subject were frequently conflated — a mistake that must have been "particularly galling" for "leaders of geography" such as Isaiah Bowman and Richard Hartshorne, both heavily involved in the American war effort.[53] Bowman's quest to differentiate (political) geography from geopolitics, and himself from geopolitical thinkers of any stripe, led him to prepare an article for the *Geographical Review* in 1942 in which he directly identified geopolitics with German aggression and older strands of German politics and philosophy. These currents of thought, he argued rather weakly, were diametrically opposed to the democratic principles of the American Constitution.[54] Such casual history, really only a minor departure from the ahistorical essences of *Geopolitik* so scorned by Bowman, was also the approach taken by his fellow geographer Derwent Whittlesey, whose *German Strategy of World Conquest* (1942) described plans for national expansion and, ultimately, global domination that were innate to a German personality.[55] Scientific geography, in Bowman's opinion, had been appropriated and applied improperly by German strategists, who had conveniently ignored the moral foundations that allowed, and required, science to be distinct from politics. Bowman spent little time delineating the principles of this science except to vaguely invoke international cooperation and democracy, where he believed scientific inquiry belonged. Like Whittlesey, Bowman was effectively using "a mix of positivism and ideology against positivism and ideology."[56]

For much of the American media, *Geopolitik* became a "hidden logic" used to explain Nazi war planning. Part of widespread "fear and fantasies about Nazi Germany," the preoccupation with *Geopolitik* was also intellectual, as Americans were instructed to take this "new form of global thinking" seriously, to educate themselves and weigh similar responses.[57] The ideas and maps of the aging Halford Mackinder had an impressive revival, topped off by his 1943 article in the influential journal *Foreign Affairs*.[58] In the United States, a number of émigrés from Central Europe were enlisted as geopolitical authorities. Media stories set their arguments against the teachings of the German strategist Karl Haushofer. In periodicals such as *Life* and *Reader's Digest*, and even in short Hollywood propaganda films, Haushofer frequently, and incorrectly, appeared as a "superbrain or scientist" at the head of a (mythical) Munich Geopolitical Institute. One magazine piece described Haushofer's compilation of a monumental Strategic Index tabulating "every phase of every nation's life," and while such a project never existed, its invention alone is indicative of the American fascination with Nazi modernity, a conspiratorial and paranoid vision of omnipotent military intelligence and unmatched social scientific knowledge.[59]

In the work of the more careful commentators, some of these lurid exaggerations were dismissed, but geopolitics — even if rephrased under different names — was still presented as essential to the war effort. For prominent writers such as the Austrian migrant Robert Strausz-Hupé, the belief that "space is power and that international politics is a struggle between different states for space" remained a timeless truth and was only rendered "degenerate" by the influence of ideology. In various discussions of *Geopolitik,* the United States was seen to lack its analytical potency, a mathematical and technical form of "instrumental reasoning." Such hand-wringing led to calls for the formation of an American Institute of Geopolitics, and although the most dramatic schemes fell through, the War Department did set up a Geopolitical section within its Military Intelligence Service in June 1942. Leading geopolitical thinkers, from Whittlesey and Spykman to Edward Mead Earle and Harold Sprout, served as consultants. At Georgetown, Edmund Walsh, the Jesuit founder of the University's School of Foreign Service, set out to turn his program into a similar Institute. According to Gearóid Ó Tuathail, Walsh, who interrogated Haushofer after the war, believed that *Geopolitik* had significant "scientific merit."[60]

For these and other American intellectuals, the appeal of German *Geopolitik* was its role as a predictive, synoptic, and global science, fascinating because of its potential for disaster if thrust into the wrong hands. In this vein, geopolitics, writes Ó Tuathail, was "*insight*, the figure of the geopolitician a seer . . . a scientist yet also a prophet, a positivist yet also a creative, envisioning artist." The spatial detail and coverage supposedly acquired by the fictional Munich Institute and its Strategic Index project were essential to the postwar world so long as they remained leashed by the correct version of national character. Similar, seemingly contradictory attributes also characterized postwar American strategic thought, which "promised the disenchantment of the surface of international affairs by reenchanting that very surface with . . . magical formulas that compelled the earth to reveal its secret strategic pathways and faultlines."[61] As another excuse for a global outlook, the slippery signifier *geopolitics* was melded onto a larger body of foreign-policy discourse, where it seemed to fit comfortably and anonymously with state-based realism. But it was more than just a representational device; the expert knowledge desired and provided by geopolitics *shaped* a world at war and the tense condition that followed.[62]

Technological Horizons

Early in September 1944, Commanding General of the Army Air Forces Henry H. "Hap" Arnold met with the aeronautical scientist Theodore von Kármán at La Guardia Field in New York to discuss the formation of a long-range science committee for the Air Force.[63] Urging the Hungarian émigré to look beyond the current conflict and forecast the shape of postwar air power, in a world where "global war must be contemplated," Arnold gave von Kármán free rein to assemble a team of scientists, study the latest in military and civilian research trends, travel abroad to interview colleagues and captured enemy researchers, and ultimately prepare a summary report.[64]

The result was *Toward New Horizons* (1945), a multivolume study produced by the Scientific Advisory Group of the Army Air Forces that directly influenced Cold War programs for air defense. In the report's key summary volume, von Kármán argued that strategy for increasingly technological forms of war must "refer to the three-dimensional space surrounding the globe." Resigning the geopolitics of land and sea to an archaic, secondary status, he set out the central philosophy behind scientific air power: a formidable air force must be able to secure "superiority over any region of the globe."[65]

While aviation, for von Kármán, was an individual effort to escape the limitations of nature, he also envisioned a future "war machine in the proper sense of the word," ignorant of weather or darkness, that would consist of "technical devices only," with a "master strategist" directing from a distance.[66] *Toward New Horizons* went on to discuss in some detail fantastic or nascent innovations such as supersonic flight, pilotless aircraft, propulsion, and radar. Von Kármán's discussion of machinic vision no doubt upset many traditionalists and was almost *chronopolitical* in its language of pure, ubiquitous war untouched by geography.[67] However, his speculations were translated and transformed into much more familiar language by his patron. In a 1946 *Air Affairs* paper, published just as he was retiring, Arnold noted simply that science and technology had "negated the concept of isolationism." The lessons of bombing campaigns over Germany and Japan, at least for Air Force officials, were clear: air power, in Arnold's opinion, had been decisive and would be integral to national security and military success in the future.[68]

Arnold's *Air Affairs* article paled in comparison to his concurrent piece in *National Geographic*, magnanimously titled "Air Power for Peace," which ran to fifty-seven pages and was accompanied by numerous photographs showing the Army Air Forces' various roles in the Second World War— including, rather ironically, scenes of urban devastation in Germany and Japan. "With present equipment," Arnold warned, "an enemy air power can, without warning, pass over all formerly visualized barriers." Not only was this a call for more effective forms of defense but Arnold additionally believed a robust air force would serve as an important deterrent to surprise attack. In both offensive and defensive strategic planning, scientific superiority was essential to national security and thus to peace. Arnold's advertisement for American air power was also a case for the continuation of military might into the uncertain postwar period, when unceasing "patrol of the entire world" would be required to maintain the precarious American atomic monopoly. (His 1949 autobiography was, after all, titled *Global Mission*.) While he acknowledged that such vigilance could be an international responsibility, Arnold remained skeptical and insisted on the supplemental monitoring of various enemies by a strong American intelligence agency.[69] From within the military, Arnold was speaking the language of many civilian strategists: the atomic bomb and its means of delivery had profoundly altered the idea and viability of security and thus the meaning of geopolitics.

Air power's destructive potential had been understood long before World War II, and its global reach was recognized not just by the military but also by the promoters of the first international airlines, mostly American, who repeatedly employed globes as corporate symbols.[70] But once saddled with an atomic bomb, a long-range plane became much more than part of a war plan. To American strategists, it was symbolic of eroding distinctions between heartlands and peripheries, whether on a national or international scale. By the 1950s, the U.S. Air Force was defining "heartland actions" as "attacks against the vital elements of a nation's war-sustaining resources," a statement that specified no particular location for such elements.[71] Air power combined with the atomic bomb signaled a concentration of the time of violence but an expansion of the reach and scope of destruction. In these and other respects, it seemed, war was becoming increasingly inhuman because the unparalleled "technological fanaticism" of the American military during the Second World War allowed for additional "physical and psychic distance" from opponents. The practical consequence, however, was the triumph of the Air Force within the War Department by the end of the war, a bureaucratic victory based on promises of maximal force at minimal expense as well as on the authority invested in air power as "the ultimate source of danger and deliverance for the United States."[72]

The Transformation of Geopolitics

It did not take long for the risks of atomic technology to be plotted over an American geography in the form of the urban disaster scenarios I discuss in chapter 5. If the bomb and the global war in which it was featured had together ushered in a new world, it was a world made suddenly insecure by science. In his 1947 treatise *The Price of Power*, the *New York Times* military correspondent Hanson Baldwin noted that a shrinking globe and a developing geopolitical "bi-polarity" meant that "choosing sides" was an urgent priority in an era of disappearing neutral ground. The United States, Baldwin argued, was now at the center of a "security zone" that stretched concentrically outward over the new maps of the air age, cartography that had discarded the Mercator projection for the accuracy of a north-polar orientation. Unlike more conservative commentators — James Burnham, for instance — Baldwin did not quite embrace the apocalyptic imagery of a necessary struggle for world domination. But he did admit that America's "*absolute* strength" was unprecedented in world history and that the trend of imperial decline meant that this status would probably not last. The only way to proceed was to combine a global foreign policy with "security measures" at home, maintaining national power as a source of "international rehabilitation and world stability."[73]

Baldwin's was a relatively crude version of diction that proliferated among American intellectuals after the Second World War. His grudging and general recommendations emphasized a division between international and national spaces and the difficulty of distinguishing one from the other in any realm beyond the discursive. The postwar "liberal moment," less a philosophy than a way of life, was also, paradoxically, the period when the question of security "became especially acute."[74] This contradiction was particularly sharp in the overlapping fields of international relations, strategic studies, and political geography.

The negative midcentury association of the term *geopolitics* with Nazi pseudoscience still figures prominently in more recent discussions of the decline or attenuation of geopolitical (and, more generally, geographic) thought in postwar America.[75] But while correct in a terminological sense, such arguments risk internalizing the very logic that they intend to critique by resigning the Cold War to a realist history and ignoring the effort required to promote a thin spatial language. As with human geography and other social sciences during the same period, postwar geopolitics was reinvented as a science — in this case a *science of strategy* — that derived much of its force from technological rather than environmental determinism and was often openly attached to the American state and to the expression of American power. In effect, an "earlier emphasis on the ontological primacy of the natural world was replaced by the epistemological primacy of the natural sciences."[76] For many strategists, this meant that the subject of study was less important than the means (scientific, or so it was claimed) with which it was approached. But science, in this instance, was a flexible expression, and its flexibility was best signaled by the predominant reluctance to abandon American affinities.

During the 1940s and 1950s, international relations, strategic studies, and political geography were united by a Cold War sensibility that treated America as the heart of a hostile world. The second half of this assertion is as important as the first: the United States was doubtlessly central to American Cold War cartography, but it was also surrounded by antagonists. For the Yale geographer Stephen Jones, by 1955 it was "safe to say that the man of affairs in any civilized land has a world-wide outlook, and, increasingly, so does the man in the street."[77] But some twelve years earlier, before the official commencement of the Cold War, Isaiah Bowman had predicted and clarified the threats accompanying America's growing responsibilities. In his 1943 presidential report to Johns Hopkins University, he wrote that "we can not safely limit our future responsibilities to narrow zones of power. No line can be established anywhere in the world that confines the interest of the United States because no line can prevent the remote from becoming the near danger."[78] Assuming this was the case, Jones, Bowman, and their counterparts were concerned with two questions: How could the dynamics of this world be accurately represented, and how could security be guaranteed, nationally *and* globally?

While acknowledging Richard Hartshorne's concern with the challenge of studying the "whole pattern of world relationships at one time," Jones was less troubled by the practical limitations of maps and globes than by "the plethora of data about the world." Given this predicament, the correct choice of "global systems" was crucial: What was the most effective way to study and display this data? In his Office of Naval Research–funded survey of these systems, Jones moved quickly through a physical grouping of latitudinal zones and climatic regions, hemispheres, and panregions and through a human category including population, race, culture, economic resources, political boundaries, state behavior, and circulation patterns. These, he summarized, were cumulative; they were ways of organizing the world that could be easily layered together.[79] In a second paper, Jones addressed "systems that are more nearly complete in themselves," the "global strategies" of Mackinder and Spykman, which, Jones noted, both shared much with George Kennan's containment theory: each could essentially be used to prevent "Soviet extensions of Soviet control in the Rimland."[80]

When compelled to choose, Jones built an "eclectic global view," but one that was nonetheless "based on the concept of national power," or the "inventory" of a state, and how the state opts to use this inventory. Inventories included variables of population, culture, and resources, while the latter, strategic component incorporated the spaces in which power, derived from these variables, was projected outward over the globe. Jones was reaffirming the classic distinction between domestic and foreign terrain. What existed inside national boundaries, for him, was fixed, or settled, and beyond was a shifting anarchic world. Maintaining a mobile military presence in this foreign realm was thus expected. Furthermore, given the simplicity of this distinction and the converse complexity of the data used to bolster a global view, it was the view's structure that was important. Once the structure was in place, additional data could be entered without much difficulty. Policy could also be enacted, because scholarly inquiry and strategic practice were, for Jones, equivalent.[81] His concurrent work on field theory was similarly all-encompassing: a bundle of ill-fitting principles behind a mask of scientific veracity and practical utility, and an attempt to supplement and support realist theories of international relations with the additional *detail* of geography.[82]

Jones was not the only geographer striving to rehabilitate and show the utility of the political side of the discipline after the Second World War. In his meandering 1950 presidential address to the Association of American Geographers, Richard Hartshorne set out a "Functional Approach in Political Geography," an approach that would move the "wayward child of the geographic family" toward "established methods of scientific analysis." The result was a remarkably depoliticized statement. Referring cryptically to the rising tide of McCarthyist anticommunism, Hartshorne described "political demagogues who find in any divergence of opinion from their own a sign of disloyalty" and added a footnote in the published version suggesting that readers "may find the pertinent connection by looking to the front pages of almost any American newspaper for any day during the month of March 1950."[83] Yet his denunciation of academic witch hunts was also an appeal to intellectual authority, an attempt to detach scientific geography from political life, or at least from certain detrimental forms of politics.

Policy-relevant realist approaches to international relations surged in popularity during the Second World War. In part because the disastrous suggestions of association with *Geopolitik* had encouraged professional geographers to shun political topics, international relations retained a limited spatial sensibility premised on a preoccupation with the state. What was novel was an increased stress on rigor, utility, and abstract conceptions of power. As expressed in the power-based writings of scholars like Hans Morgenthau of the University of Chicago, postwar realism was not yet driven purely by scientific justifications, but it was certainly not opposed to a search for empirical regularities.[84]

Influential wartime books such as Nicholas Spykman's *America's Strategy in World Politics* (1942) had advanced the understanding that global politics was a struggle for power.[85] Morgenthau and other realists borrowed the explicitly normative, national interest emphasis of Spykman's title while attempting to root the dilemmas of postwar foreign policy in the "laws, or regularities, of state behavior." The embrace of scientific method in the American social sciences encouraged Morgenthau's sweeping analyses, an acceptance that went "beyond the concern for problem-solving" to become an "operational paradigm." Again, it is important to emphasize, as Stanley Hoffman has, the role of émigré scholars, such as the German-born Morgenthau, in advancing these changes. Their philosophical training, historical sensibilities (including a sharp recognition of the catastrophes of the Second World War), and cosmopolitanism "moved them to ask far bigger questions . . . about social wholes, not just about small towns or units of government."[86] This worldliness was in turn precisely attuned to the practice of politics. A fascination with power matched America's postwar status as a global colossus. The emerging Soviet–American confrontation was unprecedented and of immense significance. To study the international system from within America was, following Spykman, to consider America's strategy. And this was not a comfortable task given the perils that beset the nation.

Spykman, a Dutch immigrant, was a prominent figure at Yale's Institute of International Studies (IIS), the most prestigious center of its kind in the United States during the 1940s.[87] Converted by his friend Arnold Wolfers, a Swiss-born student of realpolitik, from an idealist into an advocate of force, Spykman led the Institute during its formative years until his death in 1943. He is best known for a revamped version of Halford Mackinder's heartland thesis, stressing instead the strategic importance of rimland regions surrounding the Eurasian world-island and advocating an interventionist American stance in these regions and elsewhere — a worldview, in short, perfectly suited to the American Cold War.[88]

At Yale in the 1940s, international relations was a practical subject devoted to the pursuit of equilibrium over instability.[89] Spykman and Wolfers were joined by Frederick Dunn (a frequent visitor to the State Department) and later by several younger, talented scholars who sought to break from traditional military strategy, a group that included William T. R. Fox, Bernard Brodie, Gabriel Almond, and Klaus Knorr. Yale seminars drew in others with similar interests, from Hanson Baldwin to the sociologist Harold Lasswell. A mainstay in policy discussions by the end of the Second World War, and the source of large numbers of students sent to the Foreign Service each year, the IIS became a model for the *schools of strategy* that proliferated during the early years of the Cold War, and many of its members went on to distinguished careers combining academic duties with advice to state agencies. Intimate links were also established between theorists concerned with the general outlines of American foreign policy and students of specific regions targeted by that policy. The main financiers of early Cold War area studies — the Rockefeller Foundation and the Carnegie Corporation — gave generously to the IIS and similar centers as well, often from within the same funding stream.[90] Global views, after all, could contain many broadly complementary components.

If there is one conclusion that can be drawn from the vast literature on Cold War realism, it is that the term is too frequently used as a single stand-in for a much more heterogeneous philosophical constellation — often by realists themselves, who are determined to place it at the heart of international relations theory. Disciplinary debates and policing within international relations share familiar characteristics with efforts in other fields, including geography, to stereotype and exclude intellectual opponents. It does not help that the 1940s and 1950s, in the conventional terminology, included two great debates in international relations, between idealism and realism and between realism and behavioralism. But as in geography, harsh distinctions are also misleading. It makes little sense to divide Cold War realists into two groups, one motivated by the classical realism of power politics and human nature and the other by a neorealism of behavior and hypothesis testing. In both camps, a shared faith in the power of timeless laws, the utterances of canonical and caricatured figures of the past, and the preeminence of states led to a model of a directly accessible world constructed objectively. This imposition of order and control, and the naturalization of violence through a detached, expert gaze — a distinctly logocentric, masculine trait — enabled the tentative steps toward science taken by Morgenthau and others to be easily co-opted and expanded by behavioralists. A great success of the Cold War social sciences was to demonstrate "how traditional and empirical methodologies could confirm each other's conclusions about important questions of national policy." The state could be secured at the same time as international theory.[91]

I do not invoke masculinity in an offhand manner. The discourse of realist international relations, which anthropomorphized the state as an autonomous, rational actor, a *Homo securitas*, closely paralleled the rise of an "imperial brotherhood" during and after the Second World War.[92] Many of these men shared similar class, educational, and service pedigrees, including boarding schools, men's clubs, and the Office of Strategic Services or other prominent World War II units. In addition, their liberal internationalism promoted toughness, partly as a protective reaction to McCarthyist accusations of susceptibility (in the State Department, most famously): accusations that frequently employed a language of gendered and sexual failure. "Vital center" liberals, for whom John F. Kennedy was later an icon, sought to defend the boundaries of masculine virtue while dramatizing threats to an American body in "alarmist rhetoric" balanced by "patrician stoicism."[93] While employed at the RAND Corporation in the 1950s, Bernard Brodie, who was captivated by Freud and who himself underwent psychoanalysis, circulated a short memorandum comparing various war plans with sex. According to Fred Kaplan, Brodie likened one scheme that would avoid urban targets "to withdrawal before ejaculation," while the all-out destructive directives of the Strategic Air Command were treated quite differently.[94]

In the age of relevant realism, foundations frequently facilitated the narrowing of distances between Washington and academia. In addition to funding research by scholars who moved into policy-planning positions, philanthropies backed meetings where the bases of realist international relations were constructed and debated. At one such Rockefeller Foundation–sponsored gathering in 1954, Arnold Wolfers argued that the introduction of "behavioral or psychological aspects" had drawn realism away from abstract schema to "situational components." Generalization, in other words, was no longer based on "complex entities such as great powers or empires but could be derived from the study of such abundant simple elements as human demands, expectations, choices, ambitions, fears and reactions to environmental factors, the atoms, ions, and velocities of international politics."[95] Behavioralism, for Wolfers, was simply a step closer to reality and made verification of this reality easier. And although the unit of analysis may, in one respect, have shrunk drastically from the empire or state to the individual, this was compromised by the subsequent attachment of individual qualities to states, which resulted in even greater abstraction.

At the same meeting, Hans Morgenthau plainly set out the detached, masculine character of geopolitical vision shared by realist theorists. What made a singular theory of international relations possible in the face of ambiguity, he argued, was that both the "mind of the observer and the object of observation" shared the quality of rationality: "Foreign policy is pursued by rational men who pursue certain rational interests with rational means." As an equally rational observer, the strategic theorist could legitimately pretend to be one of these statesmen, whether of the past, present, or even future. The result was a "map of the international scene" — incomplete but filled with significant features "not affected by historical change." Morgenthau acknowledged that a rigorous theory of international relations could be "deflected from its rational course by errors of judgment and emotional preferences," *especially* under democratic conditions, but that this ideal was a state of mind worth seeking.[96] The striking visual and spatial language of Morgenthau's speech, and its impersonal tone, were precisely attuned to the authoritative subject-position he desired.

Despite its diversity, realism provided a consistent compass for the formulation and perpetuation of Cold War strategy. Realism stamped out lingering isolationist tendencies and justified the maintenance or expansion of power through armament and alliances. It represented the world outside America as threatening and encouraged vigilance while tempering excessive bellicosity with the principles of containment. During the 1950s, containment was translated, much to the chagrin of some, from the realm of diplomacy to the hardened discourse of military strategy. Lured by lucrative funding, and no doubt in some cases by dreams of fame and relevance, many academics moved closer to technocratic consultancy roles, deepening temporary relationships established during the Second World War. But even the most revisionist interpretations of the Cold War's origins, which argued against theories of Soviet expansionism or inevitable international rivalry, and instead placed much of the blame on American capitalism, were driven, like their realist opponents, by the postulation of an external "world of fact" without room for self-reflection or subtlety.[97]

The Case of Containment

On March 12, 1947, President Harry Truman addressed a joint session of the United States Congress, delivering the first public assertion of American Cold War strategy. The Truman Doctrine outlined a specific "situation" integral to both the "foreign policy and national security" of the United States. But a civil war in Greece was also tied firmly to a much broader moral choice between abstract and absolutist notions of freedom and totalitarianism:

> I believe that it must be the policy of the United States to support free peoples who are resisting attempted subjugation by armed minorities or by outside pressures. I believe we must assist free peoples to work out their own destinies in their own way.... The seeds of totalitarian regimes are nurtured by misery and want. They reach their full growth when the hope of a people for a better life has died. We must keep that old hope alive. The free peoples of the world look to us for support in maintaining their freedoms. If we falter in our leadership, we may endanger the peace of the world—and we shall surely endanger the welfare of our own Nation.

Truman's dogma, presenting American values as universal and implicitly connecting contemporary global events to a national history of revolution and resistance, was not unique. But it was particularly well suited to a version of the Cold War as a struggle between good and evil systems, cultures, and, not least, spaces. The binary logic of Cold War geopolitics effaced the complexities of place and employed "the abstract categories of 'the free world' and 'the enslaved world' to mentally construct a black and white map of international politics."[98] Geography was used to delineate the boundaries identifying a complete division between a primary, positive inside and a secondary, negative outside.

Discussions of Cold War geopolitics invariably begin with the writings of the diplomat George Kennan. Kennan was a product of the West, a man who constantly divided the world into two spaces, one occupied by capitalist democracies and the other by foreign despotism.[99] Both Kennan's "Long Telegram," cabled from Moscow on February 22, 1946, and his article "The Sources of Soviet Conduct," printed in the July 1947 issue of *Foreign Affairs,* stressed the expansionistic nature of the Soviet Union and advocated the strategic containment of communist influence over the globe.[100]

For Kennan, the Soviet Union occupied the space of a threatening Other. His Long Telegram described the "atmosphere of oriental secretiveness and conspiracy" pervading the Soviet government, produced partly by "the very disrespect of Russians for objective truth — indeed, their disbelief in its existence."[101] One year later, in "The Sources of Soviet Conduct," Kennan linked a determinist history and geography to "the political personality of Soviet power":

> These precepts are fortified by the lessons of Russian history: of centuries of obscure battles between nomadic forces over the stretches of a vast unfortified plain. Here, caution, circumspection, flexibility and deception are the valuable qualities, and their value finds natural appreciation in the Russian or oriental mind.

These disturbing, primordial qualities took political shape in the form of a (red) flood, "a fluid stream which moves constantly, wherever it is permitted to move, towards a given goal." What was required to limit this potent flow was a "patient but firm and vigilant containment" that might ultimately lead to a destabilizing impotence.[102]

A longtime admirer of Russian culture's "primitive vitality," Kennan bitterly addressed the faltering American–Soviet relations of the immediate postwar period in a language of psychoanalytic pathology, laced with gendered imagery, which also claimed the high ground of realism. His ordinary Russians were opposed to a cruel, monstrous Soviet leadership, a regime whose lack of reason meant that close diplomatic cooperation was unlikely. Kennan's divisions reflect the dual face of Cold War international relations: the firm application of militaristic rationality complemented by the secondary sympathies of communal sentimentalism. He was also a staunch conservative critic of American society and hoped that some of what he found, or desired, in Russia might be emulated in the emasculated United States. However, these aspects of Russian society were not in demand in America after the Second World War, when Kennan answered calls for a clear synthesis of Russian character and Soviet strategic doctrine. His conclusion that the Soviet Union was not primed for war and was weaker than the United States was bypassed by many influential readers for the more provocative picture of a threat responsive only to the "logic of force."[103]

In the binary system of Cold War geopolitics, communism explicitly suggested the Soviet Union, but it was communism more generally, as an ideology, that demanded American resistance everywhere. In its self-defined position as *the* world power, the United States could exercise containment, at least theoretically, without "clearly conceptualized geographical limitations. Its genuine space was the abstract universal isotropic plane wherein right does battle with wrong, liberty with totalitarianism and Americanism with the forces of un-Americanism."[104] Of course, the significance of containment for many scholars of the Cold War lies far beyond Kennan's theories, resting instead in specific tests of American power. Examples of such applications range from the Truman Doctrine, the Marshall Plan, bilateral aid to Japan, and the formation of the North Atlantic Treaty Organization (NATO) to specific American military interventions in Southeast Asia, Central America, and Africa.

If containment, as conceived by American politicians and strategists starting with Kennan, was originally a diplomatic, economic, or regional program, by the 1950s it had been militarized, globalized, and naturalized. The direct result was a devastating arms race and the shift from limited American vital interests to a geopolitical stage consisting of Earth itself. But this global view marks a divergence from Kennan's hierarchical interests: for him, conflict "outside the industrial formations of 'the northern temperate zone' was actually supremely uninteresting. The third world, in his view, was incomprehensible to the western mind and best left to its own no doubt tragic fate."[105]

Not surprisingly, then, by 1948, Kennan had largely disavowed his connection to the idea of containment. He had witnessed a shift from what was originally a treatise on mechanisms for dealing politically with the Soviet Union to a rapidly globalizing military position against world communism based on atomic might. From Walter Lippmann, who called the concept of containment a "strategic monstrosity," to James Burnham, the Trotskyist-turned-neoconservative who condemned it as passive and lacking in the "properly offensive qualities," the range of public responses to

Kennan was tremendous.[106] What prevailed, however, was a rigid and hostile cartography that displaced and expanded Soviet aspirations onto a world map. "After Kennan," Jim George writes, American "foreign policy analysts had a way of reading the Soviet Union that accorded them the *certainty* they craved."[107] Struggles for national self-determination were reinterpreted in terms of overarching contests for zones of power and ultimately for the globe itself. Kennan denounced the misuse of his work after the American government institutionalized containment as a policy that negated negotiation unless conducted from a position of strength, translating what he hoped would be a "temporary recharge" into a perpetual and accelerating "frozen dialectic."[108] But the resonance of his original formulation rests on its general, rather than specific, qualities: in David Campbell's words, containment was a "strategy associated with the logic of identity whereby the ethical powers of segregation that make up foreign policy constitute the identity of the agent in whose name they operate, and give rise to a geography of evil."[109]

Across the relatively narrow spectrum of Cold War commentary, a preponderance of American politicians and strategists represented the Soviet Union as markedly, irrevocably distinct from the United States—so distinct that in many cases there was little room for diplomatic compromise. Rooted in the foreign status of the Soviet Union was a vague, if potent, threat of an expansionistic communism, always on the verge of leaking out and ultimately enveloping America. Because of such related geopolitical abstractions as the domino theory, the spread of communism, no matter how distant, was *always* a danger to American security. Some cases, of course, were more crucial than others. By April 1950, these beliefs had reached their apotheosis with National Security Council document 68 (NSC-68), a paper prepared primarily by Paul Nitze, Kennan's successor as head of the State Department's Policy Planning Staff. Arguing that the Soviet Union desired and required "the dynamic extension of their authority and the ultimate elimination of any effective opposition to their authority," NSC-68 was filled with dramatic, sinister, and active language: the enemy is said to expound "a new fanatical faith, antithetical to our own."[110] By the time Truman approved it in September, events in Korea seemed to have confirmed the document's conclusions. Kennan, who had never endorsed a universalist policy in which gains for any form of communism represented an equivalent American loss, had been left behind, his original, more particular arguments distorted, to be sure, but also driven toward a logical Cold War permanence.[111] Not only was the United States mired in insecurity and vulnerability, but the passage of time through the 1950s seemed only to heighten danger, turning geopolitically specific and situated circumstances into something approaching a transhistorical state of being.

Scientific Salvation

As American Cold War strategy moved closer to a self-definition as a science, science impinged significantly on the domain of strategy to the point that national security, however defined, was understood to depend on science and technology.[112] This relationship was understood by those who manufactured it to be new. "Until the present century," the engineer, physicist, and Cold Warrior Lloyd Berkner declared in a February 1959 speech, "there has been no obvious, or even very noticeable, tie between science and politics." Berkner's amnesia, eradicating the imperial histories of botany, medicine, and meteorology, to take just three examples, was made possible by the dramatic force of the ties he had observed and had helped to forge during and after the Second World War. For him, the nature of the relationship was clear: science should be "one of the handmaidens to politics—but never a substitute." But when a political value conflicted "with a verifiable scientific conclusion," the roles were reversed, and "society would benefit most by abandonment of that value."[113] The most vexing problems, however, were those that mutually concerned science and politics. These were chiefly those of the atom, where the heavily defended boundary between C. P. Snow's "two cultures" was blurred beyond recognition.[114] This was rarely a friendly partnership. In the two decades leading up to Berkner's speech, the boundaries and purposes of scientific practice had been endlessly debated, challenged, and expanded, frequently in the context of Cold War strategic problems that were anything but objectively understandable.

I am interested here, however, in exactly the image that Berkner conjured: how could science, apparently objective and yet submissive, serve politics, which was at once more fallible and more pressing? If science was, for Berkner, amoral, then its function depended entirely on social contexts. Needless to say, this was the perfect rationale for an *American Cold*

War science—autonomous and objective, yet ready to be appropriated for the right normative purpose and set against ideological and unconvincing Soviet science. Following the sociologist Robert Merton, norms of universalism and disinterestedness could be located in science and its ideal environments, and in the next breath, science could be conflated with American democracy.[115]

Berkner's use of the handmaiden metaphor was perhaps a reformulation of a similar statement in another speech he delivered just three weeks earlier. Nature, he argued, "now gives each of us the equivalent of more than two hundred slaves in the services that energy provides at man's command. We can truly say that science has abolished slavery." How he had arrived at the number was not clear, although it was presumably the result of an unusual equation. But the implication of this statement was that man's attempt to understand his environment, to paraphrase the title of a Berkner publication, included a desire for domination.[116] This controlling impulse is a familiar historical trope. In Berkner's case, though, it was put to work not in reference to a particular site but to a global landscape. A prominent participant in the International Geophysical Year of 1957–58 and a frequent commentator on space exploration, Berkner projected his version of science onto a terrestrial whole—and beyond. But he was quite willing to step aside, or even participate, as science rushed to the aid of a more particularistic politics. If Berkner perceived an irony as American armed forces took a primary role in what he called the "assault on the secrets of the earth," using military technology that permitted a reading of natural clues with powers "far beyond the range of our simple senses," he did not mention it.[117]

My intention is not to single out Berkner, but the opposite: to position him within the booming American scientific research community of the early Cold War and the technocratic and militaristic impulses that accompanied this growth. Focusing on one individual, even one so prominent as Berkner, cannot begin to account for the complexity of relationships among scientists, government bureaucracy, the military, politicians, scientific organizations, and other factions that characterized this period of big science. Much work in the history of science has been devoted to teasing out and connecting filaments of these networks and to discussing the military's influence on period science. Some of this scholarship has a geographical aspect, although it works with a slender definition of spatiality.[118]

During the Second World War, not surprisingly, the military mobilization of American science reached a feverish pitch. In his 1941 President's Report, MIT's Karl Compton, writing from the campus with the most at stake in wartime scientific research, perceived "the outlines of an educational and research institution based upon the present ideals and objectives but incorporating a greatly magnified capacity for national service."[119] Compton's vision was realized: by the end of the war, MIT was America's largest university defense contractor, and it maintained that position throughout the Cold War. The Institute's annual reports of the 1940s and 1950s are rife with exhortations to meet the "inescapable demand... to serve the national defense and strengthen the free world."[120] It was precisely the global character of mid-twentieth-century war that had made state-sponsored, university-based defense research so necessary. A stroll up Massachusetts Avenue from MIT, Compton's counterpart at Harvard, James Conant, was similarly convinced with respect to the shape of science. He formed a Committee on the Physical Sciences at Harvard during the Second World War and joined Compton on Vannevar Bush's National Defense Research Committee. These commitments continued into the Cold War. In a 1947 address, Conant outlined a "special sense in which science is called upon to help out with national problems here in this country."[121] The following year, he endorsed Harvard's "new, more intimate association" with Washington.[122]

To be a modern nation, Gyan Prakash writes in another context, "was to be endowed with science, which had become the touchstone of rationality," and the nation-state depended heavily on "science's work as a metaphor, to its functioning beyond the boundaries of the laboratory as a grammar of modern power."[123] The Second World War lent legitimacy to the coupling of science and nationalism in the United States, particularly that done by scientific administrators and managers such as Conant, Compton, and his successor at MIT, James Killian (inaugurated in 1949), who were all frequent visitors to Washington. The remarkable list of technologies developed during and for the war — radar, proximity fuses, rockets, and atomic bombs — "formed *guiding symbols* that inspired the strategy of much postwar research" and were also the basis of much postwar strategic scrutiny.[124] These "auspicious gizmos" were matched by, and indeed demanded, new scholarly disciplines designed to manage and guide their use and evolution.[125]

In addition, the war led to the reconstitution of government science funding, notably in the creation of the Office of Naval Research (1946), the Atomic Energy Commission (1946), and slightly later, the National Science Foundation (1950). These and other agencies supported a series of large, defense-oriented research laboratories, sites for prescriptive interdisciplinary research situated between government and academia and staffed by what the sociologist William Whyte would famously call "organization men."[126] Prominent refugee scholars enlisted to work on such landmarks of big science as the Manhattan Project remained in similar positions, often with a bridge back to academic life after the war. Some of the younger, less visible technicians required for wartime research graduated from crucibles like Los Alamos to become Cold War experts and advisors. New standards for secrecy and security were also implemented; science was often understood as something done behind laboratory walls and behind national borders too.[127] Many scientists accepted these limitations and "competed for the privilege" of Atomic Energy Commission or Defense Department funding.[128] But science was also an obvious vehicle for internationalism, a prewar trend that struggled on against the constraints of national security through the 1950s. Seen in a different light, science respected no arbitrary state boundaries, and this universalism was a sensible means of overcoming cultural and political difference.

The atomic bombing of Hiroshima and Nagasaki, and the "profound ambivalence" it engendered among scientists, not least among those who had a role in the bomb's creation, spurred many of them toward politics, or at least to a "quiet diplomacy."[129] Physicists, in particular, were suddenly elevated to the status of magi, yet they had also, S. S. Schweber writes, "tasted sin," and "in the new world they had helped create, they could no longer isolate themselves in their ivory towers from the affairs of the nation."[130] Historians have ably documented this activism, the resulting persecution of many scientists (notably J. Robert Oppenheimer but many others as well), including the symbolic restriction of their travel and the failure of efforts to achieve international control of atomic energy — for a variety of reasons, from American bureaucratic intransigence and anticommunism to Soviet espionage and heavy-handed advocacy of state science.[131] These histories parallel related discussions of the relationships forged between science and government during the 1940s and the resulting constraints placed on scientific research, constraints that paradoxically signaled the authority and opportunity gained by proximity to power.[132]

Scientific internationalism carried at its core a group of common ideals: having created the modern world, science was "universally valuable," removed from cultural and political concerns, anathema to security restrictions, and as a cosmopolitan lingua franca, a source of global enlightenment.[133] For Norbert Wiener, the MIT mathematician and cyberneticist who had worked on guidance and tracking systems during the war, the use of the atomic bomb as a weapon indicated that scientists had been betrayed. As a result, he refused to provide a copy of a 1946 paper "relevant to guided missile technology to the U.S. Air Force."[134] Science, for Wiener, should not have relinquished its more benevolent authority to untrustworthy individuals and organizations with distorted agendas.[135] In the context of the United States during the 1940s, this longstanding and rather naïve conviction was also a moderate form of dissent. It presented science as a hermetic source of reason that could spread from sealed laboratories to cover the earth, eventually and benignly influencing social relations. Under inappropriate control, on the other hand, the results could be disastrous. Such was the vision advanced in the tellingly titled *One World or None,* a best-selling paperback published by the Federation of American Scientists in 1946. Inside, expert after expert, staggered by the destruction of Hiroshima and Nagasaki, detailed the dangers of atomic energy, although few could elucidate a coherent response beyond the desperate need for international control or even world government.[136] But whether the earth was headed down a path of order or disorder, it was clear in texts such as *One World or None* that the most important aspect of international relations was the role of science. The choice, precisely equivalent to the Cold War geopolitical binary, was one that would rest on the use of systems of research and knowledge production: Was science an international ideal or a national tool?

The quest for global scientific governance was not strictly that of marginal dreamers. Even policymakers such as Vannevar Bush and James Conant were convinced of the need for scientific internationalism. While they might have differed on the means, such conservatives shared a concern for global stability and a faith in the transformative power of science with intellectual colleagues who espoused alternate political values.

Conant and Bush were, after all, also prominent scientists. The communal, universal status of professional science was reaffirmed in the title and substance of Bush's 1945 report to the American president, *Science: The Endless Frontier*, despite its appeal for the national pursuit of autonomous, long-term, basic research.[137] "As early as 1937," Bush's biographer notes, "Bush had said that the quest for knowledge could replace the vanishing geographical frontier as the new source of American freedom and creativity."[138] In its inheritance of the modernist crown as a result of wartime European devastation, the United States "had become not just the center but the *sanctuary* of science."[139]

The space of inquiry for this dominant American science was literally that of the universe and, more practically, the planet. But just as *Science: The Endless Frontier* was a distinctly national vision of scientific progress, the disappearing frontier that Bush evoked was not a well-traveled globe but Frederick Jackson Turner's Western boundary zone central to American national mythology. According to the 1952 report of a national security commission chaired by Karl Compton, "the United States, and, indeed, the whole world, lives in a frontier environment," a condition of unpredictability that "*demands a frontier response.*"[140] Alongside a "shrinking of the areas of unknown territory," Compton noted similarly in an earlier speech, new forms of frontiers had been opened "for exploration, where less crude and more technical methods have been developed." These scientific spaces, which lent themselves to "quantitative measurement and technical evaluation," were also areas of battle "in which our maps and knowledge of terrain are far more scanty than in any other major conflict in which our country has been engaged."[141] Again, such language was precisely equivalent to the concurrent shift under way in geopolitical discourse. As the list of frontiers grew to encompass (and surpass) the globe, they were also abstracted, shifted from environmental or territorial bases into the safer realms of science and strategy.

James Conant's opinion of science — a singularly influential one — shared much with the principles of political realism. The immensely destructive aspects of science were accepted, Conant wrote in 1947, as "the price we must pay for health and comfort and aids to learning."[142] There was no separating the two. Moreover, in a divided world, criticism of state science policy was risky and incompatible with a stable status quo. At most, it was a challenge for administration, and criticism should not continue once a policy decision had been made, because that would imply weakness. This was a Second World War model that seemed to hold in the early years of the Cold War when theories of scientific planning and organization proliferated and when Conant, Bush, and a select few others solidified their positions as the first managers of American big science. Bush himself was fond of reminding audiences that science was decidedly antidemocratic.[143]

A key example of Conant's convictions was his role in the creation of two immediate postwar defenses of the atomic bomb's use, one written by Karl Compton and the other by former Secretary of War Henry Stimson. Along with Vannevar Bush and Paul Nitze, Conant also went on to take charge of the Committee on the Present Danger, an advocacy group of establishment figures set up to warn the public of the Soviet threat and support the implementation of NSC-68.[144] Given this background, it is not surprising that Conant was also fond of transferring military language into the domain of science. Created in 1947, his "Nat Sci 4" course at Harvard sought to instruct students in scientific "tactics and strategy."[145] He believed that the future of science—and modernity—was synchronous with the future of the United States and that, if correctly conducted, science would defend against a looming dark age.[146] As with strategic education, scientific instruction for the public was crucial to this defense, but such lessons were unidirectional.

Contrary to some pronouncements, America had not become scientific; science had become American—and strategic. Even if viewed as apolitical, science had been harnessed cleverly to Cold War policy, resulting in a poor internationalist track record notwithstanding the valiant efforts by certain American and Soviet scientists to seek out commonalities and reduce oppositional tensions. The strategic character of American scientific pronouncement, on the other hand, was epitomized by President Eisenhower's Atoms for Peace address to the United Nations General Assembly on December 8, 1953. Speaking in the "new language of atomic warfare," Eisenhower admitted that his recitation of danger was American in orientation but insisted that the subject was global in character. Following a United Nations suggestion, Eisenhower proposed the formation of an "international atomic energy agency" to which the Soviet and American governments would contribute fissionable material. The agency would then allocate this material "to serve the peaceful pursuits," or the "needs rather than the fears," of humans, such as agriculture, medicine, and energy.[147]

Peace was a word used to restrain troubled allies with whom the United States might share valuable scientific information, and it was also used to placate the anxious American public. Having indicated to the world that the United States remained superior in strength, Eisenhower hoped to get the Soviet Union — which he described, to Winston Churchill, as a "woman of the streets" who needed to be driven "off her present 'beat'" — to accede to stockpile reductions that would further America's relative position in the atomic arms race. Citing the universal benefits of science, Eisenhower actually attempted to score a psychological victory by placing the onus on Moscow. Converting the bomb into a global symbol of progress while maintaining an acceptable level of fear at home through "apocalypse management" and the perpetuation of antagonism, Eisenhower hoped to gain an upper hand in the important definition of Cold War enmity. But this was a precarious discursive construction: Eisenhower could not avoid including the United States, along with the Soviet Union, in front of what he called "the dark background of the atomic bomb."[148] As peace became war, national security became national insecurity, reflected in Eisenhower's New Look strategy, which was premised overwhelmingly on air power and nuclear deterrence in the form of massive retaliation. Eisenhower's Atoms for Peace speech and similarly duplicitous Soviet proposals ensured the perpetuation of Cold War militarization and promised diplomatic negotiations that were set up to fail. And by affirming the mythic might of atoms in the "peaceful uses of atomic energy" campaign that followed his presentation, Eisenhower also reminded listeners that "good atoms" were dialectically related to the bad. Whether utopian or dystopian, the atom carried "transmutational power."[149]

In 1950, Lloyd Berkner submitted a report to Secretary of State Dean Acheson justifying efforts to create a scientific attaché program in Acheson's department. In a secret appendix, Berkner noted that science could be, in the tense peace of the early Cold War, a very useful — and presumably more innocent — contributor to the process of intelligence gathering. His conclusions were optimistic, although by 1952, the determination that the State Department and the Central Intelligence Agency (CIA, with its recently formed Office of Scientific Intelligence) were failing at that very task had resulted in the creation of the National Security Agency. The science attaché program ran afoul of McCarthyism and was resurrected after the Soviet Union launched its *Sputnik* satellite in 1957, an event that also led Dwight Eisenhower to create a Science Advisory Committee headed by an assistant to the president for science and technology. James Killian went on leave from MIT to fill the position. While scientific internationalism was never monolithic, and while many scientists and politicians "found the abrupt integration of science into U.S. foreign policy unnerving," the power of arguments for the utility of science in a global Cold War and the willingness of many prominent scientists to support and seek what Berkner called "scientific intelligence" is historically unquestionable.[150] According to Walter McDougall, "scientific internationalism metamorphosed from the only alternative to Cold War into one of America's mightiest weapons in that Cold War."[151]

Exhibiting Culture

As with both geopolitics and science, American culture at midcentury was also frequently presented as a global, holistic force with a nationalist undercurrent. "Common universalist enthusiasms," David Hollinger writes, invariably contained "particularist biases."[152] From the Universal Declaration of Human Rights, adopted by the United Nations in New York on December 10, 1948, to Edward Steichen's much-discussed and successful *Family of Man* photographic exhibit, with its overt references to the United Nations and the horrors of war, the equality of human experience was asserted repeatedly in the years after the Second World War as a response to modern totalitarianism. Steichen described his project, first displayed in 1955 at New York's Museum of Modern Art, as "a mirror of the universal elements and emotions in the everydayness of life — as a mirror of the essential oneness of mankind throughout the world." Many of the images, taken by a variety of photographers, portrayed human–environment relations; nature and the natural quality of man was used as a constant.[153]

An early exponent of aerial photography, Steichen worked for the Army Signal Corps in World War I, and during the Second World War, he led a Navy photographic unit in the Pacific. His intention on the latter mission was to capture "the actuality of war" by documenting its humanity through images of individual soldiers. That these subjects instead behaved

as if they were on stage is one of the great ironies of military photojour-
nalism, and during World War II, Steichen's work was a small part of a
massive documentary effort. Like the concurrent proliferation of maps,
photographs of the Second World War were key vehicles for understanding unfamiliar landscapes—the stages of war. Using maps, photographs,
and other sources, domestic observers could travel with the war and in the
process arrive at an improved sense, positive or otherwise, of America's
new international responsibilities. But as the globe returned to America
in exhibits such as *The Family of Man* (which featured the work of many
Second World War photographers), an American globe was promoted
internationally. The United States Information Agency (USIA) "toured the
photographs throughout the world in five different versions for seven years
after the closing of the original display." As it was first exhibited, however, Steichen's collection targeted middle-class Americans, who were not
only driving economic growth but who were also the consumers of much
of the commentary on American culture, whether in the academic prose
of a developing American studies discipline or in the pages of influential
periodicals such as *Life*.[154] With its connection to middlebrow America
secure, *The Family of Man* could be carried abroad as another expression
of American global interest.

Henry Steele Commager's *The American Mind* (1950), one of the more
influential period texts on national character, captured the uneasy global
position of the United States by once again giving the nation an individual
human identity:

> Although still persuaded that his was the best of all countries, the Ameri-
> can of the mid-twentieth century was by no means sure that his was the
> best of times ... he was no longer prepared to insist that the good for-
> tune which he enjoyed, in a war-stricken world, was the reward of virtue
> rather than of mere geographical isolation. He knew that if there was
> indeed any such thing as progress it would continue to be illustrated by
> America, but he was less confident of the validity of the concept than at
> any previous time in his history.[155]

Such challenges to the unique and unifying American personality were
reflected in the postwar obsession, both critical and explanatory, with the
definition of America as a historical and cultural entity. The institutional-

ization of American studies, a field typified by Commager's tract, arrived at a moment of immense American power but also, as Commager, Kennan, and many others claimed vociferously, one of vulnerability. And one way to redeem the nation's timeless traits was to appeal, as Steichen did, to the universal threat of Armageddon. Choosing images with minimal social context, Steichen presented a sentimental humanism, or mass-mediated modernism, that was immediately contested from the right for its reification of the exotic and primitive and from the left for its Western superiority.[156] These critiques continued, unabated, into the 1960s and 1970s, but such was the definition of the middlebrow, which became, whether in art, jazz, or commentary, a crucial Cold War export.

The classification of popular cultural exhibitions such as *The Family of Man* was an exercise in cultural cartography and naming that lent the taxonomist an aura of authority. And while *The Family of Man,* with its relatively muted American ethos and antinuclear stance, was not a perfect propaganda vehicle, this did not stop the USIA from reconfiguring it slightly to match regional circumstances for the purpose of selling America in competitive overseas markets.[157] As a result, the exhibit became a rather innocuous Cold War vehicle, an impassioned statement of universalism that bore a "made in America" tag. Among its destinations was Moscow's Sokolniki Park, as part of the American entry in a summer 1959 exchange of exhibitions; the sprawling U.S. display followed a Soviet show held in the New York Coliseum. These exhibitions — mostly of shiny commodities and technological models that overwhelmed the increasingly timeworn photography of Steichen's display — served as temporary, timid invasions, for the goods on display were not innocent metal and plastic objects but ideological projectiles inserted with care and deliberation into hostile territory. Ideas were being presented in the form of artifacts. Using typical period diction, *Newsweek* dubbed the consumerist clash "a contest of two diverse ways of life — of modern capitalism with its ideology of political freedom and Communism."[158] Although the history of American consumerism is much older than, and distinct from, the Cold War, *Newsweek*'s description neatly captures the post–World War II emphasis in the United States on the "purchaser as citizen," supported intellectually by the hierarchy of modernization theory in which a "high mass consumption stage" was the pinnacle of progress.[159]

One major American weekly noted that the Soviet display in New York produced a comprehensive picture of "Russia's position in the world."[160] Industrial units, cars, fashions, and a model two-bedroom apartment supplemented a scale reconstruction of *Sputnik I*. Domestic media sources, from *Newsweek* to *Time*, derided the "propaganda," in the form of "representational paintings glorifying the joys of Communist life," only slightly less than the "clumsy" and dated attempts to copy Western consumer goods. Inside the apartment, "the paint easily rubs off the prefabricated walls. The furniture is frail and imitative . . . the stove is so small that the oven would cramp a large chicken."[161]

Despite these many faults, noted *Time*, "the fact was that [the New York exhibition] mirrored not life today but a combination of genuine achievements (e.g., in the sciences) and a happy dream of the future."[162] Citing similar "wishful dreams," *Newsweek*'s Henry Hazlitt also produced "real facts on output," leading him to label Russian claims of equality in living standards "ludicrous." Nonetheless, he warned,

> we should carefully distinguish between production for peace and production for war. In the latter Russia has made giant technological strides — precisely because she has put that goal first. And in propaganda she is enormously our superior. She can put on an exhibition that gives false impressions of merit, whereas our own exhibition at Brussels exhibited and apologized for our slums, and the new one at Moscow will have a painting lampooning our generals.[163]

For some in the media, the truthful projection of American culture and democratic values outside the boundaries of the state served as a hindrance in the ideological battles of the Cold War.

In Moscow to officially open the American exhibition, Vice President Richard Nixon participated in a series of remarkable confrontations with Soviet leader Nikita Khrushchev across the park grounds. Motivated by a set of active film equipment, the two politicians, surrounded by a large crowd of male aides and journalists, began to debate. Their words were shaped by the stages through which they traveled, from an impromptu television studio to the kitchen of a $14,000 model home that, according to Nixon, "any steelworker [could] afford."[164] It was in this quintessentially domestic space, framed by both the latest American appliances and a woman commissioned to demonstrate their features, that the famous Kitchen Debate transpired:

Khrushchev repeatedly predicted that the Soviet Union would soon overtake the American economy and win the brass ring of world influence. Nixon preferred to focus on the fetishism of commodities and tried to steer the conversation to color television sets. The vice-president's persistence paid off when the debaters reached the model kitchen, and washing machines became apropos. Nixon praised the freedom of choice among American housewives. Khrushchev countered that one kind of machine would be sufficient, if it worked. "Isn't it better to talk about the relative merits of washing machines than the relative strength of rockets?" Nixon inquired. "Isn't this the kind of competition you want?" The Soviet premier angrily replied that America was pursuing both types of competition.[165]

The heavily gendered conversation concluded with these words from Khrushchev: "Thank the housewife for letting us use her kitchen for our argument."[166]

The Kitchen Debate affirms two connections that must necessarily be made in any discussion of Cold War geography: the masculine culture of Cold War diplomacy and the extensive links between global diplomacy and domestic spaces — the variants associated both with the nation and with the home. Venturing into the kitchen, with its silent model, Nixon and Khrushchev reaffirmed stereotypical gendered roles and positioned themselves (and one another) as protectors of their national women and feminine domestic spaces. Later, at a separate $250,000 electronic kitchen staffed by a similar model, Nixon pointed to a panel-controlled washing machine, noting that "in America, these are designed to make things easier for our women," and stressed the universal benefits of such an "attitude toward women." That evening, searching for an agreeable subject to celebrate, the two leaders found one: "We can all drink to the ladies."[167] The "we," of course, represented an overwhelmingly male group of politicians and media.

The two types of competition identified by Khrushchev are more usefully understood as two fronts produced out of a single source. From General Electric and Goodyear to Westinghouse, a number of large American companies controlled the mechanisms of Cold War production *and* destruction, often manufacturing similar devices for the home and the military in the same industrial plant. The consumer goods praised by Nixon represented basic American tenets: the housewife's freedom from labor and the democratic opportunity to select a suitable appliance from expansive catalogs. Both geopolitics and science, staged as technology, were thus usefully expressed in the terms of cultural consumption and competition. In his remarks at the Moscow exhibition's opening ceremonies, Nixon claimed that the goods on display revealed America's proximity to "the ideal of prosperity for all in a classless society" — a clear dig at his hosts.[168] This, Scott Saul notes, conformed perfectly to what Isaiah Berlin, in an essay published a year before the Kitchen Debate, had termed "negative liberty," namely a conception of freedom as a private choice, "holding the totalitarian threat in abeyance." If NSC-68 had committed the United States to "the idea of freedom," it was primarily a freedom of enterprise.[169]

Other American entries in the Moscow exhibition were also revealing. One of Buckminster Fuller's geodesic domes, with a diameter of 200 feet, stood prominently as a symbol of American capitalism and iconoclastic ingenuity.[170] Walt Disney contributed a circular movie theater.[171] A six-room prefabricated home was similar to the structures being built for the ongoing suburban boom. Suitable middle-class families were chosen carefully as models and participants. IBM's RAMAC computer was set up to answer thousands of questions about the United States and, notably, tally the most popular choices. A variety of new automobiles were placed next to homes, camping grounds, and farm displays. Unlike the materiel of heavy industry and science brought by the Soviets to New York, the familiar flowers of ordinary American abundance were in full bloom at the Moscow grounds. But individuals and goods were secondary to a civilizational clash that had subsumed the historical anthropology of *The Family of Man* beneath the futurist outlook of consumerist capitalism and new military technologies.[172]

Timothy Mitchell has argued convincingly that nineteenth-century world exhibitions staged in European capitals were sites for the reproduction of "imperial truth and cultural difference in 'objective' form."[173] While populated with very different artifacts drawn overwhelmingly from the vital centers of two superpowers, the Cold War exhibitions of 1959, supported by impressive contingents of visiting dignitaries, created a similar version of representative certainty. As propaganda, in other words, they depended paradoxically on the claim of realism. On the exhibition circuit, the battle for ideological supremacy was frequently contested in *deferred space,* outside the territorial boundaries of either state. If the complementary Moscow and New York exhibitions represented a picture of American and Soviet life, respectively, it was by no means guaranteed that one exhibitionary order would triumph over the rest of the globe. What visitors to the New York Coliseum and Sokolniki Park experienced were two uncannily similar models of a world-as-exhibition, spectacles suggesting that, if uncontested, the way of life on display would (and should) envelop the world.

America, Modern

On January 25, 1947, Simone de Beauvoir left France for several months in the United States. Her journey was a temporal and spatial movement from old to new, a passage into an "autonomous, separate world" — into the future. Conversely, her return to Europe was a depressing one, full of old customs officers "in their crumpled uniforms," poorly dressed, "humiliated" Parisians walking the "dreary," "numbed," "dark and morose" streets. "Over there in the night, a vast continent is sparkling," she concluded, and from this contrast, it is clear that she had witnessed the beginnings of a new modernity in the bright, brash landscapes of the United States.[174]

To be in the United States after the Second World War "was not only a matter of taste, but to live where the future was unfolding: to live now what others would soon be forced to live."[175] This did not only apply to the many émigré scientists and strategists who had found new homes within military research communities. As Serge Guilbaut has documented in some detail, the "cultural center of the West" shifted from Paris to New York after the War. At the time of de Beauvoir's visit to the United States, art and other forms of culture were becoming significant factors in developing American global views. Guilbaut goes further to argue that the increasingly depoliticized creations of avant-garde abstract expressionists (the "New York school") stood as perfect symbols for the washed-out culture of the "new liberalism."[176]

The connection between art and the atomic age was made explicit in a 1946 *Fortune* article that dramatized the atomic tests at Bikini Atoll by placing photographs alongside two abstract paintings by Ralston Crawford, an artist and, during the Second World War, chief of the Visual Presentation

Unit in the Army Air Force's Weather Division.[177] The justification for this inclusion, Guilbaut argues, was that the bomb had revealed an additional individual powerlessness:

> Traditional languages, including the languages of maps and graphics, were no longer capable of giving full expression to the realities of a nuclear world. Only abstract art could communicate the new meaning of human experience, the incredible feeling of total disintegration.[178]

The *Fortune* piece was published at a moment when "political debate" was still a meaningful phrase, but, as with the implications of Hiroshima for science and scientists, the latter half of the 1940s was, for many writers and artists, a time of "deep-seated confusion," "impossible alternatives," and a retreat into empty mythological primitivism. Existentialist alienation became an American "way of being."[179]

While the midcentury embrace of a centrist liberalism was propelled by an opposition to the worst forms of totalitarianism, it also implied that an overwhelming number of Cold War intellectuals either supported or were yoked to an invigorated "containment culture" of consensus. Theirs was a serviceable elitism, which, while certainly opposed to Joseph McCarthy's xenophobic populism, also preserved the "channels of power through which intellectual authority is exercised" and limited overly threatening manifestations of direct democracy or mass culture. Abstract expressionism, for one, was staunchly backed by New York's Museum of Modern Art (itself supported by the Rockefeller family), which organized international tours of paintings by artists such as Jackson Pollock and Willem de Kooning. The universality of their explosive, seemingly apolitical canvases became an ideal statement of national culture.[180]

The narrowing of modernism's adversarial qualities after the Second World War and the incorporation of the avant-garde into a liberal geopolitical order was not limited to a select group of New York artists and writers.[181] The achievements of the CIA's cultural propaganda program in Europe, which depended heavily on a consortium of intellectuals who ran journals, exhibitions, and conferences, has been recently documented by a number of historians. Whether they were aware of their links to CIA psychological warfare or not, these intellectuals gladly cast aside their nearly ubiquitous leftist roots and went to work for America, the "new Weimar."[182]

Their influence surpassed the limited audiences of cultural criticism. In 1953, the Congress for Cultural Freedom, the CIA's main vehicle in such efforts, sponsored a substantial multinational meeting on Science and Freedom as part of the continuing attempt to "establish the disinterested search for objective truth as the distinctive epistemological posture of the Free World."[183]

The language of an American cultural citadel in an uncertain world, a threatened realm burdened with an imperial mission of incorporation and influence, was also, of course, precisely that used by Cold War geopolitics and science. The dangers of nuclear annihilation, according to Hanson Baldwin, had to be "localized" in order to solve "the problem of preserving and perpetuating American man and Western culture."[184] Modernism's slide into apolitical abstraction did not exactly require an existentialist cartography. Just as abstract art could be used to wage the Cold War, in international relations and scientific advocacy, the dizzying dreams of a harmonious atomic globe wore off quickly, and anxieties accompanying new risks were increasingly portrayed in the imagery of national militarization. What was startling was that the brush strokes remained the same.

2 REGIONAL INTELLIGENCE
The Militarization of Geographical Knowledge

> In the recent war most of the belligerents compiled encyclopedias
> on countries they were contending with or which they planned to
> occupy or otherwise swing into their orbits. These encyclopedias
> should be conceived of either as a large file of knowledge in folders
> in a filing cabinet or in some sort of finished book form....
> Their basic aim was to provide the strategic planner with enough
> knowledge of the country in question to make his over-all
> calculations on its attributes as a zone of combat.
>
> —SHERMAN KENT, *Strategic Intelligence for American World Policy*

For many Americans, the Second World War prompted tremendous interest in previously ignored parts of the world. This attention resulted in a quest for geographic knowledge that frequently rationalized and familiarized foreign landscapes in the language of military interests. Because every place, and every type of place, possessed a potential wartime purpose, the globe and its geographical parts were present in both the schematic language of grand strategy and more intricate discussions of intelligence gathering and war planning. Networks of research were therefore established for the systematic study of *regions*. These spaces—some clearly more important than others—became testing grounds where the ambitions of American militarization and social science could be localized.

Those who campaign for new intellectual approaches invariably define themselves against an earlier generation of theories and individuals. While artificial, this division is nevertheless potent. To recognize and question such differentiation is also to consider the artificiality of disciplinary boundaries. Consider that before World War II, economics textbooks were overwhelmingly descriptive. Over the second half of the twentieth century, they became crowded with graphs and equations, resembling more of a "discourse on method" for an increasingly "technical subject."[1] A similar and concurrent shift also characterized the coalescing area studies movement, whose advocates quietly borrowed certain worthwhile techniques from their prewar predecessors while disparaging these ancestors as insufficiently theoretical.

Within area studies, there was little overt appropriation from professional geographers, even though regions had long been one of their central concerns.[2] "Much of human geography before the 1960s," Tim Cresswell notes, "was devoted to specifying and describing the difference between areas of the earth's surface."[3] Regional stereotypes were certainly prominent in imperial imaginations. But the discipline of geography as of the Second World War was believed by the champions of area studies to lack suitable rigor, methods, and coverage. The resulting absence or marginalization of geographers in many area studies initiatives is reflected in the insufficient treatment of area studies by historians of geographical thought. This avoidance perpetuates narrow impressions of geography, leaving certain roles and variants of regions understudied.

The regions of concern in this chapter were not the rural *pays* or urban zones familiar to many geographers, but *strategic areas*, crucial components of the overlapping global views detailed in chapter 1. While their novelty was not absolute, since regional field experiences, in particular, were still valued, the new regions of American area studies were certainly larger, more convenient (or taken for granted), and more easily translatable into the overlapping vocabularies of social science and foreign policy. Many geographers eventually caught up to and even surpassed the sheer functionality of the world maps offered by area studies, developing a regional science that was so systematic as to be severed from the limits of local context. Intriguingly, these abstract aspirations were also compromised, wittingly and necessarily, by the Cold War. But the crucial significance of area studies was that at its most powerful it presented authoritative and yet strategically useful divisions of the human world.

Strategic regions owed their intellectual power to military imperatives in the sense that the production of spatial knowledge on the world's life is inextricable from the technologies and techniques of death.[4] As Isaiah Bowman, undoubtedly the most prominent American geographer during the Second World War, phrased it in his 1943 President's Report at Johns Hopkins University, the "'imaginative grasp of space' which science shares with poetry seemed somehow to have been impossible to attain until our Army, Navy and Air forces had taken their stations and begun their operations in almost every part of the world."[5] By suggesting this connection, I do not mean to indict individuals for their participation in various aspects of war—although stridently pursuing the opposite tack would also be foolhardy. Instead, my aim is to draw out the larger relationship between militarization and geographical knowledge. Similarly, while it would be wrong to conflate the drive for regional knowledge during World War II with Cold War area studies, it is also impossible to separate the two. The lingering longevity of categories and modes of understanding developed during a specific emergency, at the regional or any other scale, presents an intriguing historical puzzle.

A 1943 Social Science Research Council (SSRC) report, "World Regions in the Social Sciences," suggested that social scientists with regional knowledge were almost as important, and as rare, as officers familiar with combat in the same zones and that the development of effective area study was dependent on the "establishment of research and graduate instruction." Those who organized such programs, however, needed to counter a perceived absence of language abilities and cultural awareness not only among recruits and the public at large but within the academic community as well. American intellectuals apparently lacked regional knowledge, but they were also short of the skills required to produce and evaluate this knowledge. The "enlarged spatial concepts" and more comprehensive wisdom required of America could, the report claimed, be found by reducing, or focusing, the laws of the social sciences to a manageable scale. In this respect, regional study would operate like a case method, reducing the sinful "temptation toward vague generalities" and discouraging the lamented compartmentalization that beset disciplines.[6]

The SSRC report, written by staffer Earl Hamilton, was the first publication of the council's Committee on World Regions, which was established in January 1943 and continued after the war as the Committee on World Area Research. The document, distributed to intelligence leaders such as William Langer of the Office of Strategic Services (OSS), included a description of "social laws as relative to time, place, and circumstance" and noted the resulting "precision" gained by turning to intensive, comparative "study of concrete areas." In council discussions of Hamilton's paper, the psychologist Albert Poffenberger argued that the rooted nature of the social sciences distinguished them from independent physical sciences and their homogeneous laboratory environments. The social sciences, Poffenberger added, should seek a similar global generalization so as to affirm tentative regional laws. This contribution, he believed, could extend well past the immediate context of wartime problems.[7] Hamilton's brief manifesto and the debates that followed its presentation contained early versions of arguments that were echoed and expanded within the area studies community for at least the next ten years. But the calls for spatial focus *and* generalization were particularly suited to a military crisis.

The apparent lack of cosmopolitan experience among Americans unprepared "for life in the spherical world," a condition invoked endlessly by various pundits, was by 1943 already driving efforts to accumulate scraps of information, rosters of the well traveled, and most importantly, interdisciplinary teams of the few scholars judged to be competent.[8] These components were all assembled in a cluster of wartime clearinghouses, sites for the collection, dissemination, and evaluation of war-related data. The establishment of these archives, which were both physical and imaginative, was prompted by the belief that almost any photograph, article, narrative, or statistic pertaining to a particular theater of combat might just be useful — precisely the imperative that lies at the heart of the intelligence industry. The scope of the Second World War, after all, was such that the relevant organization might receive one question concerning the fireproofing of Alexandria's cotton warehouses, followed by a demand for names of people familiar with navigational conditions off Northeast Siberia.[9] "The problem," writes Mike Featherstone, is never "what to put in the archive, but what one dare leave out."[10]

The inability of an individual to keep pace with the accumulation of what Georg Simmel, in 1911, called "cultural elements" and the temporary, fleeting attempts to confine a world in a single space or picture are central to the condition of modernity.[11] Though not quite at the scale of the Library of Babel fictionalized so brilliantly by Jorge Luis Borges, the attempts to compile, coordinate, and classify geographic information during the Second World War were audacious.[12] While undoubtedly affiliated with earlier, imperial schemes of comprehensive, hierarchical knowledge, the archival efforts undertaken during World War II were at once more temporary and more advanced insofar as they were linked to an extraordinary assembly of scholars who had gathered in Washington, D.C., ready to serve the state. As Thomas Richards argues, "More than any previous conflict, the Second World War was a war of archives entailing a massive material, technical and instrumental investment in knowledge."[13] Despite the war's conclusion, the return of most academics to their respective universities, and even the closure or division of wartime institutions, much of the geographic data compiled during the war, and the systems established for its collection and use, lingered on in various Cold War guises, particularly in the area studies institutes that multiplied after 1945. Any history of the Cold War's regional intelligence must therefore consider a number of Second World War precedents.

Managed from several Washington, D.C., locations, but including many other sites, the American *war on areas* radically reshaped and empowered regional study. This conflict was premised on violent cartographies.[14] Its foreign regions were, in effect, military fields constructed by the physical activities of soldiers and espionage operatives moving across landscapes but also by more ambitious and abstract consideration conducted from the distance of domestic locations in Washington and on American campuses.[15] Long before the study of environmental security became fashionable, academics in the human and natural sciences were compiling and considering information on the world's places in the language of military danger, with an eye to overcoming natural and cultural threats through forceful methods but also through the authority of knowledge.

In the cases considered in this chapter, strategic regions, even those treated as extravagantly wild, were disciplined alongside — often before — their invasion or occupation. Such ordering was made possible not only by systematic storage and classificatory methods but also by factual selection and carefully chosen descriptive terminology. And although certain areas were more likely targets for military activity, it was crucial that *any* area could be fitted within a filing cabinet, textbook, or handbook using a singular methodology. Here we find not only an echo of earlier empires but two additional links to critical histories of geography, histories in which geography is treated as both a discourse of discovery and a more material practice of spatial articulation — accounts, in other words, connecting "histories *of* geography with historical *geographies*."[16] These links are geography's close relationship with militarism, an oft-noted but poorly elucidated affiliation, and the contribution of applied work to geography's intellectual trajectory.

The emergence of the United States as an earth-spanning power was emphatically not dependent on a "quintessentially liberal victory *over* geography." As I argued in the previous chapter, globalism was inherently geographical, its spatial limits and origins masked by certain types of abstract vision that justified the expansion of American authority around the world. The challenge is to identify this "lost geography" and then to complicate it without losing hold of its tremendous authority.[17] In the same vein, this chapter considers the resolute acquisition and construction of regional intelligence during the Second World War, but with an eye to its very specific practical potential. Geography at the Ethnogeographic Board, the OSS, and elsewhere was not so much lost as it was narrowly conceived in militaristic terms, a conception that is at once understandable and, considering the postwar epilogue to this history, unfortunate.

Surviving Wild Places

Early in 1942, the secretary of the Smithsonian Institution, Charles Greeley Abbot, appointed a War Committee in the belief that total war demanded "accurate knowledge of obscure peoples and places and other subjects chiefly of academic interest in normal times." Among the responsibilities of this committee was the preparation of a series of *War Background Studies*, which filled the "real need for authentic information" on the marginal but suddenly vital cultures and regions Abbot identified.[18] Dominated by studies of Asia and the Pacific Islands, these brief monographs also included more abstract discussions of national evolution and the inevitability of war from an anthropological perspective. A file of illustrations from Smithsonian publications and other technical journals was also compiled and made available to interested agencies. Press releases on the various people and places participating in the war were issued, and the institution set up exhibits on the Axis powers for soldiers and the public. Finally, and crucially, formal liaisons with military intelligence units were established.[19]

According to the Smithsonian's 1945 *Annual Report*, the twenty-one *War Background Studies* were the most significant publication initiative of the institution: over 600,000 copies were printed.[20] They were used to train soldiers and contributed to military rule and the various activities of the OSS. One Smithsonian employee recalled that the studies "were supposed to be advisory on how to live with people when they were liberated, or before they were liberated, explaining what the beliefs and the mores of the people were and how to conduct yourself in ways that ingratiated you and your colleagues with their ideas of living, and so on."[21] They were guidebooks to governmentality that encompassed not only those governed but also those governing. This "advice" sat comfortably alongside broader geographical claims; one study on Micronesia and Melanesia described these "distant South Pacific islands" as having "again suddenly been brought into focus."[22]

None of the individual studies, however, had the impact of a single volume produced by the Ethnogeographic Board, the organization hastily established inside the Smithsonian in June 1942 at the behest of the National Research Council (NRC), the American Council of Learned Societies, and the SSRC, and partially funded by the Carnegie Corporation and the Rockefeller Foundation.[23] Titled *Survival on Land and Sea* and distributed in the hundreds of thousands to the armed forces, the 187-page booklet, condensed enough to accompany a soldier on foreign duty, had a simple purpose: to relate, via the experiences of "men who have actually lived in jungles, deserts, and in arctic regions, . . . the main things that a man should know about living in wild countries."[24] While the gendered language is hardly surprising, the invocation of untamed places suggests a preoccupation with the control and defeat of hostile environments through the application of knowledge compiled by experts and then passed to soldiers. This was a familiar imperial obsession, but much had also changed by the Second World War.

Although the long histories of colonial travel, occupation, and the geographic limits of the Enlightenment all provided precedents for *Survival on Land and Sea,* just one example of a burgeoning military literature on endurance issued during a war of extraordinary mobility, the strategic value of certain environments was new. So was the degree of systematic knowledge mobilization that buttressed such publications and the simultaneous American presence in a dizzying array of conflict zones. These novelties were captured in a fascinating 1943 *National Geographic* article titled "Fit to Fight Anywhere," which documented not only the mental and physiological predicaments of soldiers warring "from hot tropic swamps to cold Aleutian fogs," but also the various attempts to recreate "strange conditions" in climate and fatigue laboratories.[25] Following these scattered exploits, a 1942 *New York Times* editorial conceded, required "a really strenuous effort of the imagination."[26]

Capturing the *experience* of spatial difference — what in one context has been called "tropical hermeneutics" — led to a disciplining of interpretations through intellectual categories and distinct missions.[27] While nontemperate environmental conditions were being simulated at home, "combat scientists" were descending on the spheres of war.[28] Under the leadership of the scientist and administrator Vannevar Bush and MIT president Karl Compton, the wide-ranging Office of Scientific Research and Development created an Office of Field Service (OFS) in October 1943. Soon a Pacific-focused operation, the OFS attacked the "tyrannies" of transportation, communication, climate, insects, and distance, sending scientists into jungles and between islands in search of solutions.[29] Many returned with specimens and data snatched up eagerly by the Ethnogeographic Board and other agencies.

Scientists were not the only scavengers in the Pacific. Operating in a manner akin to the British Museum during the nineteenth century, the Smithsonian Institution also published a *Field Collector's Manual in Natural History*, about the same size as *Survival on Land and Sea*, for troops with time to spare and an interest in flora and fauna. As the text noted, natural history pursuits could "provide welcome and valuable recreation."[30] But this recreation had a purpose, since servicemen properly versed in the military ramifications of biology and zoology might be of use in addressing the challenges facing postwar science. Regardless of whether or not readers had these obligations in mind, the book's detailed rules for the identification, preservation, and shipping of samples seem to have "reached a receptive audience."[31] Not only was the knowledge archived by the Ethnogeographic Board and related repositories profoundly strategic, then, but the juxtaposition of militarism and science in the generation of that knowledge reinforces the need to understand the field as both seen — an object of external comprehension — and *made*. Put differently, it makes little sense to historically separate indexing enemies from eliminating them, whether human or insect.

Survival on Land and Sea begins encouragingly: "Thousands of men whose ships have sunk or whose planes have come down in uncivilized areas of the world have made their way back to friendly territory."[32] In this spatial binary, "uncivilized areas," even if largely unpopulated, represent

U.S. Marine "Raiders" and their dogs on Bougainville in the Solomon Islands, circa November–December 1943. Photograph by Technical Sergeant J. Sarno. National Archives and Records Administration (photo 127-GR-84-68407).

the enemy. The opponent, in other words, is nature, in addition to certain cultures whose parallel wildness renders them *natural*. As the *Richmond News Leader* put it in an effusive editorial on the booklet, the "war will renew for millions of men the age of the pioneers. Those who fight in desert or jungle, in the Arctic and under blistering sun have to learn arts forgotten by the sons of comfort."[33] Offering numerous illustrations to limit the "fear of the unknown," which was "the greatest obstacle that will confront you in the wilderness or at sea," *Survival* advised downed, shipwrecked, or stranded soldiers to remain calm and to "take time to consider your plight and the best ways to go about improving it," regardless of location.[34] As I detail in chapter 5, this concern with panic and its opposite, a cool rationality, returned home during the early Cold War. It was a prominent subject of study among midcentury social scientists.

The advice offered in *Survival on Land and Sea* was thorough and in some cases probably useful. Escaping one of the book's many predicaments alive is not depicted in a language of superhuman valor but rather as a pragmatic release of ingrained skills. Yet this tone does not rule out a critical evaluation of the text. As the product of an Ethnogeographic Board dominated by anthropologists, the manual adopted a particular tone in its treatment of relations with native populations. While acknowledging the importance of local aid and "dignity," the authors of *Survival* concurrently suggested that Americans "do tricks with string" and "take out some trinket and show interest in it" as part of a successful "method of approach [that] has been used many times in many parts of the world by those going to study native peoples."[35] The intriguing component in this caricature of interaction is not the patronizing tone but the insinuation that such techniques were universally applicable from the tropics to the Arctic, particularly given the contrasting, lengthy descriptions of regional plants and wildlife preceding the brief and belated section on human relations. Environmental differences to classify and overcome were primarily physical. A pure survival was the imperative, and political and psychological questions would either assume significance later or, in the case of some regions and peoples, were irrelevant.

Because the regional categories and files of the board were created by the circulation of objects, information, and people between zones, the imaginative spaces of conflict were inextricably linked to material landscapes of warfare.[36] It was this circulation that licensed not only military government abroad but also the solidification of a particular regional map of the world. The board served in a manner similar to the First World War's various national geographic societies: it was situated in a kind of liminal space between disinterested inquiry and political advice but also deeply dependent on the activities of employees and contacts in the field.[37] A region, then, was not only a strategic but also a *physiological* construct, from scientific advice for the survival of warring bodies to the oft-invoked image of area studies as a jigsaw division of a total global anatomy. These military and intellectual approaches were complementary.

Producing and Providing Ethnogeography

In May 1943, the Ethnogeographic Board prepared a "partial list of groups and organizations working on survival techniques" that identified thirty-one agencies, from the Army Air Force's Arctic, Desert, Tropic Information Center to the OSS and Yale's Cross-Cultural Survey, engaged in the effort.[38] Whether intimately or indirectly linked to the military, all shared an interest in demystifying hostile environments, rendering them transparent, malleable, and even governable, using a measured, authoritative tone rich in experience and evidence. As the head of the Smithsonian's Bureau of American Ethnology concurrently claimed, "Travelers, fiction writers, and others have exaggerated the enchanting and the bad features of the Tropics. By placing particular stress on the latter they strive to enhance their own heroism and fortitude at the expense of the literal truth."[39]

That similar or duplicate reports and rosters were being produced in and for wartime Washington was no surprise. In the Ethnogeographic Board's official history (drafted before the end of the war), the Yale anthropologist Wendell Bennett described "fabulous confusion" in the capital, particularly at the time of the board's founding in the first half of 1942. Shortages of all types became chronic, and competition between agencies was acute, leading to contradictory and confusing classification and communication methods. While the Smithsonian personnel who staffed the board were partially removed from this chaos, they still gave precedence to military requests. Board employees additionally faced a more general predicament shared by all those whose intellectual agendas had accelerated due to war: the translation from the languages and techniques of academic, disciplinary research to the abbreviated, area-based action of policy, or at least the quick procurement and provision of information for policy.[40] This dilemma anticipated a second, related sea change, from disciplinary strictures to interdisciplinary, team-based regional approaches, that was the great impetus for the subsequent development of area studies centers. The Ethnogeographic Board was thus a clearinghouse whose regional orientation and sweeping concern with humans, their works, and the context of human life necessitated an unusual integration of disciplines, from the sciences to the humanities. Equally unusual were the questions to which the board was required to respond, inquiries that were unlikely to be answered through the standard channels of military investigation.[41]

In publications, leaflets, and advertisements placed in professional journals, the board described its concern, ethnogeography, as "the study of human and natural resources of world areas."[42] But these regions were specifically zones of warfare. Applications for financial support stressed those "areas outside the United States where military action, economic, or other action is carried on or planned." The overt use of the term *area* was both deliberate and novel. Components of the federal government, particularly the military, were structured along areal lines, while on the academic side, only the humanities-intensive American Council of Learned Societies possessed area committees before 1940. The board's adoption of a regional approach was matched by other emergency agencies, such as the OSS, the Foreign Economic Administration, and the Office of War Information.

There were some disciplines, like geography, that were, "by their very nature," concerned with the definition and study of areas, but geographers who could definitively discuss non-European regions of conflict were scarce.[43] Instead, it was anthropologists (influenced by a certain type of geography) and their ethnographic recollections of travels in "primitive" landscapes now suddenly awash with troops whose relevance was clear and who dominated the board, including both the first director and the first chairman, William Duncan Strong and Carl E. Guthe. Anthropologists, a 1943 NRC report noted, were "the only social scientists who study all aspects of a given culture." They also possessed unique experience in "native administration, resettlement and rehabilitation programs" — all indicative of their ability to jump smoothly from "knowledge into action."[44] Thomas Richards thus has it exactly right when he argues that "geography was a necessary but not sufficient tool for realizing territory. It must always be accompanied by the imperatives of state ethnography" with its emphasis on what Clifford Geertz famously called "thick description."[45]

In addition to the provision of information in response to spot inquiries, board members prepared a select number of larger reports, including confidential documents on "areas of strategic importance," particularly in the Pacific theater.[46] Featuring charts and photographs and including sections on topography and ethnography, these surveys were put together at the request of Army and Navy officials, primarily in the respective intelligence services. Some of these studies and related articles on "survival for castaways in unfamiliar environments . . . prepared by experienced scientific travelers" were published in service journals such as the *Information Bulletin* of the Arctic, Desert, Tropic Information Center.[47] Several similar inquiries, however, were apologetically rejected as too substantial, or initiated and then dropped, or passed on to larger organizations such as the OSS. Projects requiring intensive research effort were frequently turned down. The inability to address certain subjects sufficiently was compounded by limited relations with university sources beyond the mere solicitation of contact details. In the records of the Ethnogeographic Board, only Yale, the home of several board members, is prominent among academic institutions.[48]

The unusual status of the places monitored by the Ethnogeographic Board prompted the creation of a second list, a "Roster of Personnel, World Travel, and Special Knowledge," divided into regions and supplemented by catalogs of anthropologists, geographers (only a short list of Europe experts provided by Richard Hartshorne), and "Social Scientists versed in Social Analysis (for) Propaganda Purposes."[49] The area roster was essentially a huge stack of index cards gathered from more specific efforts initiated by groups such as the NRC's committees on African or Oceanic Anthropology, products of the war that were ultimately folded into the board. The roster, eventually dubbed the World File of Area and Language Specialists, held some five thousand names. In an age of *Who's Who, American Men of Science,* and the National Roster of Scientific and Specialized Personnel, it was hardly unique except for an emphasis on area and linguistic competence and a wider cast incorporating amateurs and non-Americans. As the *Washington Post* explained in a vivid 1942 article, anyone "expert on such far-ranging subjects as Timbuktu or the sleeping habits of Eskimos is fair game for the men in the first floor of the Smithsonian

Institution's West Building."[50] The varied list of seemingly suitable sources for compilation in the roster included the American Malacologists Union, the Baptist Foreign Mission, the College Art Association, and the Foreign Press Club. Thousands of generic questionnaires were also mailed to individuals, requesting such detail as length of residence; linguistic facility; and number of photographs, films, maps, and miscellany possessed. The board formally appropriated some of these objects, including Baedeker's Guides to Germany and Austria, which were then distributed to the armed services.[51]

From Classification to Government

To answer more fully the questions that flooded in from all directions, the Ethnogeographic Board also maintained reference and survival libraries. A copy of Yale's Cross-Cultural Survey (CCS) was an essential component of such repositories. The CCS, established in 1937 under the leadership of the anthropologist George Murdock, was another audacious scheme designed to assemble and organize behavioral literature on the world's "primitive peoples" — and eventually all of humanity.[52] The ideas motivating the creation of the CCS lay at the confluence of two powerful modern impulses: the urge to *classify*, according to ever more complex systems of reference and organization, and the desire to totally *integrate* disparate data. And, as with previous imperial archives merging positive and comprehensive knowledge, classification at the CCS was not just a matter of taxonomy; it was also a means of placing both types of knowledge "under the special jurisdiction of the state."[53] In this case, it was the initiative of a group of psychologists, physiologists, sociologists, and anthropologists at Yale's synthetic Institute of Human Relations, where the search for "a properly 'engineered' society with properly conditioned parts" was vigorously pursued in the late 1930s.[54] One promotional history describes "the common ground" of the CCS as "the basic assumption that all behavior, including that of people, occurs according to natural laws which ultimately are quantitatively determinable and stable by means of true equations."[55] Yet these familiar tropes were limited and given spatial dimensions by a third modern craving: to *identify* unique groups of individuals sharing certain traits. Once a sufficient sample of these distinct units was assembled, generalizations became possible, and governmental interventions became practicable. It was thus the CCS's strategic utility that closed the gap between an older, descriptive form of regional geography and contemporary, rigorous regional methods.[56]

The success of the CSS as a universal repository of cultural data owed much to its status as a team project, which in turn meant that it was easily given an explicit military and imperial cast.[57] During the Second World War, after consultation with the Navy, the CCS was revised to concentrate on Japanese possessions in the Pacific. By 1943, the Navy had adopted the project, although George Murdock, who became a Navy lieutenant commander, was still in a supervisory position. In that capacity, he sent out "additional questionnaires on strategic areas," including queries such as, *Are the natives friendly or hostile? How do you say "yes" and "no"?* and *What is the terrain like?*[58] At the time, the CCS held some 500,000 cards on more than 150 groups all over the world, data that was used in a series of Strategic Bulletins of Oceania produced by the Ethnogeographic Board, documenting meteorological conditions, food and water supplies, and the vectors of disease — all information useful to an occupying force.[59]

In 1942, Nelson Rockefeller's Office of Inter-American Affairs contracted with Yale to produce a Strategic Index of Latin America, which divided the cultures and subcultures of the area into roughly one hundred regional units. When it was discontinued just over a year later, the Index staff was estimated to have accumulated one-third of the existing "major sources on the geography and civilization of Latin America."[60] The geographer Preston James, who led the Latin American section of the OSS, praised Murdock's "encyclopedic approach" for its ruthlessly practical regionalism.[61]

The ultimate incarnation of the CCS, and perhaps the definitive instantiation of John Borneman's claim that in "mapping global categories of otherness," anthropologists engage in *foreign policy*, was the Human Relations Area Files (HRAF), launched in 1949.[62] The initiative was again under the direction of George Murdock, who used it as the basis for his landmark book *Social Structure*, published in the same year.[63] It was a wholeheartedly national project, "drawn upon by the CIA at least through 1967."[64] Also housed at Yale but distributed in component parts to interested parties, the HRAF collection was designed to provide students of human societies with a suitable statistical storehouse, a "laboratory without walls." Rather than traditional physical barriers, there would be juxtaposed, as it were, dioramas depicting life processes and cultural activities against living backgrounds of each of the world societies known to man. These data would exist in printed texts and pictures, classified by topics; each society would have its own shelf. Once the many relatively small bits of knowledge were ordered into a consistent, cross-cultural scheme, new figures could be expected to emerge from the pattern . . . [and] students of human behavior would come into this laboratory to test their generalizations against primitive, historical, and contemporary societies.[65]

The implication of this extraordinary passage was that primitive societies, the foundation of the HRAF and the CCS, were neither historical nor contemporary but *regional* — out of time and fixed in a certain space.[66] Historical, on the other hand, was a means of positioning quasi-modern societies, such as Japan, close to, but behind, those deserving the designation contemporary. "In time, as the cold war set in," writes Rebecca Lemov, the files were used to capture "the steps by which Americans would come to know the world and themselves, and the two in relation to each other." At the heart of this world picture was a scientific definition of cultural normalcy.[67]

During the Second World War, ties between the CCS and Navy Intelligence resulted in a series of Navy Civil Affairs handbooks on segments of the Pacific Ocean, to be used for military government.[68] Recognizing the value of the CCS "for both scientific and practical purposes," the Carnegie Corporation, aided by the SSRC, the Office of Naval Research (ONR), and the Rockefeller Foundation, provided the funding for the expansion of the survey into the HRAF, in the understanding that duplicates of the files would be installed at the ONR and in the libraries of partner universities.[69] At participating institutions, including the University of Chicago, Harvard University, and the University of Washington, filing cabinets held thousands of five-by-eight-inch cards.[70] These were accessed through two master indexes: an *Outline of Cultural Materials* and a geographically divided *Outline of World Cultures,* both built up and sharpened over time from the beginnings of the CCS. To maintain the growth of the files, the Army, Navy, Air Force, and Central Intelligence Agency (CIA) all contributed $50,000 per year to this unclassified operation, which focused, not surprisingly, on Southeast Asia, the Soviet Union and Eastern Europe, Northeast Asia, and the Near and Middle East. However, when the returned value was judged to be insufficient in 1954, this subsidy was cancelled and replaced by a $4 million Army grant to produce sixty-three classified and unclassified handbooks on strategic regions, both "friendly" and "hostile." These popular "plan-books," according to Clellan Ford of the HRAF, supplemented "the existing National Intelligence Surveys, particularly with respect to cultural data of interest to the psychological warfare people and others interested in human behavior, living conditions, and the like."[71]

The public versions of these handbooks are, in contrast to Murdock's strict scientific objectives, highly descriptive, expanding the sparse syntax of HRAF index cards into a more readable format. But for Murdock, there was little conflict between the conceptual aims and the concrete efficacy of his classification projects. The definitive instance of this accommodation was the use of the CCS in Micronesia during the Second World War. Once freed from Japanese control, the archipelago increasingly fell under the dominion of the U.S. Navy. Murdock was rapidly completing the CCS file on the area and had concurrently insinuated himself neatly with the upper echelons of the naval command. Together with two Yale colleagues, John W. M. Whiting and Clellan Ford, he formed Research Unit Number One for Micronesia in the Naval Office of Occupied Areas in April 1943. There, the three processed data on Japanese possessions in the Pacific according to CCS principles. Murdock also immediately circulated a memo stating that even after the war, the key role of these same islands would be as military bases and as places to stage experiments in social science. This opinion was received favorably by many Navy planners, but not by those in the State Department and elsewhere who envisioned Micronesia as one of many decolonized trusteeships that would gain independence but remain economically and politically linked to the United States.[72]

Ultimately, the geopolitical fate of Micronesia fell between these two camps. The islands were internationalized but on the condition that they remained a fortified laboratory, useful not only strategically but also for atomic testing and scientific inquiry.[73] Whether for administration or testing, a "complete knowledge of the peoples of the area" was required, and the NRC set up the Coordinated Investigation of Micronesian Anthropology (CIMA), significantly expanding the field presence of social scientists in the Pacific.[74] In addition to financial support from the ONR, the Navy also provided participating scholars with transportation, supplies from war

SAMPLE FILE SLIP, SHOWING VARIOUS FEATURES
RELATIVE TO FILING SYSTEM

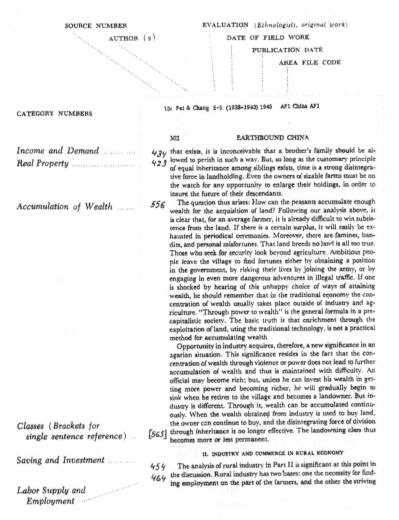

SOURCE NUMBER
     AUTHOR (s)

EVALUATION (*Ethnologists, original work*)
    DATE OF FIELD WORK
      PUBLICATION DATE
       AREA FILE CODE

CATEGORY NUMBERS

13: Fei & Chang E-5 (1938-1943) 1945   AF1 China AF1

302   EARTHBOUND CHINA

Income and Demand .... **434** that exists, it is inconceivable that a brother's family should be al-

Real Property .... **423** lowed to perish in such a way. But, so long as the customary principle of equal inheritance among siblings exists, time is a strong disintegrative force in landholding. Even the owners of sizable farms must be on the watch for any opportunity to enlarge their holdings, in order to insure the future of their descendants.

Accumulation of Wealth .... **556** The question thus arises: How can the peasants accumulate enough wealth for the acquisition of land? Following our analysis above, it is clear that, for an average farmer, it is already difficult to win subsistence from the land. If there is a certain surplus, it will easily be exhausted in periodical ceremonies. Moreover, there are famines, bandits, and personal misfortunes. That land breeds no land is all too true. Those who seek for security look beyond agriculture. Ambitious people leave the village to find fortunes either by obtaining a position in the government, or by risking their lives by joining the army, or by engaging in even more dangerous adventures in illegal traffic. If one is shocked by hearing of this unhappy choice of ways of attaining wealth, he should remember that in the traditional economy the concentration of wealth usually takes place outside of industry and agriculture. "Through power to wealth" is the general formula in a pre-capitalistic society. The basic truth is that enrichment through the exploitation of land, using the traditional technology, is not a practical method for accumulating wealth.

Opportunity in industry acquires, therefore, a new significance in an agrarian situation. This significance resides in the fact that the concentration of wealth through violence or power does not lead to further accumulation of wealth and thus is maintained with difficulty. An official may become rich; but, unless he can invest his wealth in getting more power and becoming richer, he will gradually begin to sink when he retires to the village and becomes a landowner. But industry is different. Through it, wealth can be accumulated continuously. When the wealth obtained from industry is used to buy land, the owner can continue to buy, and the disintegrating force of division

Classes (Brackets for single sentence reference) .. **[565]** through inheritance is no longer effective. The landowning class thus becomes more or less permanent.

II. INDUSTRY AND COMMERCE IN RURAL ECONOMY

Saving and Investment .... **454** The analysis of rural industry in Part II is significant at this point in the discussion. Rural industry has two bases: one the necessity for find-

464 ing employment on the part of the farmers, and the other the striving

Labor Supply and Employment ....

(*File slip reduced in size.*)

Sample human relations area file slip. From *Laboratory for the Study of Man: Report, 1949–1959* (New Haven, Conn.: Human Relations Area Files, 1959), 14. Courtesy of Human Relations Area Files, Inc.

surpluses, and other forms of assistance. For his part, Murdock returned from the Pacific in 1945 convinced, in his words, "of the need of selling social science by demonstrating its practical utility."[75] In a 1948 *Science* piece, he proudly claimed that the CIMA effort would "shortly result in the most complete, comprehensive, and up-to-date scientific coverage of the people of any cultural or geographical area of the world."[76] Murdock was clearly positioning the CIMA's researchers alongside, and at the service of, government administrators. He had no interest in aiding or attempting to speak for the administered.

During the Second World War, the Navy became concerned that in addition to the occasional visiting social scientist, a permanent cluster of officials would be needed to participate in the structures of authority emerging in the Pacific and elsewhere. In the case of Micronesia, personnel who were expected to interact with native populations required more than instruction in standardized techniques of military governance. As a result, the Navy approved the establishment of a School of Military Government and Administration within the larger Program of Training in International Administration at Columbia University. At the school, officers followed a curriculum that included education in local languages, customs, and the history of political institutions (both native and colonial), as well as the technical aspects of military rule—a course list that influenced the formation of the civil affairs training schools I discuss later in this chapter.

For Columbia faculty, the political scientist Schuyler Wallace noted, it was the Navy's intense interest in a particular area, combined with the demand for experts across the social sciences found in the Administration Program, that "raised very forcibly ... the question of the validity and the potentialities of what has come to be called the concept of area studies." Wallace, the director of both the program and the School of Military Government, was aware of the dangers that lurked within the regional approach, including the potential for superficiality and the related absence of universals found most forcibly within "that thing called Christendom." Nonetheless, grouping disciplines, he argued, might "produce an understanding on the student's part of the whole life of some particular region, rather than an understanding of some artificial segment thereof, such as is obtained through the study of comparative government or comparative literature."[77] These were deliberately chosen contrasts.

The mushroom cloud created by the underwater "Baker" detonation, part of the Operation Crossroads tests at Bikini Atoll in the Marshall Islands on July 25, 1946. National Archives and Records Administration (photo 80-G-396229).

Social scientists joining postwar expeditions to the Pacific were encouraged to proceed with the assistance of the HRAF. The HRAF's strategic value was thus complemented by a scientific motivation. Reports and guides released from the Yale headquarters audaciously deployed the rhetoric of laboratory science. This laboratory, however, was mobile, a beacon of certainty and a source of detail that could travel, in the form of jottings copied from a succinct index card, into the field. In this respect, the files were a kind of survival guide for academic voyageurs, aiding not bodily sustenance but scientific solidity far from campus.

The authors of the compact *Outline of Cultural Materials* were aware of this circularity, claiming that the manual itself had uses in the field, specifically by calling attention to a range of phenomena that were often "omitted in descriptive accounts," or scattered messily in a notebook. Because the scholarly categories had been standardized, it was easy to evaluate different groups. However, these groups and their locations were taken for granted. While Murdock and his HRAF collaborators were quick to acknowledge that the abbreviated *Outline* could by no means encapsulate the complexities of a given culture, they recommended it be consulted in advance and then kept on hand in foreign places as a means of saving time, revealing "gaps" and "inconsistencies" that could then be erased by the adept social scientist. The gesture to social science was a deliberate one. Fearing that only anthropologists would be attracted to the resources within the files,

THE HRAF LABORATORY

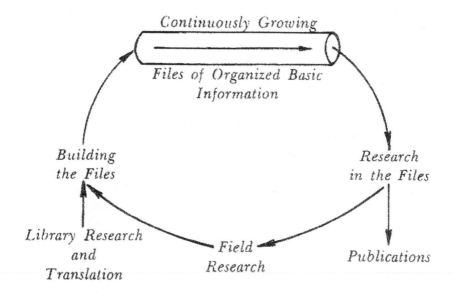

Continuously Growing

*Files of Organized Basic
Information*

*Building
the Files*

*Research
in the Files*

*Library Research
and
Translation*

*Field
Research*

Publications

Circulating regions. From *Laboratory for the Study of Man: Report, 1949–1959*, 35. Courtesy of Human Relations Area Files, Inc.

proponents asserted that the material could also serve the "universal sciences" that were "concerned with human behavior in broad perspective, i.e., with establishing valid principles not limited in time or space."[78]

To achieve these detached principles, users of the HRAF were instructed to work up from the bedrock of cross-cultural comparison. Once a sufficient level of generalization had been attained, they would be able to cast off the regional constraints built into the very architecture of the files. But only by framing the files as a single object and by practicing applied anthropology in the service of power was generalization permitted and encouraged. One document summarizing the genesis of HRAF put it best: "Much of the basic information on the peoples and areas of the world can be brought together, organized so that the facts dealing with the same topic and the same people or area lie side by side, and cast in a common language."[79] This was the detached dialect of social science but also a very useful entry into forms of neocolonial rule. It was no coincidence that the curriculum at Columbia's School of Military Government introduced officers to the social sciences of human management and control.[80] Proper governance required a larger class of trained military personnel who could be sent to certain places to make structures of purposeful regional knowledge operational.

A Map for Area Studies

The final contribution of the Ethnogeographic Board to a solidifying strategic regional geography was an analytical survey of foreign area courses offered during the war at many American universities. At a March 1944 conference organized by the Rockefeller Foundation, representatives of the board were invited to submit supplements to this survey that would be compiled in a report considering the future of area studies. Both the survey and report were largely the responsibility of William Fenton, a board research associate and an anthropologist with the American Bureau of Ethnology. The board's interest in education was reciprocated; as William Duncan Strong noted in 1943, "Universities setting up courses of regional study have written the Board or sent representatives to secure data on available personnel and certain types of information for teaching such courses."[81]

It was not the service aspect of Fenton's survey that resonated in the late 1940s. His outline of what an ideal area program might resemble arrived too late and did little to alleviate the chaos on campuses whose resources were thinned by departed students and faculty and stretched by hordes of trainees. As many of the names on the board's area roster dispersed into the Cold War market for regional experts, and as elements of *Survival on Land and Sea* were developed into more specific survival manuals and guides for the hostile environments of the Cold War, Fenton's work served as a template for increasingly frequent discussions on the subject. Nothing quite like the Ethnogeographic Board materialized during the Cold War, but numerous new sites with similar characteristics, supported through the same channels of government and foundation funding, did emerge to address another crisis. As Wendell Bennett noted shortly before the board's termination, aside from valuable contributions to military campaigns, a "unique experiment in the integration of academic research" had been completed.[82] And as early as July 1944, in a statement composed on the occasion of William Duncan Strong's resignation as director, board chairman Carl Guthe recognized the growing need for "a more informed appreciation on the part of experts of the interrelations among a large number of the special fields of investigation which relate to human activities." Guthe was looking ahead; the interdisciplinary problems that concerned him were those of the postwar world, as they became "more defined."[83]

In one of the few historical discussions of the Ethnogeographic Board, Martin Lewis and Kären Wigen note that the board's global map actually changed as military clients recognized the inadequacy of the "old continental architecture, anchored by the vast category of Asia" and the growing irrelevance of European colonial empires. The system that replaced these, one of "world regions," was largely an American invention. It was still rudimentary and in transition during the Second World War, but the modifications made later remained consistent with the initial focus on political units and cultural civilizations. The consequences of such cartography were "never given extended consideration" by board members, and Lewis and Wigen argue that the absence of geographers on the board was the reason. Primarily concerned with "physical and economic characteristics of subnational regions" and rarely venturing intellectually beyond Europe, many geographers were uninspired by the sizable areas studied and occupied by the military and its informational branches, and they lacked the linguistic training to cope with the new, relevant regions. Consequently, as geographers continued to occupy a surprisingly minor role in postwar area studies, valuable forms of geographical inquiry grew increasingly marginal in the American academy.[84]

The regional concept that owed much to the Ethnogeographic Board and its cadre of anthropologists was ultimately a subject that generated little debate because it was an instrumental means to an end, a framework that could be used to organize policy and education. An older, continental category such as Africa could thus linger on without dispute precisely because of where it sat on the geopolitical hierarchy.[85] However, the importance of the board should not be exaggerated. Other organizations were also grappling with and advocating regional approaches, suggesting that the jumbled taxonomy of world areas was less important than the intelligence it generated.

School for Soldiers

In 1944, when William Fenton returned from his visits to twenty-seven campuses hosting the Army Specialized Training Program (ASTP) and Civil Affairs Training Program (CATP), he produced a series of short reports that were quickly distributed within Washington. Although his summary of these reports was only released to a wider audience in 1947, Fenton's indictment of the disorganized state of area resources in the United States remained relevant. While the Ethnogeographic Board had become a home for academics with experience in foreign areas, they were overwhelmingly confined to the capital. The ASTP's Foreign Area and Language Curriculum, in place at 55 of the 227 sponsoring universities, was designed to improve the linguistic abilities and areal awareness of those enlisted soldiers who would potentially visit certain regions.[86] The focus was global; dozens of languages were taught. Those who participated in the ASTP not only had volunteered but were also judged to be of suitable intellectual ability; they were, two observers wrote, "a type that we are likely to see back on campuses once the war is over." At its peak in December 1943, over thirteen thousand personnel were enrolled in the Area and Language Curriculum. The CATP, on the other hand, was located at only ten universities and was designed specifically for new officers.[87]

The intention of the coordinators at the Military Government Division of the Army's Office of the Provost Marshal General was that those trained under the Area and Language component of the ASTP would serve within small police forces under the command of officers who had also received special instruction at the CATP. Planning of the language instruction (which accounted for three-fifths of the course time) was passed to the American Council of Learned Societies and its secretary, Mortimer Graves, who was also a prominent force behind the Ethnogeographic Board. Basing the program on intensive language courses already under way at several universities, Graves and his assistants advocated a concentration on colloquial fundamentals and endless, repetitive drills. Once additional components of the curriculum were added, solicitations were made to universities.

The resulting stampede of interest was not just a product of patriotism or the desire to employ faculty not already busy with war work; it also reflected the awareness on the part of academic administrators that the study of place and culture under one roof might have long-term, beneficial repercussions. Some university presidents had a dissenting opinion. Cornell's Edmund E. Day argued that small colleges would be swamped for the duration of the war by soldiers who had little need of a liberal arts education — which was not how the Army and Navy "make killers" — and who would instead demand extensive "technical and professional training."[88] This was, as it turned out, an incorrect view on several counts, but it was devastatingly accurate on another. Both military education and even later instruction in area studies centers created a class of *technicians* who were distinct from their detached social scientist teachers.

Unlike the Ethnogeographic Board, ASTP area initiatives were primarily concerned with Europe and Asia, although for comparative purposes, allusions to "parallel developments in the United States" were made frequently.[89] This was an indication of the integral but shadowy presence of America in a globe of regions. The study of foreign places and cultures encouraged comparison but also a parallel appreciation for distinct American values. In this respect, it made sense that the dual foundations of the ASTP course of study on areas were geography and anthropology bolstered by a splash of history. The result, wrote Fenton, was "an experiment which attempted rather uniquely to prepare soldiers for field work of sorts in the civilizations (or cultures) of great areas."[90]

That this would be an uncommonly violent form of fieldwork did not merit mention. But Fenton was aware of the threat that such an integrated program would pose once it gained a foothold in American universities. Grasping the "total civilization of a region" meant an innovative blend of the anthropological focus on culture with a specific physical geographic area.[91] And while the two elements of this synthesis were, in the terms of anthropology and geography, not always sufficient for Cold War social science, they were still essential to the growth of area studies programs after World War II. Also prominent, however, was a third component that received far less scrutiny: instruction in law enforcement. Not all personnel received policing education. But the military pretense of the course was consistent for all: to minimize the difficulties of social and spatial adjustment by preparing "the individual to act efficiently in a new environment."[92]

In searching for suitable geographers, the Army encountered the same problem faced by the Ethnogeographic Board: a dearth of teachers who were "attached to man and his culture" and a corresponding preponderance of technical specialists, often from combined departments of geology and geography, who were preoccupied with the instruction of Air Corps cadets in meteorology and map reading.[93] As a result, scholars from disciplines such as classics and economics were occasionally enlisted to lecture on geography. Given the limited educational experience of the trainees, the emphasis shifted from explanation to description. Scientific elements were kept to a minimum. And geographers whose understanding of regional work was the intensive scrutiny of a diminutive territory were suddenly forced to broaden their horizons. Yet this same background meant that the generic approach of geographers suited the ASTP's aims quite nicely, even if it did not succeed in the more focused, authoritative circles of postwar area studies. As two sociologists who participated in the ASTP at Vanderbilt University opined, the Army "does not want its soldiers to be acquainted with the problems of population, stratification, labor, and the family as such but with all of these as part and parcel of an entire civilization."[94] Still, Fenton reported that the skeptical opinion of geography held by other social scientists was reinforced when those from a range of alternate disciplines stepped in and "did a fair job of teaching geography" when necessary.[95]

The ASTP also provided instruction in several other fields, including engineering and medicine. It was believed, not surprisingly, that as many soldiers as possible should possess an understanding of military equipment and its mechanisms, but because this equipment "embodie[d] abstract principles of science," such principles needed to be taught as well.[96] For participants in the geography component of the ASTP, a vital piece of equipment was the map, and key principles included those of basic cartography. Soldiers were provided with a world atlas featuring maps prepared exclusively for the Basic portion of the two-part program, a three-term "phase" that totaled roughly nine months. Holders of the atlas were urged to use it in conjunction with a globe "in order to keep in mind continually the true picture of the global relation of the land and water areas of the world—and especially of the world that matters." Credit for many of the maps, none of which were at any scale smaller than the globe, was given to the Department of State and its geographer, Samuel Boggs, and the Map Division of OSS, led by Arthur Robinson.[97]

Boggs and Robinson were not the only geographers with a hand in the ASTP curriculum. Alongside the atlas, students read a substantial textbook titled *Geographical Foundations of National Power*. Planned by Isaiah

Bowman, Richard Hartshorne, Derwent Whittlesey, and Charles Colby, and written by Whittlesey, Colby, John Kirkland Wright, Dorothy Good (of the American Geographical Society), and Harold Sprout, it was effectively the combined product of the most prominent geographers in the United States. The tome established a connection between a soldier's "duty and the kind of world in which he will live after the war." It was thus a contribution to military geography and geopolitics, a statement of disciplinary principles that could not escape its context: a conflict that represented the "most destructive form" of the "competitive struggle for existence," a struggle whose "basis and . . . nature" was the subject of geographical study. Competition, for the authors of the volume, applied at all scales. The individual struggles of the recruit-soldier were telescoped to an international arena where the role and potential of a state was determined by considerations of power.[98] But this connection between bodies and states, suggesting that both shared an antagonistic relationship with external enemies, was not a temporary association; the textbook implied that it was rooted in the principles of geographical science.

Anthropology's relevance to the ASTP was signaled by the discipline's treatment of the "whole man" and the ease with which it could shift to accommodate the rudimentary "universal cultural pattern" favored by curriculum planners. Even more crucial was the experience anthropologists possessed in contact situations and field methods. As Fenton put it, anthropologists held "an informant's view of culture," a particularly relevant approach for those soldiers who would be engaging in "social control at the local level."[99] After all, informants were always useful under conditions of military government. These were practical matters of social understanding and occupation and were certainly not as complex as contemporary anthropological theory. In this rushed program, the "total picture of an area" sought was profoundly elementary.[100] But it is this exhibitory aspect — the stubborn quest for some form of complete and bounded perspective — and the military orientation of this view that are important here.

The exigencies of wartime area education were emphasized by its administrators and teachers and marked the difference between the ASTP and more advanced, graduate-oriented area programs of the Cold War. But like the interactive suggestions of *Survival on Land and Sea*, the assumptions within this popular version of social science were closer to those of its more sophisticated descendent than many cared to acknowledge. And despite the basic level of instruction, the ASTP revealed that the challenge of a "more inclusive reality" — a global worldview divided into manageable regions — could "only be met by an integrating effort of all the social sciences."[101] The combination of social science, global ambition (realized in language training), and, crucially, policing and governance, was not available at American universities before the Second World War. It would have to be invented.

Character, Culture, and Nation

In its impact on American area studies, the educational experiment of the ASTP was comparable to the emergence of mental testing in the First World War and the subsequent growth of psychology at universities. Indeed, social science during the Second World War was led by similar, albeit more advanced and ambitious, research in psychology. A pamphlet about the ASTP claimed that the most powerful weapon of the soldier was "his brain," and as soldiers progressed through courses on campuses across the country, they and members of other training programs were being surveyed and monitored by teams of experts.[102]

While experimentalists conducted "man–machine" research on sensation and perception, and another set of social scientists sought to comprehend the "national character" of German and Japanese foes, a third group combined extensive, positivist survey methods with an interest in personality — both individual and national — to analyze American combatants. All three of these efforts advanced psychological techniques and interdisciplinary approaches while concurrently articulating the relevance of social scientific authority within a society engaged in prolonged, global conflict. If nations possessed characters, cultures could then be diagnosed and classified as disordered, and psychology could lead to policy. This was obviously a narrow perspective on culture. Militarization allowed psychology to "operate as a weapon system" of the imagination against "epidemics of irrational emotion and flawed national characters in need of containment or reconstruction."[103]

Studies of national character are commonly associated with Margaret Mead and Ruth Benedict, both students of Franz Boas, a towering figure in twentieth-century American anthropology and no stranger to geographers either. Even before the Second World War, Mead's and Benedict's works had focused on the relationship between psychological factors and cultural conditions, but with the commencement of combat, they, along with Mead's husband Gregory Bateson and Geoffrey Gorer of Yale, shifted from the study of small, "primitive" societies to the psychocultural condition of "modern," *national* units such as Germany, Japan, and the United States itself. In turning to the home front and to the emancipatory aspects of social science, the anthropological traditions of traveling abroad in order to understand social patterns and avoiding explicit questions of policy and social change were cast aside.

The imaginative location of the primitive—the inversion of the national—is also the terminus of the national character thesis, where the veneer of cultural difference can hardly disguise a proximity to biological theories of race. Yet these primitive geographies—and the philosophies, rich in moralistic tenor, that create them—are not empty fantasy or the stuff of innocent academic curiosity. The performative discourses of national character, which are "simultaneously descriptive and normative," should be considered as part of a larger, powerful apparatus. That apparatus includes statements and practices of nationalism, citizenship initiatives, and individual acts in contexts in which the language of national character can be employed "as an explanation, justification, or rationalization" for such behavior. All of these elements are as applicable to the United States as they are to a more typical target of national caricature.[104] And although primitive societies were of strategic interest only for their *presence in territory,* if these alien cultures should make progressive strides toward a contemporary status, an alternative treatment was required, one which by the 1950s required the attention of modernization theory. Culture and personality studies, after all, were molded from anthropological work on tribal peoples. When the theory supporting these studies traveled, savage characteristics could be found anywhere, and when mixed with the seeds of modernity, industrialization, and political organization, these characteristics were particularly threatening.

Given the Boasian critique of scientific racism that defined early twentieth-century anthropology, it was hardly surprising that many American anthropologists, beginning with Boas himself (although he died in 1942), would take issue with Nazism. Several formed a Committee for

National Morale (which included Mead, Bateson, Gorer, Benedict, Clyde Kluckhohn, and George Murdock) to promote the utility of interdisciplinary behavioral science even before Pearl Harbor. They and others then joined the rush of social scientists to war-related posts. Benedict, for instance, replaced Gorer as head analyst in the Office of War Information's (OWI) Overseas Intelligence Bureau in June 1943. This division prepared propaganda for use against enemy nations and distributed more innocuous information to allies and neutral states. A year later, having produced a series of cultural profiles on various war-torn countries, Benedict was asked by Alexander Leighton—a psychiatrist who, along with Clyde Kluckhohn, headed the OWI's Foreign Morale Analysis Division—to write a study of Japan.[105] Despite her inability to read Japanese or conduct fieldwork in Japan, Benedict wrote the report, which was revised and expanded after the war and published as the best-selling *The Chrysanthemum and the Sword: Patterns of Japanese Culture* (1946). Margaret Mead's work with the NRC's Committee on Food Habits, meanwhile, led her to ruminate more generally on American character and, later, to expand her thoughts into a comparative project on various types of national ethos. Her consideration of democracy and culture in the United States also led to a significant work of popular anthropology, *And Keep Your Powder Dry* (1942).[106]

For Mead and other liberal intellectuals, the American democracy pressingly in need of defense during the Second World War was a *normative* form of culture, set contrapuntally against Nazi, and later Soviet, totalitarianism but also against all cultures not defined by "freedom-as-autonomy." That the state of American character or morale occasionally seemed on shaky ground only increased the importance of neutral and detached social scientific contributions to the war effort. This absolute defense of democracy countered the relativism encouraged by Boasian anthropology, subsuming it beneath the pragmatic decision to embrace patriotism, whether cautiously or not. It also challenged Boas himself, since the American Anthropological Association had censured him for criticizing anthropological espionage in World War I and for his pacifist leanings. Benedict, by contrast, was aware of the OWI's role in formulating psychological warfare doctrine, but she also perceived this to be a natural entry point into the compilation of information on national character, which in turn would be critical to postwar peace—a peace to be achieved, it was hoped, through a United Nations in which the uniqueness and independence of cultures could be fully respected.[107]

The research spearheaded by Mead, Benedict, and others in the early stages of the Second World War eventually spread from a focus on the Axis powers to cover nations such as Greece, Thailand, China, and, later still, the Soviet Union. Given the hasty assemblage of many studies, the obvious constraints placed on travel to Axis countries, and the dependence on existing secondary sources and interviews, the work that fell under the banner of national character was recognized as "only approximately accurate," although it certainly retained academic pretensions.[108] Critics, including one associated with the OWI, immediately pounced on what was perceived to be the lost objectivity, caricatured phrasing, and moralistic ethnocentrism of some national character scholarship, called a "curious doctrine for the heirs of Franz Boas."[109] But proponents of this approach successfully infiltrated various wartime agencies, from the OWI to the OSS, where Bateson found a home along with dozens of other anthropologists. National psychological traits were also common subjects of discussion in the Army Specialized Training Program.[110]

The support of the War Department and the general context of wartime meant that studies of other cultures, positioned next to ongoing consideration of America's unique attributes, contributed substantially to the solidification of a globe of regions with the United States at its center. Crucially, the cultural foundations heralded in culture and personality research were consistently those of "America," and not "the West." This made sense given Europe's wartime horrors and the role that America was beginning to play as the locus of a new modernity, but also for practical reasons of geopolitics made manifest by militarization. The psychoanthropological work conducted during the war advocated a global geography that was hierarchically and discordantly divided by culture and politics. It emphasized differences and the inherent flaws within other regions — and not only those occupied by American enemies. Mead and others were beginning to walk the theoretical and practical tightrope that defined postwar area studies: an opposition to authoritarian forms of rule and persuasion next to a desire or need to nudge populations toward democratic values, even if undemocratic techniques were required.

In the OWI's Far East section, under the direction of George Taylor, a historian of China, a group of anthropologists attempted the dual task of devising methods for Japanese surrender and convincing American military leaders of the complexity of Japanese culture, a nuance that would, contrary to the opinions of certain decision makers, permit this same surrender. The conclusion that Japanese morale was faltering was certainly unorthodox and contradicted subsequent arguments in favor of atomic bombing. Whether Taylor and his staff were able to moderate opinion of Japanese character—particularly concerning the role of Emperor Hirohito in discussions of surrender—is a matter of significant dispute. For Clyde Kluckhohn, the appraisal of the emperor's symbolic value was the most important contribution of the OWI's anthropologists. He deemed it far more valid and far more practical than vulgar psychoanalytic gestures to lingering childhood traumas. Yet these gestures may have ultimately had more resonance, if only because they reinforced popular prejudices within policy circles and beyond and did not do much to confront them.[111]

American behavioral scientists working in intelligence analysis and psychological warfare had, by the final year of the war, collectively determined that Japanese culture was submissive but malleable, childishly uncertain, and ethically situational. As such, it was capable of oscillating between "fanatical militarism and some form of qualified democracy." This latter state could be achieved by maintaining the emperor in place, shifting his role from the "embodiment of ultranationalism" to a democratic symbol, and by combining "authority, example, and symbolic manipulation." While this view was more accommodating than that held by conservative Asia hands, who were repeatedly stifling discussion of policy options by stressing the incompatibility of self-government with obedient and tradition-bound Japanese masses, the psychological perspective remained condescending and cautious.[112] It was also directly tied to military strategy, as mental and moral weaknesses were revealed for the benefit of propaganda operations, diplomatic negotiations, and strategic choices such as the use of atomic bombs.

Yet in stressing the possibility of Japanese democracy, the behavioral science approach, enfolding even the more humanistic scholars like Benedict, also suggested that Japan's national culture was, like all others, unique *and* adaptable, full of potential for either improvement or degeneration. Here lay the key difference between theories of culture and those of race. However influenced by inheritance, cultures could, and often should, be

changed. The appropriate, measured solution was to retain certain traditions and symbols as part of a cosmopolitan aesthetic, provided they could
be detached from and made inferior to a more "ascetic," responsible approach to questions of policy, particularly in the foreign sphere. This was
the same doctrine of simultaneous cultural tolerance and strategic realism advocated by Cold War liberals. *The Chrysanthemum and the Sword*,
with its deliberately contradictory title, concluded, as Christopher Shannon puts it, that "peace of the world depends not on a liberal intellectual
elite controlling world events but on the peoples of the world controlling
themselves in accord with the values of that elite."[113]

National character studies — with their appeal to tolerance and diversity balanced by suggestions for manipulation, and with their abstract categories of populations and partially autonomous individuals — can clearly
be linked to broader discussions of governmental reason. It is thus relevant
that the Second World War and its aftermath were described by Pendleton Herring (ironically, an administrator) as an "administered age," a high
modernist epoch of planning and ordered human relationships. These relationships could be found, according to Herring, in totalitarian societies
but were also prompted within new international organizations, such as
the United Nations or the World Bank, or the blooming bureaucracy and
global commitments of the United States itself.[114] Herring's diagnostic
overview did not dwell on the American occupation of Japan, where military control was tied closely to techniques of psychological manipulation
and economic liberalization.[115]

The role of the social scientists in the OWI was unambiguous. They
were to provide regional intelligence, composed after a good deal of basic
fact compilation, to those who were in a position to use it. Whether naïve,
overconfident, or moved by the extremity of the situation, these scholars
frequently failed to question the interpretation of their studies. Instead, they
believed that their more sophisticated analyses were a distinct improvement over impoverished military logic. Yet their analyses could not be *too*
sophisticated or populated with professorial prose. As was the case at the
Ethnogeographic Board, the OWI's superiors demanded rapid syntheses,
a call that surely influenced the cast of certain claims. Those involved with
OWI who were not social scientists — journalists, advertising executives,
public relations consultants, and others — had little tolerance for pursuits

that were not seen to be practical. Relevant sources were not always on hand, research had to be marketed intensively to secure an audience, and other agencies, competing for this attention, produced overlapping or contradictory reports. One academic veteran of the Overseas Branch was certain of the resulting subjectivity but questioned it from another direction, bemoaning the infrequency of systematic, scientific analysis that pooled "the available data and information of all experts."[116] The centralization efforts undertaken during the war were clearly insufficient for some.

It was still assumed by most social scientists, regardless of their workplace, that improving understanding of unknown places would reduce ignorance and thereby improve the prospect of cooperation with the United States not only during but also after liberation. Knowing your enemies and allies also required and made possible further study of one's own strengths and weaknesses — the preoccupation of texts such as *And Keep Your Powder Dry*. This was the central dialectic of national character studies that is too often hidden or is treated as a dichotomy: the uncertain distinction between social engineering, or the applied aspect of anthropology and related disciplines directed at other cultures, and a cultural analysis directed toward the critic's own nation. These are undoubtedly dependent on one another in the manner that all discourse on a world beyond American borders was linked intrinsically to America, simultaneously calling into question the strength and legitimacy of those borders themselves. Regardless of this paradox, nations and cultures as objects of analysis remained unassailable vessels that could be filled with the facts and conclusions generated by social science.[117]

In its most simplistic variants, national character study documented the acquisition of a culture's singular personality. To create unity and pattern in a world of shifting alliances and heterogeneous populations, an integrated whole had to be created and bounded, at both cultural and global scales, and history had to be deemphasized in favor of timeless categories. Advocates of this approach focused on cultural anthropology and psychology; Benedict, for her part, was, like Geoffrey Gorer, obsessed by childrearing habits, though in a less psychoanalytic register. A largely unstated geographical element also influenced the study of national character. Even if certain qualities might cut across regional and other differences, such speculation was halted by the strategic preference for blocs and states. In this respect, Japan — widely viewed as completely homogeneous — was the perfect case study because it was set apart from the United States through practices of demonization to a greater degree than any other foe. Such habits were not always lessened by the judgments of national character projects. Benedict's interest in disputing stereotypes did not prevent her from repeatedly contrasting Japanese and American values in plainly ori-entalist tones. Although she and other innovative theorists such as Bateson were keenly aware that "characters" were constructed markers of differ-ence depending on stereotypes, they defended these formations as none-theless significant. Characters were norms from which one could speak of alterity and without fully acknowledging complicity in the continuation of essentialism.[118]

Such long-range scrutiny, aided by the compilation of various data sources and *not* by personal experience, necessarily adopted a detached perspective that was more prone to generalization and abstraction than much colonial ethnography. Grounded observation had been replaced by library research, interviews, study of popular cultural sources, and even statistical analysis. Individuals lacked nuance and agency. These were all common themes in the study of totalitarianism that dominated Cold War scholarship on the equally inaccessible Soviet Union, its satellites and, later, China.

The destruction of Hiroshima and Nagasaki did not diminish Mead's interest in national cultures or in associations with the military. Shortly after the war, she and Benedict received a grant from the Human Resources Division of the Office of Naval Research and, perhaps to limit association with wartime propaganda efforts, set aside the term *national character* for *cultures at a distance.*[119] The Research in Contemporary Cultures initiative at Columbia University continued well after Benedict's untimely death in 1948 and was extended into several other projects on Soviet culture (for the RAND Corporation) and contemporary cultures (for the ONR and, later, MIT's Center for International Studies). Attention had shifted from Japan to the complex terrain of postwar Europe. The ultimate result was a "manual," presented first to the ONR in the autumn of 1951 and then published for popular consumption in 1953.[120]

In a section of this book titled "Political Applications," Mead noted that the methodology outlined in the manual had already aided military occupations by facilitating interaction with allies and partisan groups in enemy countries, estimating the capabilities of opponents, and preparing foreign policy documents. All of these tasks required a diagnosis of "cultural regularities in the behavior of a particular group or groups of people that are relevant to the proposed action." Whether issuing propaganda, offering threats of reprisal, or announcing a new regulation, a "specific plan or policy" was consistently the anchor for cultural study, which would be used to predict the success of such plans and policies. For Mead, regularities were most valuable when applicable to large groups of individuals and long stretches of time. It was only after an interval that "patterns of reaction" to the bombing of German cities could be determined and set against preexisting predictions and panic typologies, an argument echoed in the concurrent predictive study of American urban disasters.[121] Not only did work on national character fail to fade after its flowering in the Second World War, then, but it also proved easily translatable — after shedding certain "softer" cultural and psychological elements and occasionally shifting its spatial boundaries — to a broad swath of social science. Perhaps the best example of this was the Soviet Refugee Project initiated at Harvard's Russian Research Center, an initiative I discuss in chapter 3. But before the Soviet Union occupied center stage in a world of studied regions, national character scholarship was already in debt to the discourses and institutions of modern militarization.

Scholars, Soldiers, and Spies

While the Navy was using CCS material for handbooks, it was also producing, in a joint initiative with the Army, a more basic form of regional intelligence. These were the Joint Army-Navy Intelligence Studies (JANIS), anonymous and confidential volumes about various strategic regions, each packed with photographs and comprehensive detail on the resources, terrain, infrastructure, and other geographic features.[122] Contributions arrived from over twenty sources, including the Board on Geographical Names, the Coast and Geodetic Survey, and the OSS. The intention, according to a weighty tome on Korea, was to "make available, subject to limitations of time and material, one publication containing all the necessary detailed topographic information upon which may be based a plan for military operations." Most of the chapters, including those on general military geography, were descriptive in style, focusing on physical landforms and impediments to an invasion.[123] However, these catalogs were also supported by significant theoretical work. The model on "sea, swell, and surf forecasting," for example, was developed for the Navy by the prominent oceanographer Harald Sverdrup.[124] According to one immediate retrospective, the JANIS catalog included the most important and reputable of the Second World War's "detailed regional surveys."[125] Reviewing the state of American military geography in 1954, Joseph Russell wrote that JANIS represented "the finest example of wartime area reports ... the range of topics covered, the variety of sources tapped, and the high quality of the writing in the cartographic work placed them among the major geographic achievements of recent decades."[126]

Like the CCS and its postwar progeny, the JANIS project, as the list of contributors indicates, was initiated to clarify and integrate the muddle of imbricated information analysis produced by various branches of the American government. A meeting of intelligence heads led to the formation of a Joint Intelligence Study Publishing Board that oversaw the production of thirty-four comprehensive studies between April 1943 and July 1947. JANIS was also the template for the CIA's National Intelligence Surveys (NIS) initiated just four months after the CIA opened in September 1947. The NIS program required improved geographic gazetteers and better maps; the Department of the Interior produced the former with help from the Board on Geographical Names, while the CIA took care of the cartography. The NIS was also supplemented by the classified country surveys produced for the CIA and the Department of Defense by the HRAF.[127]

The geographer Kirk Stone was the first OSS representative on the JANIS Publishing Board. He was frustrated by the lack of competence of fellow members, calling them "dead-heads" and suggesting to Richard Hartshorne only a few weeks after being appointed that "this Board should be dissolved."[128] Edward Ullman, who headed the Transport Section of the OSS's Research and Analysis (R&A) Branch, followed Stone and ultimately became board director.[129] He was less concerned with board membership than with how it defined the notion of a strategic area. In a December 1944 memo on "topographical intelligence," he complained that too many JANIS reports had been focused strictly on regional description:

> Specialized knowledge of a subject is more important than knowledge
> of an area. . . . When the Research and Analysis Branch of OSS was first
> organized, it was set up primarily on a regional basis . . . [but] most of
> the product was poorly organized, unbalanced and of preliminary value.
> Later that branch was reorganized and some functional sections were set
> up . . . the result was a better, more useful product.[130]

Ullman's support for "functional" organization prefaced the scientific turn
in postwar human geography, but he also recognized the need for a related
"breed of *intelligence* cats" to bring discussions back to a regional scale
at any time.[131] Inside the OSS, more than any other wartime agency, the
definition and value of *area* or *region* was contested, caught between the
need for basic strategic information, of the type found in the ASTP, and
the more ambitious aims of the social scientists tasked with providing this
information. Cold War area studies resolved this tension.

R&A was the heart of the OSS, the agency responsible for collecting
and, more importantly, for evaluating foreign intelligence — a centralized
capacity that the United States lacked before the war.[132] And given the linguistic and research abilities and the regional awareness required of intelligence analysis, the academic community was an obvious place to hunt
for talent. In advance of the OSS's secret intelligence or covert operations
divisions, the R&A branch was in place, charged more with the study and
"finishing" of raw, compiled information than with explicit policy recommendations. It was home to a remarkable group of scholars. Led by the
Harvard historian William Langer, R&A attracted to its staff such diverse
individuals as Wassily Leontief, Herbert Marcuse, Franz Neumann, Walt
Rostow, Arthur Schlesinger Jr., Carl Schorske, Edward Shils, Paul Sweezy,
and, in addition to Ullman and Stone, geographers such as Edward Ackerman, Chauncy Harris, Richard Hartshorne, and Preston James. Frequently
working in teams, they compiled and interpreted source material relevant
to the operations of war.[133]

As was the case within the Manhattan Project, military officials at the
OSS, beginning with its head, General William "Wild Bill" Donovan, were
mostly tolerant of left-leaning intellectuals deemed useful to the American
cause. R&A's list of luminaries alone enabled the OSS to carve out a niche
in the crowded field of Washington information clearinghouses. Early on,
the core of R&A's elite community was drawn from an Ivy League cadre
of young gentleman scholars whose education had instilled the merits of

what Schlesinger called "thoroughly objective and neutral" research techniques in the humanities and social sciences.[134] Schlesinger's objectivity
was more moral than methodological; he was no proponent of abstract
empiricism. But those recruited also "possessed expert knowledge of particular regions or localities."[135] This knowledge fit well with Donovan's
desire for centralization and calculation:

> We have, scattered throughout the various departments of our govern-
> ment, documents and memoranda concerning military and naval and
> air and economic potentials of the Axis which, if gathered together and
> studied in detail by carefully selected trained minds, with a knowledge
> both of the related languages and technique, would yield valuable and
> often decisive results.[136]

As their ranks diversified, R&A scholar-analysts remained united by
an opposition to totalitarianism and a disdain for relativism.[137] The search
for certainty in social science reflected a broader cultural shift during the
build-up for war. But within the walls of the OSS, a combination of the
"wide use of sources, avoidance of the first person singular, and an absence
of overt political partisanship" was usually sufficient to earn the stamp of
scientific scholarship. When formal reports authored by more than one researcher and containing summaries and conclusions were prepared, they
were often heavily edited, usually within Hartshorne's influential Projects
Committee, and stripped of identification and emotion.[138] This collaborative, interdisciplinary, and surprisingly efficient system was new to many
on staff, but it was essential to the preparation of integrated studies — the
work found so useful by R&A clients.

In an August 23, 1945, memo to Donovan summing up the service
and "assets" of R&A, William Langer noted that "the specialized knowledge as well as the training in research of American universities was for the
first time made the core of government service."[139] The consequences were
significant: Langer wrote three years later that some R&A analysts were
"shocked . . . to find how narrow much of our specialization had become
and how difficult it was to get people from the various disciplines to work
together."[140] Shattering the "artificial barriers separating one approach from
another," in Langer's words, was the solution.[141] This was done in early
1943 by dismantling disciplinary units and replacing them with divisions
based on military theaters, a distressing if momentous decision.[142] Regions

were thus paramount, but only functionally so, as manageable containers for interdisciplinary collaboration—precisely the organization that characterized subsequent area studies initiatives. The connection was unsurprising, since according to the consummate Cold War intellectual McGeorge Bundy, numerous postwar area studies programs were "manned, directed, or stimulated by graduates of the OSS."[143] Back on their respective campuses, these graduates forcefully advocated the collaborative methods and objective postures adopted by R&A. Many of their students, in turn, wound up in Langley, Virginia, at the CIA.[144]

During the Second World War, the connection between academia and intelligence was "hardly a shady plot." Still, it was not at all clear whether this clarity would survive in the more murky terrain of the postwar period. And yet, many R&A leaders had concluded that a more lasting connection between universities and intelligence agencies for research on regions was a worthwhile enterprise, a judgment shared by State Department planners such as George Kennan. As early as the summer of 1943, Langer had explicitly directed his various outposts to consider intelligence from a "long-range" perspective. As the next chapter shows, a number of OSS veterans, as well as other social scientists who worked in similar wartime capacities, continued to enthusiastically consult for the state while sending their best graduates to its various branches and advocating for interdisciplinary area studies at their respective universities.[145]

Wartime intelligence work also obviously required cartographic support. R&A's Map Division—headed, as mentioned earlier, by Arthur Robinson—was initially supplied by the Library of Congress, the Department of State, and the Army Map Service.[146] The Map Division then supplemented its collection after a national radio appeal by OSS Director Donovan, and by the war's end, R&A possessed over two million maps. In four years, the Division also answered fifty thousand cartographic requests and produced eighty-two hundred new maps.[147] Wartime cartography was aided by the 1943 reorganization of the Board on Geographical Names, which sought to regularize the proliferation of unfamiliar place designations. These titles had become "'fighting' words, tools of war."[148]

As the OSS grew, Map Division offices were established in Algeria, Egypt, India, and China, resulting in a number of audacious map-procurement expeditions. The London branch of the Division solidified agreements with the exiled governments of several European states to produce maps made from "delivered" geographic data. Other maps and models were passed on to Washington from field offices, and OSS map teams examined German cartographic collections and interrogated Nazi military geographers. Admitting that "the overwhelming majority of American geographers had had little training in the use of maps, particularly foreign maps," the Division's Deputy Chief suggested that the subject of map information might be included in any reevaluation of geographical inquiry focusing on "systematic" elements. He was, on the other hand, impressed with the regional achievements of the Division, suggesting that they stretched far beyond "military purposes."[149]

The piles of items accumulated by the Map Division and the lack of preexisting information about the collection necessitated the development of expanded classification and cataloging standards. After casting around for suitable precedents, the decision was made to "make a fresh start." The primary partition of materials was by area. While this was not radical, the choice of appropriate areas does merit mention. Aware that political factors were central to "most maps," Division researchers identified an important cartographic class that did not conform to global, hemispheric, or continental scales: "parts of the world which are smaller than continents, but contain more than a single political unit." The solution was the partition of the world into twenty-two "primary regions," including oceans and the poles, broken down further by secondary and tertiary filing categories. Primary regions, it was assumed, were large enough to accommodate political change. This system was more flexible than segmentations dependent on exact limits, but only insofar as political boundaries shifted. It could be made to fit an altered atlas. And the OSS cataloging was, obviously, undertaken during a period of tremendous changes "in the political control of territory," not just as a result of military victories but also due to widespread decolonization. There was also the question of subject classification, or the branches of "systematic geography," that appeared, in various combinations, on the same maps. These, however, presented a far less significant problem, since it was simply a matter of identifying appropriate aspects of the physical or cultural landscape and attaching them to an area.[150]

The exploits of the Map Division, of course, paled in comparison to other OSS groups, particularly those such as the Special Operations Branch that engaged in covert activities to support military campaigns and resistance movements. It is difficult to avoid the narratives of skullduggery and derring-do that proliferate in discussions of these groups. The anthropologist Carleton Coon, who operated under diplomatic cover for the OSS in North Africa, summed up these themes by stating that "it is probably the secret ambition of every boy to travel in strange mountains, stir up tribes, and destroy the enemy by secret and unorthodox means."[151] These sentiments, which are part of a long tradition of masculine adventure travel, were neither novel nor remarkable and can be found in an array of literature relating to the Second World War, from the accounts of scholars such as Coon to manuals such as *Survival on Land and Sea* and the "Fighting Forces" handbooks distributed to voyaging soldiers.[152] The 1943 *Pocket Guide to Alaska,* prepared by the Army with the assistance of the OSS, advertised the "close-up of America's last frontier in action" to those posted north.[153] Compact manuals and guides consulted and carried by soldiers and spies were twentieth-century versions of the "hints" for colonial travelers, establishing the premises of a war on regions and their populations in advance of departure. "Distant dangers," according to one aid tellingly titled *On Your Own,* "seem most hazardous. Remember this when you go into strange country."[154]

Another anthropologist sent overseas on behalf of the OSS was Gregory Bateson, who spent much of 1944 and 1945 in Burma, Thailand, Ceylon, India, and China. Among his duties was the creation of false, exaggerated Japanese propaganda broadcasts in Thailand and Burma. Unlike his wife, Margaret Mead, Bateson, who "took part in the communications systems which he helped to create or hoped to disrupt," grew disillusioned with applied anthropology after the war, perhaps because he had participated in much more visceral operations.[155] Yet he seemed to throw himself headlong into his OSS tasks, which also included analyses of raw intelligence, compositions on the conduct of intelligence, and even secret rescue missions. Nor did he hesitate to comment on the potential postwar role of the OSS in Asia, arguing, in the language of culture and personality studies, that an American neocolonial order could, and should, be maintained by altering local attitudes and dissolving older institutions of empire. As Bateson wrote in a 1944 OSS memo, by studying, encouraging, and shaping "native achievements" rather than imposing an external cultural model and risking the rise of "nativistic cults," the United States could ensure that nations such as India progressed in the appropriate direction.[156]

While still in the field, Bateson managed to compose a sobering memo to Donovan just days after the destruction of Hiroshima and Nagasaki — incidents that eventually propelled Bateson further toward the clarity of cybernetics. He wrote that the atomic bomb altered the "relation between *attack* and *defense*," increasing the likelihood of psychological and economic warfare because those with and without the technology would collectively seek to avoid atomic confrontation at all costs. Agencies responsible for the indirect, "peaceful" methods of combat would thus be even more pervasive and powerful.[157] This irregular or shadow activity, which fascinated OSS chief Donovan, inspired the psychological warfare techniques standardized by the CIA in the 1950s. Donovan himself acknowledged that these unorthodox approaches were neatly applicable to the struggle with a conspiratorial communism.[158]

After the war, the OSS developed a reputation as an organization that had produced extraordinary achievements in the field of espionage. Part of this mythology resulted from the deliberate efforts of William Langer, the director of R&A, who later took an additional hiatus from Harvard to establish the R&A-like Office of National Estimates at the CIA in 1950. He also assisted in the creation of the CIA's Office of the Historian, which subsequently churned out a series of histories that lambasted prewar American unpreparedness and lionized the prophetic Donovan. One insider's account describes Washington officials

> astounded by erudite government reports in the language of Harvard, presented in *Reader's Digest–Life* style. The maps, the topographic models, and, in particular, some five-foot floating globes — supported on hidden ballbearings — brought distinction as well as occasional envy to the [OSS].[159]

Responding to these exaggerations, revisionist historians have described the OSS as "the product of foggy or even nonexistent reasoning allied to a large dose of bureaucratic opportunism boosted by effective hyperbole," its achievements overshadowed by the more practical successes of its British and American military counterparts, who were privy to better intelligence.[160] A shrewd public relations tactician, Donovan realized that the minds of R&A, however impressive to some, were hardly as magnetic as covert operations. A desire to remain in Donovan's favor led Langer to reluctantly move R&A, dubbed the "Chairborne Division" in Washington,

closer "toward the battle zone."[161] But Langer's preference for pure strategic intelligence also made possible the intricate regional structure of
R&A. His efforts were only partially successful at reducing the distinction
between R&A and the rest of the OSS, and they were not aided by the
reluctance of many R&A scholars, whether because of their idealism or
squeamishness, to work in "compromised" places such as China. Overall,
the OSS became a "sideshow," if an important one, since the obsessive bureaucratic battles fought by Donovan and others contributed to the emergence of the security state and one of its key components, the CIA.[162]

Given this last contribution, to dismiss the OSS as insignificant is to
follow an overly instrumental route. The security state so frequently associated with Cold War America was a complex entity. It encapsulated all
manner of intelligence production and analysis, both informal and formal.
The members of R&A were often several degrees from policy decisions.
When they did move into these spheres, near the end of the war, ideological tussles became increasingly common, as in the case of a legendary dispute over the fate of the German Socialist Party between Richard Hartshorne's Projects Committee and Herbert Marcuse, Carl Schorske, and the
Yale historian Sherman Kent, chief of the Europe–Africa Division. This
was a disagreement that reflected the strains of interpretive and explanatory methodologies pushed together.[163] The key outcome, however, was
that R&A staff not only dispersed to new centers for the regional study of
a global world but also amassed resources and experience on areas that
roughly conformed to these same regions, most notably the Soviet Union.
While Langer transferred a similar framework, and a few personnel, to the
CIA, the "decisive impact" of R&A, according to Barry Katz, was in that
"ultimate de-centralized intelligence agency," the university.[164]

Despite his frustrations, meanwhile, Edward Ullman continued to
contribute to the JANIS effort after the war. In March 1946, as studies
on the "European USSR" and Manchuria were prepared, he drafted a report titled "The Future of JANIS." This document noted the use of JANIS
across the armed forces, including "the military training system," and,
once declassified, within other government departments. The relevance
of these "studies on foreign areas," Ullman argued, would linger "because
of the static nature of much of [their] content," an inert quality that presumably included both physical and human geography. However, some of
the contributors could simply not keep up with the increased demand of

the postwar period. Ullman wrote that as a result of "the widespread nature of American interests, more of the world needs to be covered now
than during the war when efforts were concentrated on theaters of operation." The intent was to provide "handbook material" for traveling American diplomats and politicians, to prepare the military for minor "policing" duties, and to contribute to intelligence tasks that grappled with new
methods of warfare, particularly those involving atomic weapons.[165]

Wither Geography?

Less than two months after Japanese officials signed an instrument of surrender in Tokyo Bay, the Social Science Research Council's Committee on
Problems and Policy met to survey the "changed situation brought about
by the war." A key topic of discussion was the accelerated "trend toward
regional specialization." While the products of wartime regional work did
not all match the "best research standards," important advances were made
in the "compilation and organized presentation of factual material." The
publications singled out for praise included the JANIS volumes, Navy Civil
Affairs handbooks, and OSS reports. The study of geography, as a result of
this accumulation of resources, had "advanced farther in this brief period of
wartime research than it would have in fifty years of normal endeavor."[166]
Those within geography departments felt similarly and invoked this new
status, using the experiences of war, to advocate for their discipline.[167]

A postwar report of the NRC's Committee on Training and Standards in the Geographic Profession identified six fields where American
geographers had made important wartime contributions: regional surveys,
cartography, topographic models, map intelligence, place names, and administrative duties. But the same report identified several tasks fulfilled,
inappropriately, by nongeographers, including considerations of terrain and
water supply, climate and weather, port facilities, commodity studies, and
major policy decisions (with the noted exception of Isaiah Bowman). As
Syracuse University's Preston James summarized in 1946, the readjustments
required of geographers were dependent on finding ways to "fit with the
work of other social scientists" and do so while maintaining a distinctive
"geographic point of view."[168]

In one of the few critical histories of academic geography and the Second World War, Andrew Kirby cites Owen Lattimore's observation that

government service was, for geographers and others, a seductive and corrupting enterprise, offering access to classified material, guaranteed funding, and an entrance (however partial) into the halls of decision making and influence.[169] Like other social sciences, this was a type of experience that had begun during World War I, when calls for an improved global geography led to an increased focus on *human* concerns. Scholars were thus, as Harvard's Derwent Whittlesey put it in a 1941 editorial, "ready to speak and move promptly and to the point" at the advent of another conflict.[170] Commentaries proliferated on the contributions "which the science of geography could make to the conduct of war and to subsequent reconstruction."[171]

A 1947 survey found that from 1942 to 1945, the American government employed two out of every five geographers who were members of the three national associations: the American Society for Professional Geographers, the Association of American Geographers, and the National Council of Geography Teachers. Geographers traveled and resided in countries outside of the United States to an unprecedented degree and for various purposes.[172] Washington, where many geographers were based, became, according to a 1942 *Newsweek* piece, a "city of maps" where it was "considered a *faux pas* to be caught without your Pacific arena."[173] The status of cartographic images as fashionable icons of knowledge and power had been solidified. Geographic representation and observation was also seen as key to the production of a postwar peace and reaffirmed, for Stephen Jones, the "need for field work" to "look down the vistas that spread before political geographers" firsthand.[174] In sum, as Kirk Stone famously declared many years later, "World War II was the best thing that has happened to geography since the birth of Strabo."[175] The veracity of this claim is less important than the suggestion of change; it is certainly the case that the war, and more specifically the demands of global war, hastened a significant shift in American geographical thought.

The interlude spent by geographers in government agencies during the Second World War, Kirby writes, was "not a diversion: it helped redefine their subsequent intellectual positions."[176] Many supported the trend of cooperative research that would bear directly on political problems and foresaw the continued relevance of intelligence work and area specialization. But in the opinion of Edward Ackerman, among others, despite geography's "unquestionably... wider recognition" in the United

States, its practitioners had entered service positions with poor training, and they were unable to provide a useful body of facts to the war effort. Two deficiencies, for Ackerman, were particularly damning: the "inability to handle foreign languages and lack of competence in topical or systematic subjects." Geography's regionalism, with individual exceptions, lacked international coverage, or a rigorous pattern, as well as social scientific principles that sprang from and encouraged collaborative work. "If our literature is to be composed of anything more than a series of pleasant cultural essays, and if our graduates are to hope for anything more than teaching positions," Ackerman stated derisively, "we shall do well to consider a more specialized, or less diffuse approach."[177]

Ackerman believed that geographers were insufficiently worldly, and they had additionally made the wrong choice by dividing regional and systematic methodologies and then overwhelmingly selecting the former as more suitable. This was a telling clue to the mystery of geography's absence from the Ethnogeographic Board. At R&A, where Ackerman led the Geographic Reports section and later moved to the Europe–Africa Division as assistant chief, the regional approach was not chosen because of its proximity to reality but rather for its strict utility. There was little attempt to identify the variation between regions. Instead, it was suggested that the same forms of inquiry could be carried out within *any* region and that this inquiry could be cumulative. According to Trevor Barnes, the 1943 reorganization of R&A "should have been one in which geographers flourished given the new regional geographical focus. Instead, the reorganization made the inadequacies of geographical training seem that much starker."[178] Leonard Wilson captured this embarrassment bluntly: "Wherever they trained those geographers found themselves confronted by other social scientists, with earlier and better developed claims."[179]

Ackerman was not willing to discard areal differentiation. He maintained that it was the heart of geography. But he did find it ironic that, despite the prewar preoccupation with regions, the resulting scholarship could not provide "adequate data for wartime geographic research on few, if any, parts of the world." This data, Ackerman suggested, covered such a "wide range of subjects" that "no matter how long we had worked," it could not have been gathered using older techniques but instead required the skills of "systematic specialists" who could more successfully attempt correlation. Those "technicians" whose skill with detail was offset by poor interpretative abilities should, like laboratory employees in physics or biology, be put to use on "mechanical work of the mind and eye" — an equivalent status to those educated within the ASTP.[180] Like many contemporary writers, Ackerman believed that the globe as an object of investigation was too complex and interrelated for an individual regionalist. His concern was not with the provision of data — the "place-specific information" that wartime geographers were adept at providing — but with its detail and integration.[181] In a vivid analogy that became popular in Cold War area studies, Ackerman likened regional geography to the medical concern for a single body part, whereas analyzing "functional units" such as skeletal structure could lead to more profitable discoveries. This advice did not apply only to the special circumstances of war. Adopting the standards of rigor and interdisciplinarity held by intelligence agencies would improve all aspects of peacetime geography, as would the abandonment of the idea that the world was a "mosaic of localities, districts and regions, with a potential student assigned to describing each tile." Geographers, in short, needed to measure up to the hardened social sciences whereby a more thorough understanding of regions would become the ultimate goal of a unified systematic geography.[182]

Many American geographers no doubt read Ackerman's important call for practical knowledge that could contribute to social and political problem solving. But his demand for additional systematic scholarship, so parallel to the promotion of area studies, was not exclusive to geography. It was a call that swept across the social sciences at the end of the Second World War. And while the laws generated as part of the response are a key part of this intellectually tumultuous period, we would do well to consider also the *data* that Ackerman claimed was lacking in prewar geography. This data, "never raw," was used interpretatively by those seeking to formulate laws.[183] Its collection and study led to the regional intelligence of the Second World War.

The case, however, is incomplete without a Cold War sequel. Robert Matthew's 1947 evaluation of the ASTP's area and language courses recognized the proliferation of regional knowledge — in the form of "language guides, pocket guides, war background studies, and the civil-affairs handbooks which were distributed in great numbers" — that had occurred since Pearl Harbor. The stressing of "little-known regions," the contemporary focus, and the interdisciplinary methodologies found in these and other texts were all novel, bearing little resemblance to what counted as foreign-area programs before 1941.[184] But in addition to an awareness of the language, culture, or geography of a particular area, those who deserved the important title of regional experts in the postwar period would additionally be what a 1944 Rockefeller Foundation memo called "subject-matter specialists," preferably in one of the authoritative social sciences.[185] Matthew was also clear on a "possible priorities arrangement" for future area studies in civilian institutions: it should begin with "the Slavs."[186]

3 ILLUMINATING THE TERRAIN
Social Science Finds Its Targets

> At the *end* of this road of increasing frequency and specificity
> of the islands of theoretical knowledge lies the ideal state,
> scientifically speaking, where *most* actual operational hypotheses
> of empirical research are directly derived from a general system
> of theory. On any broad front, to my knowledge, only in physics
> has this state been attained in *any* science. *We* cannot expect to be
> anywhere nearly in sight of it. But it does not follow that, distant
> as we are from that goal, steps in that *direction* are futile. Quite the
> contrary, *any* real step in that direction is an advance. Only at this
> *end* point do the islands merge into a continental land mass.
>
> —TALCOTT PARSONS, *American Sociological Review,* 1950

By the end of the 1950s, the habit of partitioning the world into three parts
was commonplace across disciplines, political perspectives, and even
states. The simplistic "three worlds" framework was firmly anchored, if
not unequivocally tailored, to the *metageography* of the Cold War and simultaneous decolonization.[1] Those who invoked a tripartite globe were
keenly aware that the third, developing sector was the object of a political and economic competition between the first and second worlds. In
the United States, modernization theory was the most significant scholarly outgrowth of the Cold War's stark divisions. Modernization theorists
mandated that in the titanic struggle between first and second worlds,
with their equally generic characteristics of freedom and totalitarianism,
the third world had to "choose" the correct path. This adage could be used
to advocate or justify the overthrow of unfavorable governments or be
held up as an example of a new liberal pluralism. In this respect, American
social scientists, with few exceptions, overwhelmingly embraced national
interests during the 1940s and 1950s, continuing a trend launched by patriotic contributions to the Second World War. To cite just one example,

this smooth transition is apparent in the essays—and the title—of the 1952 volume *Current Trends: Psychology in the World Emergency,* a collection that was both intellectually and geopolitically current.[2]

If there was a single thread running through the American social sciences during the early Cold War, it was that the nonideological character of the United States distinguished it from the tainted technocracy of the second world. Yet research on communist societies was not segregated from the rest of the social sciences. The information gleaned from studies of the second and third worlds was ultimately "suitable to the formation or modification of general laws."[3] Forays into other spaces under the sign of area studies thus enabled the globalization of social scientific theories built first within an American environment.

In the case of the third world, that such expeditions be both intellectual and physical, conducted from the library and within the field, was crucial. The same could not be said about the study of the communist bloc, especially before Joseph Stalin's death in 1953, although intriguing efforts were made to compensate for the overwhelming absence of local observation.[4] Meanwhile, academics, foundation employees, and intelligence professionals argued together that the Soviet Union was taking advantage of the open society of the United States to skim plenty of valuable information. In contrast, American knowledge of the Soviet Union was felt to be unacceptably slender. Area studies scholarship was therefore very much aligned with intelligence estimations of Soviet capabilities and intentions, assessments that were the very fabric of Cold War rivalry. Given its expansive title and geographic reach, area studies has long been a realm of contention, and some of that diversity is acknowledged here. But I am interested in the particularly useful role played by area-based institutions as practical vehicles for the generation of strategic knowledge.

Much social scientific research in the United States during and after the Second World War was not just collaborative but cumulative: it bridged and combined disciplines and advanced quests for total theories and total archives of knowledge. The latter were especially important. Gathering data propelled theorizing ahead and simultaneously proved the utility of social science in the most challenging of circumstances. A new stratum of intellectual integration had been reached. If the "fundamental aim" of the Social Science Research Council (SSRC) was to improve "knowledge concerning human relations," as Penn's Donald Young wrote in 1949, then benefiting human welfare through the control and use of this knowledge could not be far behind. But these benefits would be reduced without appropriate "organization for research."[5]

What Henry Kissinger called the world's "grey areas" were used, in a colonial echo, as laboratories for the testing of methods and theories originally conceived on American campuses.[6] Universities were crucial institutions for the training of *national subjects* who possessed global knowledge, and university area studies institutes were the seedbeds of modernization theory and related developments across the social sciences.[7] At sites such as MIT's Center for International Studies (CIS), a slightly more nuanced version of the three-worlds framework was prevalent, but it still permitted a region to be summarized in a single descriptive word — *traditional,* for instance. And because the growth of area studies in the United States was intimately tied to military and intelligence agencies, as well as prominent philanthropic foundations, the spatial knowledge produced by area studies, on "areas of concern," was by definition strategic.[8]

Bruce Cumings has argued that despite the prominent role of state-centered economic and political power in the shaping of area scholarship, the most intriguing effects of this power actually occur at the various intermediary locations where it is spread, filtered, and transported among more perceptible nodes.[9] Such institutions are crucial because they produce "not only forms of representation but also mediations among representations, behaviors (both personal and social modes of subjectivity), and larger social processes."[10] This chapter travels to several sites of Cold War social science, setting down in laboratories, libraries, and boardrooms. But it is also concerned with the relationship between these places and the solidification of the strategic regional framework discussed in chapter 2. Cold War considerations, of course, led to the identification of new areas and new methodologies for their study. Equally, the ambitions of social scientists strained increasingly at the regional structures that had received so much investment. Even in the most abstract of models, however, geographical elements lingered, unacknowledged or buried beneath clean surfaces. The contours of the American Cold War were also often the limits of rationality, and abstraction could be used to make certain strategic spaces *reasonable.* While not always perfectly reconciled, intellectual advancement and service to the state certainly made a comfortable match for many of the era's most prominent scholars.

In his 1949 book *Strategic Intelligence for American World Policy,* the Office of Strategic Services (OSS) veteran and Yale historian Sherman Kent posed a question: Should "the basic pattern of intelligence organization be regional or functional?" His answer was a compromise: the globe must be broken down into "four or five major geographical areas" and these areas then divided further into smaller regions. Intelligence work, Kent observed, was frequently national or regional, as was the data flowing into agencies or already available on file from the Second World War. Still, efforts that prevented national or regional grids from becoming static were also essential, because espionage did not always respect such boundaries. This chapter does not directly address the early history of agencies such as the successor to the OSS, the Central Intelligence Agency (CIA), where Kent occupied a key role at the Office of National Estimates and where the Sherman Kent School for Intelligence Analysis was dedicated in 2000. Instead, I focus once more on the generation of regional intelligence — a type of geographical knowledge typically overlooked in conventional disciplinary histories — at precisely those area programs that Kent believed would "produce exactly the kind of expert" he envisioned for collaboration between the CIA and universities. Scholars of this variety, he went on, would simplify the "administrative problems of intelligence organizations," because certain classes of analysis could be farmed out and hidden behind a shroud of dispassionate academic research.[11] These chains, as I show, were pervasive, but the Cold War and the social sciences were also intertwined in much more insidious ways.

Scientific Aspirations

Even as it was overshadowed by the achievements of wartime scientific research, social science was seen as a sphere of tremendous opportunity after the Second World War. The same methods used to develop radar or the atomic bomb could, it was believed, be deployed to reaffirm and understand "reason, security, and social peace under the umbrella of the United States."[12] As a result, the definition of *applied* scholarship was broadened precisely as government funding and facilities allowed for data collection and hypothesis testing "on a scale previously unimagined."[13] Historical accounts casting back to the years between 1945 and 1960 document "substantial advancement" in the social sciences, from numerous landmark publications and an "unprecedented volume of work" to a new generation of academics trained in appropriate research practices.[14] Support that trickled in from a reluctant federal government — reluctant at least compared to the investment in national "big science" projects — was supplemented by the generosity of philanthropic foundations. Across the social sciences, money that did come from government was overwhelmingly provided by the military; one 1952 National Science Foundation report put this figure at 96 percent of the total.[15] But these were not "one-way conduit[s] of influence." Some intellectual approaches were easily compatible with military funding and worldviews, while others resulted in much more hesitant affiliations with Cold War priorities.[16]

Much of the history of American social science during the 1940s and 1950s can be characterized as a collective attempt to pull theories and methodologies in line with the perceived successes of the physical sciences.[17] One result of this drive for additional respect was a paradoxical blurring *and* reassertion of disciplinary boundaries. Projects, centers, and grants were set up that endorsed cooperation but inevitably excluded one discipline or another, and more often than not, geography was among the ostracized. Following the standards set by both the OSS and the Manhattan Project, interdisciplinary teams were seen as necessary in the face of an increasingly complex reality. Neither geopolitical nor intellectual isolationism was permissible. However, the extension of scientific status could only proceed cautiously, after all, and thus it was best to begin with fields that had already begun to infiltrate the temple — such as economics and psychology — and with measured, verifiable empirical studies rather than theoretical flights of fancy. But as the Talcott Parsons quote that opened this chapter indicates, it was equally important to set each discipline on the same general path.

Within the net of the leading behavioral sciences (as they began to be called), and also in apparently cruder fields such as geography, quantitative methods were encouraged as a means of standardizing phenomena. This was an approach signaled by the formation, in 1952, of the SSRC's Committee on the Mathematical Training of Social Scientists. Scientists such as the engineer and physicist Lloyd Berkner, a member of numerous Cold War initiatives and projects, demanded as late as 1960 that social science find elementary, fundamental, and independent concepts or parameters, whose coefficients can be determined numerically, and which combined in suitable mathematical formulations could predict analytically something about the ultimate capacities of the individual.[18]

If groups were merely aggregates of individuals, the body could be mapped in tandem with the nation, which could in turn be simplified as a collection of "social facts, political publics, and economic markets" or characterized bluntly as either backward or enlightened.[19]

Such a framework, typified by an unambiguous faith in numbers as "nature's own language," and the excuses of precision, verification, and testability trotted out in its defense, was certainly not just a Cold War phenomenon.[20] When Michel Foucault described the quest for "man as an empirical entity" as characteristic of the human sciences, he did not locate this search in the middle of the twentieth century.[21] Yet notwithstanding various international intellectual debts, the United States during this period was uniquely marked by pervasive aspirations to scholarly detachment, from economics to philosophy and literary criticism, which have been given the suitable catchall label of *rigorism*.[22] But these ambitions, however radical or novel, were complemented by conservative suspicions of change, uncertainty, and social criticism, misgivings resting nicely behind a shield of supposed neutrality held up to guard against charges of ideological depravity.

A second condition provided additional spatial context: the argument that the triumphs of the sciences had "won the war" was crucial for calculating social scientists, because it meant that any quest for universality required an accompanying practical application (an awareness gained in the prewar New Deal period as well). The aversion to excessive complexity was characterized by the virtual elimination of "the distinction between the technical and the practical" as one was enclosed "within the categories of the other."[23] This move was justified by the organization of defense research — not just because of the questions that were asked, or the answers sought, but owing to a *security problematic* that demanded cultural and geographic certainty and also encouraged difference rather than similarity in the study of the "remote and the strange." In this sense, the demon of relativism was discarded while social and political discrepancy was not.[24]

To respond to the demand for utility, social scientists were forced to confront the peculiarly human aspects of their research. What was the meaning of a controlled experiment in urban, national, regional, and global societies that seemed, more than ever, to be fluid? The answer was not the abandonment of a laboratory sensibility, so it was argued, but rather a redefinition of what constituted a laboratory, as in the case of the Human Relations Area Files (HRAF). This task required the creation of consistent tools to maintain levels of generalized authority and legibility under mobile conditions. Similarly, as a Rockefeller Foundation official put it in 1948, the worth of "relatively simple societies for the intensive study of specific problems" was substantial and was one way of building up to the challenges of "modern societies."[25] But given a backdrop of geopolitical antagonism, social scientists could ill afford to be preoccupied by merely cultural distinctions between primitive and modern spaces. Certain environments were more strategic than others, and the definition of *scientific* was, as a result, in some instances extremely malleable.

Although demands for advances in social engineering were already common in the early twentieth century, social science was seen as increasingly relevant after the Second World War precisely because of the uncertainty brought about by rapid technological and geopolitical change as well as the cultural contact and psychological insecurity accompanying these alterations in the fabric of modernity. A "revolution in our physical environment," as Rockefeller Foundation president Raymond Fosdick described it in 1946, was destabilizing the very foundations of the ground beneath human feet. For all of its recent triumphs, American science had also prompted a new set of ethical debates. Rockefeller's Annual Report for 1945 stated plainly that humans were "discovering the right things but in the wrong order."[26] Social science might smooth some of these concerns, but it could not do so at the cost of progress, and its role was therefore more one of management and adaptation — rationalizing the human factors that went into scientific research and application, observing the conditions of life, and promoting "orderly adjustment to technological advance."[27]

In a speech titled "Sociology and the Strategy of Social Science" to the American Association for the Advancement of Science in 1948, Samuel Stouffer diagrammed the route to "orderly adjustment":

> By developing limited theories, testable and tested empirically, by being
> modest about them and tentative, we can, I think, make a small but
> effective contribution toward an ultimate science of society whose engi-
> neering applications will help regulate the complex civilization wrought
> by physical science and society.[28]

This was not an unusual comment; by the time of Stouffer's address, many
social scientists, including some geographers, "had begun to see themselves
as engineers, designing and evaluating complex systems."[29] And one of the
most effective ways to restrain Stouffer's "limited theories" was to do so
geographically — to attach them to a particular region. Such spatial sepa-
ratism also produced ideal opportunities to practice the regulatory aspects
of social science.

Asserting the scientific legitimacy of the social sciences while maintaining their exceptionality was a difficult task. Much of this discussion
was characterized by equivocation. For instance, in a 1950 booklet on
research for the federal services, two SSRC representatives argued that
while the preoccupation of the social sciences was "man, rather than the
substances and forces that surround man or the lower orders of organic
life," humans were still "subject to physical laws" that could presumably
be clarified and inspected. Although a "nation, a community, a family, cannot readily be put in a test tube," this did not mean that the search for
"uniformities of behavior" should be discarded. More impressive than
these general ruminations were the examples selected to illustrate "new
approaches to the study of social behavior": they included the interview
methods of the Office of Strategic Services; the community studies of the
War Relocation Authority; the HRAF; and the propaganda, psychological
warfare, and administrative techniques aided by culture and personality
studies. The militarization of social science was a point of pride. During
World War II, the SSRC authors noted, many detached scholars had been
"converted into social practitioners."[30] What was on display in such advertisements was not just a sensible synchronicity but also an intentional one.
By distinguishing, and then entangling, the dialects of social science and
social practice, entry to the realm of Cold War strategy was made more
certain.

During the Second World War, Samuel Stouffer was the director of
the Research Branch of the War Department's Information and Education
Division. The military was an ideal organization for the testing of social

scientific theories; it was an institution where knowledge of the self remained as important as the determination of an enemy. For Stouffer and dozens of colleagues, the massive and influential study *The American Soldier* (1949) was the result. The scale of the book's background research was unprecedented and was encouraged by Stouffer's government sponsors. Over two hundred questionnaires were handed to more than a half million soldiers, and the results of these tests, in addition to the analysis of "operational statistics," were compiled in a four-volume set of books.[31] The Research Branch, Stouffer wrote, "existed to do a practical engineering job, not a scientific job," but in the softer light of the postwar period, its work could be added to the cumulative achievements hastening "the development of a science of man." Initial results of the research were published in December 1942 as *What the Soldier Thinks* and were used in the planning of the GI Bill and demobilization programs.[32] But the broader influence of *The American Soldier* derived from its balance of quantitative methods with regulatory recommendations.

Stouffer was a key member of Harvard's Department of Social Relations after the war, and continued to do contract research on leadership and behavior for the military. The SSRC and the Carnegie Corporation generously funded his research for *The American Soldier* and other projects.[33] Some social scientists were quick to distinguish their craft from that of less meticulous inquiry, but the opposite complaint was also common, and several of the scholars charged with reviewing *The American Soldier* intensely questioned Stouffer's military sociology. Nathan Glazer wrote in *Commentary* that while Stouffer's aim was undoubtedly to "create sciences" on the model of physics, the result was "the mechanical and formal confining of knowledge, not the increase of it."[34] Arthur Schlesinger Jr. went further, denouncing the study as a "ponderous demonstration in Newspeak" and claimed, strikingly, that sociology had "whored after the natural sciences from the start."[35] In a *New Republic* essay, "The Science of Inhuman Relations," Robert Lynd argued that the books depicted "science being used with great skill to sort out and control men for purposes not of their own willing."[36] However much these critics clung to a model of pure science, their charges should not be dismissed. Stouffer's project, after all, attacked a certain kind of grand theory, trumping speculative, humanist inquiry with a more orderly methodology. And in the influential hands of Paul Lazarsfeld and Robert Merton, and aided by a massive public relations campaign, *The American Soldier* was ultimately redeemed as a model of social science that eliminated "guesswork and conjecture." But it also pioneered a brand of applied behavioral research that the military found quite useful (and supported handsomely) precisely because of the benefits for techniques of human manipulation.[37]

The 1947 passage of the influential National Security Act was a milestone in the identification and funding of suitable Cold War research. Among other novelties (including the CIA and the National Security Council), the Act created a Research and Development Board (RDB) within the remodeled Defense Department. The RDB was not wholly new; it shared much, including the same first chair (Vannevar Bush), with the wartime Office of Scientific Research and Development. But unlike its predecessor, the RDB did not actually distribute funds for research contracts or supervise research. Instead, it was composed of a series of coordinating committees. Within these, panels, sometimes of limited duration, were created to tackle more specific subjects, such as radar and Arctic environments. Although most RDB committees were scientific in character, one directly addressed the social sciences. Chaired initially by Donald Marquis of the University of Michigan, the Committee on Human Resources included military and foundation representatives and many of the most prominent social scientists in the United States, all scattered across its panels. Attention was concentrated on four related themes: psychophysiology, personnel and training, manpower, and human relations and morale. Similar fields were also prioritized by the Office of Naval Research (ONR) and by social science research centers within the military, from the Air Force's Human Resources Research Institute to the Army's Operations Research Office at Johns Hopkins University.[38] Not only did all of this activity suggest that the organization of social scientific research for defense, to paraphrase I. I. Rabi, had continued after the Second World War, but it had proliferated.[39]

Foundations and the Relevance of Area Study

Writing for the Carnegie Corporation in 1952, William Marvel took stock of the explosion of area studies programs in the United States, arguing that the phenomenon was "closely related to the changed position of the United States in world affairs—as contrasted with the situation prior to

World War II." Prewar areas, Marvel went on, were sites investigated by individual scholars, and a global conflict had turned these regions into "segment[s] of humanity" with a direct bearing on American interests. Marvel did not mention that these same locations had also, in many cases, become battlegrounds. But following the war, and in the midst of a very different set of hostilities, all aspects of foreign cultures were of potential significance. Apparently without an imperial history to draw from, the United States was forced to "improvise" in the form of language courses such as those built into the Army Specialized Training Program (ASTP).[40]

Marvel's laudatory sketch arrived at a precipitous time. The Carnegie Corporation and its friendly rival, the Rockefeller Foundation, were on the verge of eclipse at the hands of the upstart and staggeringly wealthy Ford Foundation, the third member of the philanthropic big three. After 1952, both Carnegie and Rockefeller support for area studies became more specialized, and both turned increasingly to other pursuits. But the same two organizations—eminently qualified to serve as intermediaries between government and universities, and affiliated closely with related bodies such as the SSRC—were chiefly responsible for the explosion of interest in area studies following the Second World War. By identifying appropriate funding lines, coordinating interdisciplinary meetings of interested academics, and above all, contributing to discussions concerning the approaches that area studies scholarship should take with regard to the places under scrutiny, Carnegie and Rockefeller trustees, executives, and staff were undoubtedly essential to the shaping of area studies as a Cold War *episteme*, a "field of scientificity."[41]

The identification of specific areas as strategically relevant was accompanied and reinforced by a belief that comprehensive and comparative study of "another society, nation, or cultural area" was also a "means of achieving a more profound and more valid grasp of the field of international relations." John Gardner, president of the Carnegie Corporation between 1955 and 1967, wrote those words in a 1958 memo. Earlier, following a tour of duty in the OSS, as a Carnegie Corporation executive associate he was crucial to the funding and promotion of a regional intellectual framework. In the winter of 1946–47, new to the corporation and working under the tutelage of President Charles Dollard, Gardner conducted a review of Carnegie's area studies program and concluded that few of the centers in operation, or even "in the planning stage," were adequately concerned with "possible contributions from the fields of social psychology, sociology and anthropology." Trained as a psychologist, Gardner was personally affronted by this elision. In the summer of 1947, seeking suitable sites for over one million dollars in area studies funding, he corresponded with numerous scholars in precisely the three fields he felt were neglected by the corporation. By 1958, when he drafted the memo extolling the promise of area studies for international relations, these disciplines were synonymous with Cold War social science.[42]

The end of World War II and the development of the Cold War only sharpened Gardner's perception that Americans of all stripes lacked an understanding of the world's strategic regions. This was not only an educational deficit but also a profoundly practical, military problem. In a 1952 letter supporting the HRAF to John W. Macmillan of the Office of Naval Research, Gardner could not have made these sentiments more clear. Military globalization, in the form of the "concept of total war," had "produced an enormous increase in the *range* of information which must be brought under the category of intelligence." But it was not just the scope of Cold War intelligence that troubled him:

> If one encircles on the map those areas of the world in which we may have to carry on intensive military operations, one finds that many of these are areas in which we have a minimum of readily accessible information. Yet in most cases useful information exists. The problem is to dig it out, to sift it, and to put it in readily usable form.

This triple predicament of research, organization, and classification required scholarly competence such as that found at the HRAF's Yale home and various area studies institutes. From its conception in 1949, the HRAF had been explicitly designed to aid "universities which are interested in area-investigation and which are developing integrated social science research programs that require factual information from all over the world."[43] The range cited by Gardner could be covered by the interdisciplinary staffs of these programs, fed by clearinghouses such as the HRAF, and made more manageable by a focus on a single region. After all, the very title Human Relations Area Files combined two "promising" routes to a potential "integrated science of behavior": social scientific theory, especially that which drew on the scientific principles of physiology and psychology, and "the concentration of attention on a particular area or ethnic group by scien-

tists of different disciplines." The files were the data, represented in convenient printed form, that permitted both regional examination and the
testing of theoretical hypotheses.[44]

In a rare published statement alluding to his work at the Carnegie
Corporation, Gardner used a specific, significant case from the Second
World War to illustrate the value of social scientific proficiency under conditions of geopolitical duress: the aerial bombing of Germany. Before such
operations could be launched, alternatives had to be weighed, which required the collection and sorting of large amounts of information, from
studies of industry and transportation to speculations on morale and character. Once sufficient data had been gathered, vulnerability could be estimated and targets could be selected. But the key was Gardner's assertion
that the demands of *postwar* international relations were no different from
this use of area information and expertise during an intensive military
campaign.[45] His choice of examples suggested not only the role of the city
as a strategic space but also the development of a specific form of geographical vision, one that could read landscapes from a distance, whether
in statistical tables or through the sights of a bomber.

The initial grants of the Carnegie Corporation to area studies centers were part of a larger funding initiative on global knowledge, including support for related work in strategic studies. As the corporation's 1946
Annual Report declared, assistance was necessary to make "this country
more literate and more emotionally mature in international affairs." Given
that military funds appeared to be widely available for the physical sciences, the officers of the corporation saw a need for the support of both
national heritage and the principles of "a sound and stable society."[46] By
the following year, grants in this sector had tripled in value, and "the
extensive study of geographic areas," as part of the international affairs
scheme, was "a particular object of invigoration."[47] The double concern
with stability at home and global awareness was one and the same, encapsulated by the ideal of a known, orderly world with a liberal America in a
position of dominance.

For both the Carnegie and Rockefeller philanthropies, the elitist obsession with adult thought on matters of foreign policy was brought to the
fore during the Second World War. By deliberately subsidizing groups such
as the Council on Foreign Relations (closely aligned with the State Department during the war) and the Foreign Policy Association, the foundations

targeted a group of professionals devoted to American globalism. This internationalist perspective was invariably accompanied by tough, masculine realism, built on the blocks of national security, forceful military strategy, and a stable balance of power—precisely the stance applauded by Carnegie officials. While encouraging intellectual innovation, foundation officials were also determined to maintain an existing order of authority, preferably through leadership rather than blunt coercion. A concern for stability, gatekeeping, and moderation, of course, sat well not only with liberalism at home and abroad during the early Cold War but also with concurrent academic doctrines such as structural-functional sociology.[48]

As the Second World War was ending, foundation officials conceived proposals for sound postwar programs in the social sciences. At Carnegie, the decision was made to transfer support from established, successful projects to those in their initial stages. In addition to direct grants, both Carnegie and Rockefeller also began to funnel significant sums through the SSRC, mostly for the hands-on aspects of area research, from conferences and travel stipends to graduate fellowships. These initiatives, run through the SSRC's Committee on World Area Research, were part of the attempt by philanthropies to supplement immediate forms of area study with a lasting base of more considered scholarship.[49]

Foundations, Rockefeller's Norman Buchanan reflected in 1955, were not in the business of competing with the CIA or the *New York Times* by supporting the "compilation and dissemination of area intelligence of an ephemeral or current events character."[50] But this was a flawed distinction. It was the tremendous volume of material built up during the Second World War that would be evaluated more meticulously, and one explicit impulse for widespread graduate training was to educate future intelligence workers—not only academic researchers. This was a question not simply of labor but also of epistemology. For the anthropologist Julian Steward, writing under SSRC contract in 1950, just as nascent intelligence agencies such as the CIA were obsessed with prediction, the needs of area studies were "better served by better science," which would similarly reduce uncertainty, in the ultimate hope that Americans might "understand the nations in foreign areas so thoroughly that we could know what to expect of them." The only meaningful difference was that pure scientific research was conducted on a longer time scale, but that did not stop the repeated raids on area institutes by various government branches.[51]

At midcentury, the SSRC oversaw and selected members from seven professional American societies of anthropology, economics, history, political science, psychology, sociology, and statistics. The council also published several of the most prominent surveys of area research in the immediate postwar period. University of Michigan geographer and Japan specialist Robert Hall, an OSS and Ethnogeographic Board veteran and the head of the Committee on World Area Research, wrote the first survey. Published in 1947, it was based on travels to twenty-four American universities in the spring and summer of 1946. After summarizing the dearth of area experts in World War II and the resulting launch of centers for regional study, Hall explained in familiar language why the SSRC was so interested in these developments:

> Here was a possible means of bringing about cross-fertilization within the social sciences and of bridging the gaps between the social and the natural and the humanistic disciplines. Here might be a way of working toward the fundamental totality of all knowledge. Here might lie means by which research in the social sciences could be made more cumulative and comprehensive.[52]

Hall went on to spell out the multiple motivations for interest in area studies, from the "sterility" of disciplinary structures, which perpetuated "twilight zones and vales of ignorance," to the latent "provincialism" of work that passed as universal, as well as the increasingly global role of the United States. The problem was that enthusiasm for the area approach, as epitomized in wartime training programs, was, for Hall, "makeshift" and quite distant from models of liberal education or sustained scholarship.[53] However haphazard, though, the instruction of soldiers did share a similarity with Hall's expansive update of classical pedagogy: the aim to span *space* as well as time. But the introduction of social science into this model meant that regardless of the area under consideration, regional intelligence would be contemporary and more precise.

One of the most common rationales for regional research formations was the assertion of provincialism—the argument that universality in the social sciences and humanities was impossible unless the limitations of reliance on the "civilizations of the western European and North American world" were overcome.[54] While nominally an admission to the prevalence of culture-bound Eurocentrism and a vote for liberal cosmopolitanism, it was also an affirmation of the West's distinctiveness. As Hall put it, "We need to know all other areas: we need the data of other areas to check our assumptions." Hall advocated the maintenance of cross-cultural study in graduate area programs, "especially between the particular area of study and the United States." But neither America nor the basis of civilizations was really up for question. A shift to "unusual areas" was all that was required for balance.[55] Key area studies texts rarely granted foreign contacts the recognition they deserved and invariably treated them as mere informants or research assistants. The field was much more often viewed as an American laboratory, especially for graduate students, who could spend a requisite time inside this space and return to receive accreditation.[56]

The claim that area studies would bring social scientific disciplines into contact with an intricate globe was filtered through a version of pluralism leashing diversity to expert authority. A similar dependency characterized the relationship between area studies and conventional disciplinary formations. The latter would continue to dominate while being integrated and supplemented by the former. "An area program," Charles Fahs wrote in 1949, was "a focus for the practical application of methods and concepts in the established disciplines, not a substitute or an alternative to these disciplines."[57] In the imaginative world of integrated yet autonomous regions, differences could thus be both *systematized* and *segregated*. And this uneasy relationship could be monitored by the invigorated social sciences whose aim was Hall's total knowledge. But such grand theory was a distant objective, to be preceded by endless, patient observations and empirical methods.[58]

Hall also contributed the preface to Charles Wagley's report on a national "world areas" conference, organized by the SSRC, funded by the Carnegie Corporation, and held in the Men's Faculty Club of Columbia University in November 1947. Among the speakers were two prominent members of the Harvard professoriate: Pendleton Herring, a political scientist who became president of the SSRC the following year, and Talcott Parsons, a sociologist nearing the peak of his fame. Both "drew an analogy between area study and the science of medicine" in the sense that while there was "no single science of medicine," it was generally devoted to the "practical problems of the total human organism, to the whole man."[59] This was no coincidence. Parsons was prone to drawing on medical practice to demonstrate that social science could effectively control deviance and shape the postwar "national purpose."[60] When extended beyond American borders, however, he believed that social anthropology and institutional sociology were essential to grasping "the total system of an area" as well as to an "understanding of the *differences* between societies." Herring added that any parochialism in the social sciences would quickly be revealed upon the application of certain theories to "alien cultures." Integration was possible circuitously by understanding regional divergence.[61]

Neither Parsons nor Herring was directly equating social systems with biological organisms. Rather, they were interested in the shared idea of an organizational whole. Yet the image of an area studies center as a "clinic" diagnosing and healing the ailments of a regional body is a provocative one, because the social sciences had, for several decades, been frequently drawn together with biological and medical fields as part of a "comprehensive explanatory and applied framework of social control," or human engineering.[62] Just as this supertheoretical discourse, which Michel Foucault would later call biopower, was configured toward the body, it could also be transferred to a regional *population* differentiated by brute geography.[63] In another venue, Herring was careful to stress that social science emphasized "analysis rather than force," understanding over manipulation, and that "command of data" could lead to greater precision in policy.[64] However, this was an unsustainable distinction. The Cold War social sciences were less concerned with the direct application of a separate power than with a more subtle influence over conditions of conduct and human potential.

Many of the typical statements from the search for "a universal and general science of society and of human behavior" are present in Wagley's synopsis. But given the conference's 1947 date, it is not surprising that attendees were also preoccupied with geopolitical matters. Apologetic yet enthused, Hall noted that it was "perhaps fortunate for the continued development of area studies, if for nothing else, that our world has remained even after the war in a state of critical uncertainty."[65] In his address, "Objectives of Area Study in Colleges," Herring asserted that the knowledge generated by area studies, "while acute during wartime, can now be regarded as crucial for the post-war period."[66] The postwar relationship between area researchers and strategy had been raised even earlier, in October 1945, at a meeting of the SSRC's Committee on Problems and Policy. There, Hall "recognized the need for cooperation between scholars and government but advised against any official connection, in order to protect the position of the American scholar working abroad."[67] The focus on the researcher's safety, rather than on ethical principles, suggests that Hall was not averse to *unofficial* links—exactly the secretive connections that haunted area studies.

Recognizing the continued merits of economic and psychological warfare in 1945 or 1947 was part of the perpetuation of American militarization. But in 1947, Herring was more direct than most in his assessment of the Cold War role and responsibilities of area studies. Additional understanding of the Soviet Union and other regions, he argued, was fundamental to "a cool and calculated execution of the Truman Doctrine or its equivalent." Once it was acknowledged that the "great population masses of the world are simultaneously in a state of unrest that is unparalleled" and that the United States and Soviet Union both faced outward toward these zones of disintegration and disorder, the true cast of the conflict became clear:

> The struggle, in other words, is rather a competition to win adherents friendly to the United States and more disposed to accept our values than to follow the course of Russian leadership. . . . To the extent that we are able to exert our influence upon these areas and win their adherence through our understanding of their problems and, in turn, through their understanding of our objectives, we shall be able to win out in our competition with the Soviets.[68]

Similar comments were made innumerable times during the Cold War. But Herring's remarks were not only predictive. They were also offered by an individual who, as president of the SSRC (1948–1968), later oversaw the committees on Political Behavior and Comparative Politics, in which the third world "competition to win adherents" was given social scientific muscle in the form of modernization theory.[69] For Herring and many others, Cold War competition was precisely compatible with a scientific and systematic social science. As foundations came under conservative attack for funding "subversive" inquiry in the 1950s, Herring went before a congressional committee, according to Mark Solovey, and "vouched for the integrity of the social sciences and their private patrons, taking the opportunity to explain that an objective, empirical approach to social research was, in fact, particularly congenial to the American way of life."[70]

The Committee on World Area Research held a second national conference in May 1950 that was again financed by the Carnegie Corporation and again held in New York. The roster was a veritable who's who of Cold War social science: scholars, bureaucrats, and foundation employees who regularly attended such gatherings, where "manly interdisciplinary camaraderie" was the tonic for the emasculation of disciplinary cells.[71] Area studies had, by midcentury, become a genuine academic vocation, as a 1951 SSRC survey by Wendell Bennett indicates.[72] But notwithstanding this growth, the 1950 postconference report is notable only for its tone of heightened urgency — "areas deemed exotic and peripheral more than ever impinge upon our daily life" — and a continued shift away from European topics to Asian concerns. Several geographers present at the first conference, including Richard Hartshorne and Preston James, did not attend the second, and the peripheral status of the discipline in area studies was made clear by the comment that the contributions of geographers were "still largely of a reconnaissance role," immaturely concerned with identification and description.[73]

The same year also saw the publication of Julian Steward's dense consideration of the theory and practice of area research: an SSRC bulletin that drew upon his own experiences in Puerto Rico and at the Smithsonian Institution. His essay was an example of the SSRC's relentless appeals in the first few years after the Second World War for "clarification of current thinking" in the social sciences, a task that could be accomplished in the form of "systematic surveys."[74] Summarizing the objectives of regional study as practical knowledge provision, awareness of cultural relativity and wholeness, and the furthering of universal science, Steward sifted through the various meanings of areas, from spaces of global significance to subnational regions such as the American southeast. To his credit, he acknowledged that the impetus for area programs rarely derived from scientific or cultural theories but rather from strategic imperatives and institutional conditions. Steward believed that in some cases, such as the Soviet Union, seemingly nonpolitical criteria could be made belatedly to fit a space suddenly targeted for its role in international affairs.[75] But an emphasis on culture was nonetheless becoming secondary as more *functional* economic and ideological factors took precedence in the identification of areas worthy of scholarly consideration.

Steward recognized that once "structural-functional" typologies such as value systems were introduced, the definition of a region became "infinitely complex." The solution, at least as it emerged in the 1950s, was to naturalize or downplay the geographic divisions between areas and to concentrate on the array of structures and functions that were found within an area. This hierarchy, with its debts to the sociology of Talcott Parsons, including a stress on factors of social control, was perfectly suited to Cold War concerns. As Steward wrote, area studies were, in practice, "dictated by international relations, by the necessity of formulating United States foreign policy."[76] But given the vast size of these regions, and the need, during and after the Second World War, for more detailed forms and sums of knowledge, interdisciplinary methods were still required to achieve "a total picture."[77] And while both area studies and the behavioral sciences were in many ways already well established by 1945, their intellectual prominence crystallized in the study of *the* Cold War strategic region.

Sovietology as Strategic Social Science

In 1947, the Carnegie Corporation gave $740,000 to Harvard for the support of a new Russian Research Center (RRC). The justification for this substantial grant, the corporation's 1948 Annual Report explained, was the emergence of a new, *non-European* world power "which has its roots in a culture vastly different from ours." This very difference rendered the unfamiliar Soviet Union a challenge to those concerned with the global influence of the United States. It became "of utmost importance that we achieve systematic and full understanding of Russian culture and history, and of the habits, beliefs, motivations, fears and loyalties of the Russian people." But while the RRC would "bring expert knowledge to the aid of those officials who must conduct day-to-day negotiations with the Russians," excitement accompanied foreboding. Despite the politically charged context, the expertise and resources of the RRC also promised "development of new facts and methods in the social sciences."[78]

Sovietology was the centerpiece of Cold War area studies, a substantial, multifaceted enterprise on its own. In the late 1940s and 1950s, as Americans discovered a strange new foe, Soviet scholarship was a hobby, or a fervent obsession, for many of the most prominent intellectuals of the period. It was, by the standards of the social sciences, lavishly funded. Partly as a result, it became controversial. Attacks first arrived from conservatives concerned with the sums of money spent on pursuits that appeared dangerously proximate to communism itself. These campaigns, which lay at the heart of McCarthyism's assault on universities, claimed several victims and certainly influenced the direction and tenor of some research. But they now appear to be feebly, if frighteningly, ironic given the deep bonds forged between Sovietologists and those in government responsible for fighting a Cold War.

The earliest significant entry in Soviet studies was not Harvard's Center but Columbia's Russian Institute. Although this site opened its doors to students in September 1946, it was conceived as early as October 1943, when Geroid Robinson, a Columbia historian on leave in Washington, sent a letter to his home university predicting a postwar expansion of Russian studies corresponding with "a change that is rapidly developing in the distribution of world power." Robinson, who led the USSR Division in the Research and Analysis Branch of the Office of Strategic Services, followed this missive with another in April 1944 that sketched a suitable program of teaching and research, and a third that suggested Russia's "powerful place" in the postwar world would lead to "considerable expansion" in American study of the region.[79] Building on but also departing from European precedents, he argued that the proposed institute at Columbia should include a combination of geographic specificity, disciplinary integration, multiple ties to other parts of the university, and a blending of the humanities and social sciences. The Rockefeller Foundation provided a five-year grant; a space was carved out within Columbia's new School of International Affairs; and five staff members, all (with one exception) fresh from government service, were appointed in the fields of economics, law and government, international relations, history, and literature.[80] Graduate students were to concurrently specialize in one of these five fields inside and outside the institute, a coupling of regional and functional "citizenship" that became the hallmark of area studies.[81] In supporting Robinson's appeal, the Columbia administration recognized the appropriateness of his OSS duties. One committee report noted that he had already "in fact been operating a Regional Institute on the Soviet Union" and had gained "a wholly new sense and concept of the multiple values of experience on concrete, practical problems and especially of group attack."[82]

The institute's faculty members were also very active in the postwar promotion of area studies and Cold War social science, dominating the early roster of the SSRC's Slavic Studies Committee (formed in 1948) and working with the RAND Corporation, for instance.[83] From the outset, Columbia's training mandate in Soviet studies was broad. As Schuyler Wallace, the head of the School of International Affairs, wrote in 1946, the purpose of the institute was to prepare "regional specialists to do work of authority and influence in business, in finance, in journalism, [and] in various branches of government service," as well as in academic circles.[84] This was no mere boast. The institute's "national character" was proudly demonstrated by the presence of government employees, including Army, Navy, and State Department foreign service personnel, some of whom went on to important positions in diplomacy and Soviet policy.[85] Prominent visiting fellows, such as Herbert Marcuse, stopped by for regular guest lectures.

The international relations scholar appointed to the Russian Institute was Philip Mosely, who during the Second World War was chief of the State Department's Division of Territorial Studies. Mosely succeeded Robinson as institute director in early 1951 and continued to consult with the State and Justice departments, the CIA, the Air Force, and a number of additional federal agencies. Mosely repeatedly scheduled CIA job interviews for Russian Institute graduate students, testified against the Communist Party of the United States before the Subversive Activities Control Board in 1953, and aided the Ford Foundation in a program of country studies that led directly to influential work in comparative politics and modernization theory.[86]

According to John Hazard, the first Russian Institute professor in law and government, Robinson encouraged the formation of a competing center at Harvard because he was concerned about a potential backlash against Columbia for its interests in Soviet studies.[87] Another account describes an inability at Columbia to "take care of the demand."[88] At the 1947 SSRC Conference on World Areas, Mosely suggested that it made sense to have several centers for Soviet studies "'strategically' located throughout the country."[89] And the Columbia Institute—to Robinson's eventual dismay—consistently emphasized training over research.[90] All of these factors were probably influential, and Robinson found willing partners in Harvard's William Langer (formerly of the OSS, and at the time linked to the fledgling CIA) and Clyde Kluckhohn, as well as Charles Dollard and John Gardner of the Carnegie Corporation. As a result of his 1947 investigation of area studies research, Gardner believed American scholarship on Russia to be "in deplorable shape."[91] In a contemporaneous discussion with Mosely, Gardner grew perturbed when the lack of Russian Institute students in sociology, anthropology, and social psychology was mentioned.[92] He also consulted closely with Kluckhohn, the Harvard center's first head, on the merit of certain faculty prospects. Gardner did not comment on this role when he wrote in June 1948 that Kluckhohn had "done a brilliant job of gathering talent for the program at Harvard" and that the RRC would shortly "be the equal of the group at Columbia in terms of brains and ability." But in discussions with his Harvard contacts and others, Gardner emphasized that the research to be conducted at the proposed center should follow hard on the heels of George Kennan's recent "X" article in *Foreign Affairs*, World War II analysis of Japan, and the work of the OSS.[93]

In his June 1947 conversation with Gardner, Mosely mentioned that Soviet studies lacked a "pattern of coordination for the country as a whole, and particularly between government and the academic world."[94] The Russian Institute's training of military and diplomatic personnel was one aspect of the relationship between government and academic Sovietology. But Mosely was clearly gesturing toward a deeper bond, and it was left to the Harvard center, ostensibly a research institute, to forge it. As Sigmund Diamond observed in his landmark exhumation, "the interests of the intelligence agencies—the CIA and the State Department, in addition to the FBI—and those of Harvard were most likely to intersect at the Russian Research Center—and they did."[95] From the beginning of operations in February 1948, the results of research at the RRC were made available to these contacts before they were released to the public. And although the university as a whole declined to conduct classified or secret research, the public versions of certain RRC publications were missing sections pertaining to the direct conduct of war, in its various forms, against the Soviet Union.[96]

The image of the RRC presented to a wider audience was, in Kluckhohn's words, "the study of Russian institutions and behavior in an effort to determine the mainsprings of the international actions and policy of the Soviet Union." By emphasizing interdisciplinary cooperation and curiosity across fields "neglected" in Soviet studies—not coincidentally, those fields leading the charge toward social *science*—and inviting an array of visiting specialists (from Isaiah Berlin to Margaret Mead) and graduate students into its seminars and projects, the Harvard center was designed as a place where a "coherent whole" could be produced, with additional repercussions for general questions of methodology, theory, and comparative study.[97] As one of its researchers, Alex Inkeles, put it in 1951, the "primary task is not that of making our regional methods more adequate for the study of foreign societies, but of improving our conceptual tools and methodological equipment to make us more effective in the study of *any* society."[98] He was echoed in the summation of the center's first ten years, where area studies was described as "not merely the study of a particular geographical subdivision, but a grounding in a number of disciplines whose convergence upon the same area produces a more rounded appreciation of the societies involved."[99]

For Russian Research Center scholars working on social relations, since the observations made possible by fieldwork were unavailable, refugees and former citizens of the Soviet Union became subjects of tremendous scrutiny. The center used many of these individuals in a vast, team-based Harvard Project on the Soviet Social System, dominated by researchers with substantial intelligence experience and organized with support from the Air Force's Human Resources Research Institute (HRRI). Interviews (many conducted in the Munich area) and questionnaires were employed to capture the conditions of daily life in the Soviet Union. The "ultimate goal," Kluckhohn wrote near the beginning of this initiative, was to build "a provisional working model of the Soviet Social System."[100] To do so, "special scales" were formulated that adjusted for various types of distortion and cultural difference. This work was assisted by quantitative sociologists at Columbia's Bureau of Applied Social Research (BASR), who were acknowledged for their "technical collaboration" in the project's final report.[101]

Although Kluckhohn described the social system work as unclassified, at least one participant, Raymond Bauer, used materials from his German fieldwork in a classified project at MIT's CIS. The Air Force also reviewed the manuscript of the most significant public social system report, deleted some names and references, and demanded the omission of any "assumption of peace."[102] Two versions, in other words, were produced: one that was applicable to military planning and one that presented an image of academic objectivity. This difference was strikingly reflected in the respective titles of the documents: *Strategic Psychological and Sociological Strengths and Vulnerabilities of the Soviet Social System* became *How the Soviet System Works*, published first as an RRC study and then as a monograph in 1956.[103]

Established in 1949 within Air University at Alabama's Maxwell Air Force Base, the HRRI was designed to conduct human resources research within the Air Force, but also to "assist the Air Force in making use of social science in general."[104] It came into its own during the Korean War, when its dual interests of strategic intelligence and "psykewar" took on added urgency. But those military officials interested in such contentious subjects as target selection, the intricacies of targeted populations, and the results of atomic targeting on humans (or the "social fabric as a target") lacked data on the Soviet Union.[105] Given a puzzle — the reactions of certain groups to hypothetical bombing and invasion — Soviet studies provided one solution.

Social science such as that practiced and produced at the RRC was useful to the HRRI's mandate in two mutually reinforcing respects: reports containing sociological and psychological data would be sharpened but so would the process of data collection. It was believed that these "two tasks [could] be accomplished concurrently." Advances in intelligence methodology would accompany, and in turn improve, the "human factors intelligence" so desired by the Air Force.[106] The Harvard center therefore contributed to both the *reconnaissance* and *research* operations that are at the heart of intelligence work.[107] As a supplement, the HRRI also contracted with Yale's HRAF to create a "file of knowledge" on Siberian populations, "a direct reflection of Air Force interest in this strategic region."[108] This file, HRRI staff believed, was flexible and could be employed to create psychological warfare plans, intelligence reports, and manuals outlining proper and improper conduct with respect to certain social groups.

According to HRRI director Raymond Bowers, the Air Force was interested in Kluckhohn's "working model" because it could be subjected to "various strains and stresses."[109] While this model was not as mathematical as those produced by the RAND Corporation, it did provide military planners with novel information that could be used in psychological warfare campaigns or, more ominously, in an atomic attack on the Soviet Union. Here the RRC's work dovetailed with that of the Bureau of Applied Social Research at Columbia, where the sociologist and demographer

Kingsley Davis was leading an HRRI-funded initiative on global urban data that would permit comparative study and statistics for the selection of air targets. But this Urban Analysis Project, which I discuss in chapter 5, was only a component of the bureau's work for the Air Force. A related group, also headed by Davis, examined interview methodologies and provided advice to RRC staff setting out for Germany. These interviewers were also given an Air Force *Air Interrogation Guide* reminding them to collect information on Soviet cities that would be useful in the production of maps displaying the location of industry and other important strategic details.[110]

The HRRI–RRC–BASR axis was undoubtedly one of the most intimate collaborations established during the Cold War across university, intelligence, and military lines, particularly in terms of the parity of aims and theories. It solidified a pattern that lasted into the 1960s: military requirements were molded into "operable research designs."[111] This transformation was mutual. However awkwardly, the results of the interviews were made to fit the Air Force's generous understanding of strategic intelligence — a kind of partisan basic research that, although not directed to particular missions, nonetheless remained spatially anchored.

That the Harvard Refugee Project shared a title with Talcott Parsons' *The Social System* (1951) was no coincidence.[112] Not only was Parsons a featured speaker at the 1947 Columbia conference inaugurating coordinated area studies and heavily involved with the RRC, but he was also working at the time on a similar total model of society. Although his model deprived modernity of a historical geography and instead, in the words of Jurgen Habermas, "stylize[d] it into a spatially-temporally neutral model for processes of social development in general," it certainly took the United States as its foundation.[113] As two of the original members of the RRC's executive committee, Parsons and Samuel Stouffer were frequent attendees at center seminars. Parsons' Department of Social Relations at Harvard (where Stouffer ran the Laboratory of Social Relations) started almost concurrently with the RRC and was also fueled by Carnegie funding. The Department of Social Relations was a divisive creation precisely because its members sought the foundations of a synthesis that flowed beneath artificially divided disciplinary compounds, although early on the individual idiosyncrasies of Parsons, Kluckhohn, and other members prevailed.[114]

Parsons derived "neither his theories nor his hubris from the Cold War."[115] But he responded favorably to John Gardner's initial inquiries concerning a meeting of behavioral science and Russian area studies at Harvard. And in the summer of 1948, Parsons traveled to Europe to lay the foundations for the interview project. Sigmund Diamond devastatingly sums up this role:

> Parsons approved attaching universities to the intelligence apparatus of government—covertly; bringing persons accused of collaboration with the Nazis to the United States—covertly; using Harvard connections to influence government officials to ease their entry to the United States—covertly; breaking down the distinction between research and intelligence.[116]

Dogged by the FBI because of his proximity to both intelligence circles and Soviet studies, Parsons was demonstrably an opponent of McCarthyism, which he saw as a political anomaly.[117] But he was even more aggrieved by leftist radicalism.

Universities, in Parsons' view, should be above the ideological fray, and scientific inquiry should be disinterested and aspire to universalism. Although a lengthy debate could be had on the severity of the Soviet Union's threat to Western Europe during the late 1940s, this is not the issue here. Parsons clearly believed that this threat was imminent, but more importantly, he understood his geopolitical vision to be rational and thus singularly appropriate. It is thus doubly fair to describe his sociology as a "strategic vocation," an activist attempt to shape a social and political reality rather than merely describing it. Parsons, whose scholarly output was never more prodigious or more influential than in the 1940s and 1950s, was a proponent of the *stability* provided by capitalist democracy. However, his advocacy of a neutral, professional social science as a "progressive, rationalizing force" directly paralleled the narrowed field of Cold War intellectual expression. And because of his work for the RRC, these two lines are explicitly connected.[118] It just so happened that the destabilizing forces present in American society and threatening to impinge upon it further from the outside were identifiable under the category of national security.

In books such as *Toward a General Theory of Action* (1951), produced with Carnegie Corporation support under the auspices of the Department of Social Relations and edited with his University of Chicago colleague

Edward Shils, Parsons set out to piece together a "unified conceptual scheme for theory and research in the social sciences."[119] It was this quest that led C. Wright Mills to lampoon Parsons's exemplary "Grand Theory," composed in dense, imprecise prose requiring translation "into English," which even after conversion "would not be very impressive." Although Mills skewered the Grand Theory *and* the abstracted empiricism of Parsons, the preponderance of "high generalities" did not prevent others from borrowing Parsonian concepts to serve a social science premised on merciless observation.[120]

As early as 1944, Geroid Robinson had written that what distinguished the Soviet Union was its exceptional integration of the "elements" comprising its singularity, characteristics "of its life and thought" that combined to "form the Soviet unit of power and policy with which Americans must deal."[121] This view of the Cold War enemy, a perspective that coalesced around the single word *totalitarianism* (already inherited from another foe), became orthodoxy in America's elite institutions of higher education in the 1940s and 1950s, when Sovietology spawned numerous programs and institutes. The absorption and preservation of a rumor "transformed a speculative vision of the enemy into a powerful working hypothesis," one that could be put to work globally, according to a spatial taxonomy of good and evil. Cold War behavioral science may have been ruthless by nature, but it was also marked by militant ambition, a weakness for contemporary "cultural mores," and a trenchant distaste for circular arguments.[122]

Led by political scientists who conceived of Soviet politics as a unidirectional, omnipotent rule over "a passive, frozen society," and finding influential recruits in anthropology, history, and sociology (at the very least), Soviet studies, while growing rapidly, was also a scene of "self-impoverishment." Its practitioners denied adverse national influences while reasserting familiar axioms and marketing the weight and urgency of their scholarship to the CIA and the State Department. (Ironically, this stultification occurred in large part during the mid-1950s after Stalin's death had signaled the beginnings of profound changes in the Soviet system.) The "pattern of coordination" sought by Philip Mosely may have been found, but it was also crude, defied by only a few rebels at least until the mid-1960s.[123] This was the case, at least, in the social sciences. In the humanities, where Soviet studies was also booming, the idea of Russian culture as similar or at least connected to others — not least the United States — was a recurring and prominent claim.[124] But with the practical exception of language training, the humanities were not crucial to the sponsored growth of area studies or its enthusiastic adoption as a form of strategic knowledge.

For obvious if bizarre reasons, the movement that came to be known as McCarthyism scrutinized students of the Soviet Union. John Hazard of Columbia's Russian Institute, for instance, was forced to file an affidavit stating that he was not a Communist in order to receive a passport. He was also called before the House Un-American Affairs Committee.[125] Such trials undoubtedly frightened Sovietologists, particularly older scholars, and there was little refuge to be found in the arms of the foundations, which were also targeted by various conservative congressional investigations, or within the confines of prestigious universities such as Harvard, where the FBI was active. These contexts, and less personal constraints, certainly discouraged critical reflection and furthered specific methods. Nor was it easy, given the state of the Soviet Union in the early 1950s, to contest standardized narratives. But the pervasiveness of "pseudointerpretation" was also due to the norms and concerns of scholars invigorated by wartime service and the even stronger passions of the crusading ex-Communists or refugees from Communism who joined them.[126] Some of these writers were driven by a commendable idealism, and a smaller group made lasting contributions to the field of Russian research. Rather than blindly impugn a substantial body of work for its politicization, then, this same partiality should be acknowledged and its implications considered carefully.

The FBI investigation of RRC employees included director Clyde Kluckhohn. An anthropologist who had worked extensively on Navaho culture, he also possessed a top-secret clearance as a result of his participation on the Department of Defense's RDB. In a 1951 FBI report uncovered by Sigmund Diamond, Kluckhohn's loyalty was confirmed. According to the bureau's Boston representative, the

> State Department would communicate with him to suggest they were short in a certain aspect of Soviet activity. Kluckhohn would then suggest to a graduate student at the school that he might do a thesis on this particular problem, making no mention to him of the fact that the State Department was also interested. Subsequently the results of the individual research could be brought to the attention of the State Department.[127]

In July 1954, William Langer moved back from the CIA, where he had been summoned in 1950 to lead the Office of National Estimates, to replace Kluckhohn as RRC director, further cementing a relationship that was already very profound.[128]

Communicating Modernity

When the Ford Foundation dramatically raised the stakes of social science funding in the early 1950s, it did so with a program markedly similar to those of the Carnegie and Rockefeller philanthropies. Ford's major innovation was a formal initiative in the behavioral sciences — so named because of the potentially "socialist" implications of "social" science — incorporating psychology, anthropology, and sociology in an attempt to further advance the scientific aspects of human study already confronted by Carnegie and Rockefeller schemes. (Geography's own "behavioral" turn was still several years away.) The heavily quantitative analysis of observable actions, which led to testable models of political systems and their group or individual machinery, was a cautious reproach to alternate theories of power elites, most obviously the work of C. Wright Mills. Though not without successes, the Behavioral Science division's esoteric, ambitiously integrationist aims were received poorly by Ford's conservative trustees, and it was terminated in 1957. But the slack had already been taken up by military and intelligence agencies.[129] And the tally of Ford's broader contribution to strategic social science by the mid-1960s was astounding: some 138 million dollars directed "to a limited number of universities for the training of foreign-area and international-affairs specialists."[130]

In political science, behavioralism may have gained its greatest triumph in the field of domestic politics. The presence of a totalitarian menace led to an affirmation of American institutions, backed by a "consensus school" of history and American studies that positioned a national creed above petty European struggles over class and ideology. To be sure, a reinvigorated liberal curriculum in the humanities concurrently stressed the timeless values of Western civilization.[131] But the United States was perceived to be the torchbearer for these values. While the consensus label is largely a myth fashioned in hindsight, the assertion that America was unique resonated with scholars who applied it internationally in discussions of modernization and economic development.

Ford's Board of Overseas Training and Research followed the lead of Carnegie and Rockefeller initiatives, organizing conferences and disbursing fellowships, but on a vastly greater financial scale. This funding stream, supplemented intellectually and institutionally by the SSRC's prominent Committee on Comparative Politics, was explicitly designed to address instability and the dangers of communist advance in the developing world. Established in 1954, the Committee on Comparative Politics was an outgrowth of another Council Committee on Political Behavior. A conference at Princeton University late in 1953 led directly to the formation of the new body, and Princeton's Gabriel Almond became the first chair.

For Almond, the fact that he and his committee's young members studied Southeast Asia, Africa, India, and China was a stand against conservativism. Their intention was to "bring foreign area studies up to 'state-of-the-art,' encouraging scholars to delve into the political infrastructure of European and other foreign countries."[132] The key to this upgrade was intense fieldwork in "'laboratory' situations," followed by cross-cultural comparison that included, but moved past, the depthless anthropological indicators of the HRAF.[133] Almond's assertion of radicalism now appears dubious. While the location of interest might have shifted, the methodologies of comparative politics were still premised on a kind of categorical imperialism, which in turn encouraged and justified more material forms of neocolonial activity, such as military intervention or support for coups. Over the next fifteen years, the committee, its members, and their approach became intellectually hegemonic, especially through the Studies in Political Development book series.[134] "From the beginning," MIT's Lucien Pye, another prominent committee member and one of Almond's best students, wrote in 1959, "the committee has sought . . . the means to differentiate political systems as *wholes*." This goal was achieved by recourse to Parsonian sociology, to the identification of "certain universal functions of all political processes" or institutional structures.[135] Included among these was modernity itself, which according to Daniel Lerner was "felt as a *consistent whole* among people who live by its rules."[136]

Comparative politics encouraged rankings of regional systems and spaces, producing broader judgments by setting areas against one another, with the Western Hemisphere, and more specifically the United States, as the "final citadel" of political perfection, just as it was for modernist culture. While disciplinary boundaries were transgressed, geographic and moral divisions and hierarchies were actually reinforced so as to pursue useful comparisons.[137] Despite a gloss of conceptual innovation, the result was the continuation of a very familiar imperial distinction between traditional and modern societies, even if the possibility of transition from traditional to modern had been "unthinkable in the colonial era."[138]

Fears of foreign volatility—from political unrest to tropical diseases—prompted further research on the practical aspects of the behavioral sciences, and policy was used to directly test hypotheses. Ford Foundation grants under this rubric were overwhelmingly directed to prominent institutions: Columbia, MIT, Harvard, Chicago, Stanford, Berkeley, Princeton, and Washington. The resulting modernization theory, perhaps best epitomized by Walt Rostow's brazenly titled *The Stages of Economic Growth: A Non-Communist Manifesto* (1960), held that "underdeveloped nations," threatened by the "blandishments and temptations of Communism," could instead be more suitably directed up through an arc of progress.[139] If the "ultimate aim" of Rostow and his collaborators was "the production of an alternative to Marxism," this was ironic, because modernization theory, according to Odd Arne Westad, shared with Soviet doctrine "many of the same positivist traits. . . . Indeed, it could be argued that both constitute a form of 'high modernism.'"[140] Not surprisingly, then, the historical certainty of modernization theorists was seductive for "policy makers groping for an explanation of the United States' place and responsibilities" in a confusing Cold War world and also well suited to the military focus on command and control.[141] Senator John F. Kennedy used the appropriate language in 1956 in reference to South Vietnam: the United States was "directly responsible for this experiment—it is playing an important role in the laboratory where it is being conducted."[142]

The terminus of Rostow's curve was a norm of complete American modernity, a social structure whose needs for reform were minimal. This *development* could be achieved by instituting techniques of gradual, evolutionary economic change run by a well-educated class of elites who shared certain values with their American patrons. Modernization theory was usefully multiscalar in that it could apply to cities, states, and regions, but groups of these elites would, it was hoped, ultimately meld together to constitute a global culture of singular faith in order to manage what Joseph Willets of the Rockefeller Foundation's Social Science Division described, in 1949, as "the orderly evolution of the unindustrialized countries."[143] The

most attractive evolutionary path was one that proceeded "least convul-
sively"; upsetting the already fraught process of development was seen as a
potential rationale for military intervention in the name of stability.[144] Despite its ostensible universalism — in the sense that the Family of Man could
all be made to act like Americans — such an outlook also shared much with
earlier anthropological scholarship composed in the service of empire. Indeed, development was a response to crises of decolonization in the mid-
twentieth century, and many anthropologists trained in a cultural relativist
tradition rushed to embrace Rostow's hierarchical evolutionism.[145]

Numerous studies exist on the local consequences (by no means
unidirectional) of the America-centric modernization framework, consequences that were only just becoming apparent in the late 1950s, especially
in Southeast Asia, where the limits of regional intelligence were also rapidly becoming clear. Earlier in that decade, however, the quintessentially
strategic temperament of modernization theory was already perceptible.

Long before he and his cohort of social scientists migrated into the
Kennedy and Johnson administrations, Walt Rostow was blending his intellectual gifts with the practice of strategy. During the Second World War,
he worked as a bombing analyst for the Enemy Objectives Unit of the
OSS — a team of economists sent to London by the Research and Analysis
Branch. After brief stints at Oxford, the Economic Commission for Europe, and Cambridge, Rostow moved to Cambridge, Massachusetts, to a
university with exceptionally close ties to defense and intelligence sources.
Inside MIT's CIS, Rostow, with outside assistance from the likes of Philip
Mosely and McGeorge Bundy, ran a classified initiative on the "dynamics of Soviet society" for the CIA.[146] Founded "to apply social science to
problems bearing on the peace and development of the world community," during the 1950s, the Center was covered with CIA fingerprints.[147]
The CIS director from 1952 to 1969, Max Millikan, arrived directly from
the CIA, where he had briefly managed the new Office of Research and
Reports, which "wrote in-depth studies on geographic characteristics of
foreign areas, especially on foreign economic developments in the USSR
and China."[148] Straddling area studies and international relations, the eminently practical MIT center had its origins in a State Department exercise
appropriately dubbed Project Troy, an interdisciplinary study group that
brought together scientists, social scientists, and historians to study the
role of communications in strategy and diplomacy. More specifically, Troy

participants were interested in "getting the truth behind the Iron Curtain," an activity that had been frustrated by Soviet measures to jam Voice of America broadcasting.[149] Several of them, including Clyde Kluckhohn and RAND's Hans Speier, were also consultants for the Pentagon's RDB.

In a period of atomic angst, the promise of psychological warfare (and its political and economic cousins) was tantalizing because it was believed to be enlightened and relatively bloodless. The consequences of this impression were felt most in the third world, where rival American and Soviet programs escalated minor conflicts and subjected large populations to repeated manipulation attempts.[150] What is intriguing about psykewar, in addition to this explicit focus on civilians, is that it was as much a problem for historians and social scientists as it was for those interested in the technical aspects of broadcasting and propaganda. Convened in October 1950, with the enthusiastic backing of MIT president James Killian, the

An international communications seminar in 1960 at MIT's Center for International Studies, led by Ithiel de Sola Pool *(head of table)* and Daniel Lerner *(right, arms folded)*. Courtesy of MIT Museum.

Troy group met at the institute's Lexington Field Station and journeyed to Washington to meet with Secretary of State Dean Acheson. Although Troy's final report cited the Marshall Plan and other economic initiatives as important precedents, it also called these measures "essentially defensive" and recommended a more aggressive and comprehensive psykewar program that could use social science to identify and distinguish what Allan Needell dubs "target populations" in the Soviet Union, Europe, and China.[151] By November 1951, MIT's vice president and provost Julius Stratton had alerted Charles Dollard to the new CIS, which already held "some large Government contracts for work on propaganda."[152]

Rostow's analysis of Soviet social dynamics, Harvard's interview project, and the establishment of a permanent institute for the social scientific study of psychological warfare all directly followed Project Troy. The last was MIT's CIS. Funded by the CIA and the Ford Foundation, through the 1950s it "was a place where a wide variety of academic specialists could come together in academic surroundings to participate full- or part-time in classified research and discussions."[153] This abandonment of academic freedom at MIT led to significant campus protests in the 1960s, but not before a model of military–university cooperation had migrated from the institute's scientific laboratories into those of social science. Secrecy and classification restricted discussions regarding funding arrangements and teaching or research methods. These concerns had been anticipated by some members of Troy, who were, as authorities on communications and propaganda, in a position to struggle with the contradictions of "democracy in a garrison," as the MIT historian Elting Morison put it. But not surprisingly, calls for "simultaneous research on the impact of the Cold War on American society and on ways of mitigating its negative effects" went unanswered.[154]

Unlike the strategists who plotted an impending Armageddon, the behavioral scientists at CIS and related sites were more interested in mundane forms of conflict. This preoccupation was pushed to the forefront of social scientific research during the Korean War, when all but the most feverish planners resisted talk of atomic applications. Using the premise of an enemy operational code of political decision making, teams of advisors from the HRRI and the RAND Corporation traveled to the Korean peninsula, where they found a formidable study set in the thousands of Chinese and Korean prisoners of war housed in United Nations compounds. This was historical research, but it could also be used to provide intelligence to future psykewar campaigns requiring data on target areas and their vulnerabilities.[155]

Another purveyor of psykewar advice was the Army's Special Operations Research Office (SORO), housed at American University in Washington, D.C., and set up in 1957 to assist the HRAF in the preparation of country studies. SORO went on to achieve infamy for its role in the disastrous Project Camelot during the 1960s, but in 1958, it also began to produce Psychological Operations Handbooks providing "appeals and symbols of tested persuasiveness for communicating messages to specific audiences in a given country." These audiences comprised various social groups—ethnic, economic, and geographic—who were understood to be differentially susceptible to certain messages and techniques and diversely opinionated on the subject of American influence.[156] The SORO and its older relative, the Operations Research Office, were essentially attempting to provide *maps* of a region's communications networks, both technical and cultural, that identified locations of weakness to be exploited by American propaganda campaigns.

Psychology is central to the formation of governable subjects, and this is a particularly crucial imperative in wartime. But wars are fought domestically, too, and the premises of psychological warfare spilled beyond its foreign objects to infiltrate American life, as I show in chapter 5. According to Nikolas Rose, the human sciences, including psychology, "embody a particular way in which human beings have tried to understand *themselves*—to make themselves the subjects, objects, targets of a truthful knowledge."[157] Psykewar can therefore be understood as a kind of banal terrorism, relegitimizing a state monopoly on violence and the militarization of daily life, licensing one form of collateral damage while preparing the ground, if necessary, for another.[158]

The crystallization of communication research into a distinct academic field was directly attributable to government-funded work on psychological warfare and related subjects such as persuasion, surveys, and human behavior. This field was a generation removed from Second World War studies of national character but was no less concerned with personality. Psychological warfare, according to one survey published for the Operations Research Office, was defined as the *planned* use of *propaganda* and *other actions* designed to influence the opinions, emotions, attitudes, and behavior of enemy, neutral, and foreign groups in such a way as to support the accomplishment of national aims and objectives.[159]

Because psykewar's methodologies dovetailed with those of advertising, a domestic national body was not immune either — a parallel clearest in the novel field of disaster research.

Psychological warriors sought to create new, modern persons in an alien space. But the Cold War practitioners of imperial social science were not only devotees of a distant, technological approach; they were also concerned with molding humans en masse. This was a program of improvement that appealed to many leaders of newly independent countries as well, partly because much of the initiative could come from local administrators. For modernization theorists, however, the threat of communism meant that young states had to be closely parented. As he expanded his BASR Urban Resources project to global proportions, Kingsley Davis also became a prominent figure in population control and argued that demographic trends in underdeveloped countries made these nations vulnerable to communism. The solution, for Davis and others, was controlled birth rates: the American nuclear family's role as a bulwark against subversive behavior could be replicated in the third world.[160]

In a 1956 memo to MIT chancellor and provost Julius Stratton, CIS's Ithiel de Sola Pool claimed that university research on the "human implications" of scientific problems was minimal in the United States; only think tanks such as the RAND Corporation were shouldering this important burden. Heralding the rising relevance of game theory, organizational theory, and similar perspectives, Pool added that at MIT, there was "little temptation . . . to build irrelevant models for their own sake." Rather, "the development of mathematical and other rigorous approaches tends to be infused by a continuing focus on the major issues of national importance."[161] The CIS was not alone in these pursuits, nor was Pool's blithe acceptance of the "infusion" and the geographic divisions it necessarily inscribed unique. But the work of the center symbolized a new form — perhaps the deadly terminus — of Cold War area studies. It was comparative, multiregional, and activist, and, as with similar institutes at Harvard and Princeton, area studies overlapped increasingly with strategic studies.

One early expression of CIS principles offered an impressive if caricatured distinction of the center's work from area studies:

> The usual area program is not policy-oriented; our whole effort is focused on the illumination of policy problems. Area programs as a rule look backward; we are primarily concerned with implications for the future. Area programs tend to be purely descriptive, interested in a country or culture *per se;* we are concerned with what can be learned about a set of generalized problems from a selective study of the ways in which they appear in specific locales.... It can be said that the area program constitutes a starting point for our kind of inquiry.[162]

The MIT center combined the latest in social scientific theory with a kind of military anthropology directed toward hostile environments. This mixture was a key intellectual component of the Cold War and a blend that deepened and distorted Second World War trends.

Abstraction and Application

The surveys conducted by the Bureau of Applied Social Research and similar institutes sought what Jean Converse calls the "large shapes of social geography, movements of populations, flows of information, opinion, and feeling."[163] This concern was equally characteristic of the "spatial science" that transformed geographical study beginning in the early 1950s. The imperative behind these changes was to move past limited theories and seek holistic understandings of the world approximating those offered by science.

Alternatives to this ambitious upstart certainly lingered, but were they, in comparison, poorly organized? With the use of a telling image, Carl Schorske, who is no aficionado of positivism, makes a similar argument concerning philosophy:

> It would be too much to say that before the creation of the new, rigorous analytic schools that acquired salience in the 1950s, darkness lay on the face of the deep; however, the lights that hovered above it were many-hued and scattered, lacking the power to illuminate the terrain with strong, focused beams.[164]

That one source of the improved beams was the light of explosions illuminating darkened German and Japanese cities during Second World

War bombing runs is not anecdotal speculation but documented opinion. Enlisting in the Army Air Forces during the Second World War, the iconoclastic geographer William Warntz found the maps he had "always enjoyed" to have suddenly taken on "'life or death' aspects" and began to deduce a new spatial order from the vantage of his planes.[165] As the Carnegie Corporation's John Gardner also suggested, this airman's view, both transcendent and limited to crude contours, could be translated into a popular conceptual outlook for social science.

The Second World War democratized professional geography in the United States: the elitist Association of American Geographers (AAG) was contested by an alternative American Society for Professional Geographers (ASPG). The exclusivity of the AAG was embarrassingly exposed in wartime Washington, where nonmembers, mostly younger scholars, worked alongside and often outranked older, more established colleagues. The growth of geography during and after the war meant that the elected system adopted by the AAG was no longer suitable, and the two organizations merged in 1948.[166] But the generational schism was also methodological. In one of the earliest ASPG missives, William Van Royen bemoaned the "agonizing detail" of "microchorography," which "in its search for minor facts . . . has often ignored major problems which are staring at us in the face." This was an entry into the widening discussion over the respective merits of regional and systematic approaches and whether the two could be amalgamated. But the ASPG's *Bulletin,* which later became *The Professional Geographer,* was also devoted substantially to the "practical applications of geography," including extended discussions of the discipline's role in defense work.[167] George Kish reflected that after the 1948 merger, "nothing was ever quite the same," and while subsequent changes occurred gradually and even hesitantly, a decade later human geographers were rushing to adopt scientific methods. The region, the object at the heart of geographical study, was consequently reimagined, less as a physical realm for description than an abstract tool for instrumental objectives.[168]

In the previous chapter, I noted Edward Ackerman's strident call for precisely the same scientific geography, an appeal that derived from his wartime experience with the OSS. Ackerman's challenge had particular relevance for the most prominent geographer at the OSS, Richard Hartshorne. Two years before the United States entered the war, Hartshorne had published *The Nature of Geography,* a formalistic study surveying, as its

An American bombing raid over a German city, circa 1942. National Archives and Records Administration (Franklin D. Roosevelt Library Public Domain Photographs, 1882–1962).

subtitle indicated, "current thought in the light of the past." This landmark text, unsuited to its time of composition in several respects, later became a symbol of desecration for younger geographers.[169] As Eugene Van Cleef wrote in a 1952 issue of *Science*, Hartshorne's cherished areal differentiation, or the description and interpretation of regional differences, was "untenable if it is not based upon certain fundamental and established principles, which may be utilized as standards of reference." Although he recognized that geographers, unlike physicists, did not always enjoy the benefits of a controlled laboratory, Van Cleef argued that with a "vast number of observations in which similar conditions occur, we may be able to generalize," producing a true science of geography.[170] Statistical methods, Preston James later wrote innocuously, "offer the equivalent of a geographic laboratory."[171]

Van Cleef's comments bore a striking resemblance to broader statements on the social sciences that included the use of Cold War areas and other geopolitical divisions as anchors for the pursuit of laws and uniformities. The same can be said for geography's "quantitative revolution," which "placed human geography for the first time within the social sciences."[172] The outlines of this episode, which was transnational in origin but preponderantly American, have their share of raconteurs, and I do not wish to travel some of the same paradigmatic paths.

As Ackerman and others since have argued, in *The Nature of Geography*, and more importantly, in his work with the OSS, Hartshorne was by no means opposed to inquiry that fell under the labels systematic, scientific, or objective. The Research and Analysis Branch of the OSS, where Hartshorne occupied a significant position and took his work quite seriously, was the site of extensive collaborative, deskbound and fact-based work.[173] In addition to his R&A experience and his continued interest in postwar political geography, Hartshorne's participation in some of the early meetings to coordinate American area studies, his appointment to the staff at the National War College for the fall of 1949 (while president of the AAG), and his concurrent consulting duties with the Department of Defense collectively indicate that his "geography" is not best understood in the tones of internalist history.[174] As he stated in his capacity as a member of the Air Force Planning Board, "Any specific operation of the Air Force is concerned with particular countries as regions of the world. A good working knowledge of the major differences among the different areas of the world is therefore essential to intelligent operation."[175] In the new postwar strategic order, specific sections of the planet—whether states or regions—were significant only if they could be fitted into a larger global picture. If Hartshorne did not understand this before the war, he certainly did after his time with the OSS.

During the 1950s, appropriating the accoutrements of "hard" and "tough" sciences such as physics and mathematics—or the work of economists and some sociologists—was a practical act at a time when the mantles of science, social science, and overall rigorism provided additional protection from the crusades of anticommunist politicians.[176] Belief in the power of mathematics to change or fix the world meant that some quantifiers were far from politically conservative. But while their intellectual approach "may have seemed radical to those satisfied with an inferior status," Richard Morrill recalled, it "was actually conservative in the sense that we wanted to save geography as a field of study and to join the mainstream of science."[177] Equally, for a discipline that was just beginning to gain entrance to the social sciences, government demands for technocratic, quantitative research and the promise of federal funding were extremely enticing. A depoliticized policy-driven geography, prescribing "the *optimum means* of achieving a *given set* of social objectives," was tailored to attract corporations and the state.[178]

"For several years after the war ONR was a major sponsor of geographical study," Kirk Stone, a former naval ensign himself, noted blithely in 1979.[179] But while it supported an important 1959–60 quantitative symposium, the ONR, along with the SSRC, the National Research Council, and the Army, had also funded the meetings that produced *American Geography: Inventory and Prospect* (1954), a weighty tome featuring mostly statements of unabashed regionalism.[180] Quantification was thus not the key, nor the only, determinant for military patronage. By 1950, when Louis Quam was hired to head ONR's new Geography Branch, the Navy's research mission was global, but it nonetheless decided "what kind of scientific work was done, by whom, and with what equipment."[181] It chose its funding targets strategically, stressing subjects related to national security, tightly controlling clearances and publication, and maintaining "listening posts" within the academy to prepare for military appropriation of cutting-edge scientific laboratories.[182] Even when, by the end of the 1950s, the Geography Branch began to focus on specific "problem" themes, these were a meaningful mixture: "the Arctic, the coastal zones of the world, fundamental methodology and foreign areas, and the interpretation of the results of airborne sensing devices."[183] Thus, as the Navy encouraged the "objective, precise, and penetrating" inquiries of Brian Berry, William Garrison, Waldo Tobler, Edward Ullman, and William Warntz, among other individuals central to the quantitative revolution, it sent dozens of geographers overseas under the auspices of the Foreign Field Research Program, which ran from 1955 to 1966. Concurrent attempts to produce large compendia on countries ranging from Finland to Thailand, as part of an effort to resuscitate classic variants of regional geography, were less successful.[184]

Some of the most adventurous geographic work of the early Cold War was conducted under the sign of "social physics" (later called macrogeography), an approach pioneered by Princeton's John Q. Stewart. "There is no longer an excuse," he wrote in 1947, "for anyone to ignore the fact that human beings on the average and at least in certain circumstances, obey mathematical rules resembling in a general way some of the 'primitive' laws of physics."[185] At a 1949 conference sponsored by the Rockefeller Foundation, Stewart and likeminded attendees invoked cybernetics, operations research, and other holistic products of the Second World War to justify the extension of methods from the study of "physical nature" to the analysis of its human counterpart.[186] Stewart was also interested in geopolitics, and in an audacious 1954 article, he attempted to read "natural law factors" such as gravity and "social mass" into American foreign policy. The national limitations were significant. According to Stewart, precisely the same laws were built into the American Constitution, whose structure of checks and balances was set up to "forestall the natural tendency of leaders towards self-aggrandizement and of followers towards mass hysteria."[187]

Among Stewart's collaborators in the 1950s was William Warntz, who worked at Princeton while employed with the American Geographical Society (AGS). Warntz wrote against "purely verbal and descriptive methods," ignoring his own conceits in a search for regularities and "working hypotheses."[188] Against "the tendency of American geographers to be preoccupied with the unique, the exceptional, the immediate, the microscopic, the utilitarian," Warntz sought "general laws" that unified "individual, apparently unique, isolated facts."[189] Like aerial photography, the god's-eye perspective adopted by spatial scientists reduced terrain to "a set of coded topographic features, 'grounded' by the digital logic of the grid."[190] In his advocacy of macrogeography, Warntz placed legitimacy on the deep structure of space, not the interpreters of that space: "Geography recognizes what geographers may not."[191] The aim of detachment is inevitably brought down to earth by the particularities of perspective; while at the AGS, Warntz's inquiries were supported by the Office of Naval Research.[192]

At midcentury, the AGS was clearly at a crossroads. In addition to the work of Warntz, its director, the British-born George Kimble, had published a significant critique of regional geography, an essay including the oft-quoted line, "Regional geographers may perhaps be trying to put boundaries that do not exist around areas that do not matter."[193] And yet in summary publications the society was still attempting to contend that geographic "tools and techniques have such universal application that it is viewed by some as the common denominator, or, by others, as the catalyst of the sciences."[194] This was an argument that did not sit well with the leaders of the Carnegie Corporation, who rejected a substantial number of AGS funding applications at the height of philanthropy's support for area studies. A clue to the corporation's rationale can be found in two 1949 conversations between the cartographer Richard Edes Harrison and Carnegie's John Gardner. In the first, Harrison bemoaned the state of cartographic education in the United States, opining that there was no one institution able to "provide a man with really adequate broad-gauge training." This included the military, where he had been "doing some work." In the second exchange, Gardner explained to Harrison that the corporation was not willing to support initiatives in geography, owing to "a whole series of problems facing the field." Harrison, who was not a professional geographer, and who was on poor terms with a number of geographers, concurred that this was a "reasonable position."[195] The contrast with the corporation's concurrent enthusiasm for area studies was stark.

Cold War area studies, moreover, was already dividing the world into suitable regions linked only by the vaguest of connections, such as a propensity for communist disruption. Geography's first responsibility, at least as it was traditionally understood, had been preempted. An older generation of regional geographers was ill equipped to do much more than act in a secondary, descriptive role because the regional framework of area studies was merely a vehicle for social science. While geography departments in many public universities were expanding after the Second World War, at a number of influential institutions, the impression of inferiority was devastating: in 1948, at the moment that his school effectively ended the discipline's formal presence on campus, Harvard president James Conant opined that "geography is not a university subject."[196] In this sense, Richard Hartshorne's pedagogical role at the National War College can be compared to the Army Specialized Training Program of the Second World War. Regions as vehicles for description were living on as units of strategic intelligence and military geography, largely outside the academy.

The gurus of spatial science, on the other hand, traveled in the opposite direction, dismissing regions in favor of distributions and correlations that did not always respect the compartmentalization of the planet. If regions existed, they did so as purely functional units useful for scholars en route to universal conclusions.[197] But this was also a derivative movement, borrowing from economic models that had already arrived in area studies programs or that were useless for the comparative politics of the modernization theorists. Too many geographers, one Arctic specialist complained to another in 1948, were already "more interested in pure economics than in geography," and fieldwork, an important premise of area studies, was "almost an unknown activity in geography departments."[198]

While historians of geography have repeatedly emphasized the divisions between regional description and the systematic spatial science that largely superseded it, the middle ground — that of a practical *and* scientific Cold War geography — has been ignored. Regardless of theoretical or methodological differences, American geography of the 1940s and 1950s still appears to have been overwhelmingly and profoundly *national*, a "service discipline," in the scathing judgment of David Harvey.[199] But the conditions of geographic employment and support at midcentury should not be detached from a wider context. These nuances are clear in the comments of John Kerr Rose, surveying the roles of geographers working directly for the federal government in the 1950s:

> Certainly the lot of the systematic geographer would seem to be somewhat happier. But there *are* regional geographers and graduating students who wish to specialize in a region. More than that, there are positions in Washington, not a few of which call for regional experts.[200]

The detachment claimed by spatial scientists rested on the ability to erase or thin humanity, to dematerialize information from the substrates that carried it. This was the language of the computer revolution; life was reduced to models and equations that possessed a truth above and beyond the limits of the individual. Moreover, the elimination of earthly, emotional humans is precisely what was also sought by postwar synthetic subjects such as cybernetics and also by the high modernism of Cold War urban planning that featured Olympian views of nuclear destruction. At the height of quantification in geography, according to David Mercer, urban scholars focused on the city's "hardware," reinforcing "the technocratic engineering view of the city as a *machine*" without passing judgment on its failings or limitations or on their own.[201] If during the Cold War information lost its body, then, for a time, it also lost its spaces, in any rich sense of the term.[202]

In the 1950s, many American geographers were turning to the "advancing front" of scientific progress. The troops of science were taking possession of the social sciences, and geographers were encouraged to desert the resistance immediately. Reflecting on these changes, Edward Ackerman remarked in 1963 that geography, in its determination to declare independence, had missed participation on the "forward salients in science." The previous decade had been spent just catching up. But he was prepared to state the priority for a scientific geography: to understand the "system of humanity and its natural environment on the surface of the earth." The mention of systems was deliberate: Ackerman's world had become a cybernetic grid for design and engineering, a "revolution in rationalism," he acknowledged, that had profoundly altered "the nation's defense program." But regions had not vanished. They had become subsystems that were part of a complex whole. Geographers had to ensure, however, that they selected the proper subsystems for study. And while dismissive of "the old concept of a 'geographic' region," Ackerman was equally quick to acknowledge the significance of *strategic* regions. Political scholarship "within the systems framework," he claimed, "is concerned with regions that have true functional significance in the great man-land system."[203] It is to this scholarship, and to a particularly significant subsystem, with its own very material frontier, that I now turn. Talcott Parsons was correct; at least *one* continent was coalescing.

4 THE CYBERNETIC CONTINENT
North America as Defense Laboratory

> There is no geographical approach to U.S. strategy which does not
> wind up finally in the laboratory.
>
> —WILLIAM BORDEN, *There Will Be No Time*

Coined by MIT's Norbert Wiener in 1947, the word *cybernetics* referred to
"the entire field of control and communication theory, whether in the machine or in the animal."[1] During the Second World War, Wiener worked
extensively for the military on a unified human–machine system that could
target an enemy plane and launch antiaircraft fire. After the war, he and
a diverse group of intellectuals generalized this cybernetic vision to encompass human behavior. Humans were, as Alan Turing "proved" in
1950, not so different from machines.[2] Computers operated like human
minds—and vice versa, as cultural archetypes from organization men to
brainwashed communist subversives seemed to indicate. The "fascination
with information-based feedback systems" spread far beyond mathematics and engineering.[3] Despite Wiener's principled postwar abandonment
of defense affiliations and the widespread concerns accompanying an age
of tremendous power over life and death, cybernetics quickly became a
heavily militarized "universal discipline," blurring human and nonhuman
most successfully "in the agonistic field, if not the battlefield itself."[4] The
military–industrial complex diagnosed by the departing Dwight Eisenhower was itself a cybernetic entity.

In *An Introduction to Cybernetics* (1956), W. Ross Ashby positioned
his subject next to the "real machine—electronic, mechanical, neural or
economic—much as geometry stands to a real object in our terrestrial
space."[5] But as conceived by advocates such as Wiener and Gregory Bateson, it was also a synthesis built up from the results of finite experiments
and "exact theorems." Ambitious cyberneticists attempted to extend these
experiments and theories into other fields of inquiry, including the social

sciences.[6] In this sense, cybernetics shared with the Cold War social science detailed in the previous chapter the traits of certainty and fearlessness; the frontiers of science were not fragile or susceptible to deconstructive impulses but were instead heroic zones of confident endeavor. In his 1948 manifesto *Cybernetics*, Wiener described these interdisciplinary "boundary regions of science" as offering "the richest opportunities to the qualified investigator." Proper investigation of such blank spaces required a team of specialists who could depend on one another's expertise. If not managed, steered, or *governed* (akin to the Greek derivation of the term cybernetics), these explorations would, for Wiener, resemble "what occurred when the Oregon country was being simultaneously invaded" by many competing groups: "an inextricable tangle" of laws and nomenclature.[7] But the idea of control at the heart of cybernetics was self-regulatory rather than coercive, meaning that the human subjects who were part of a cybernetic system — such as the continental defense network detailed in this chapter — would ideally govern themselves.[8]

The Cold War spread of cybernetic ideas and images was also exciting for the popular press. Once removed from the analog register of instrumentation and placed in a digital sphere, computers of the postwar period were increasingly referred to as "electronic brains" — a term that delighted some, including Wiener, and frustrated others.[9] The speed of new computers meant that they could be usefully compared to organisms, but a specifically midcentury fascination with human engineering and organizational behavior meant that humans could also be equated with machines. For the first time, cyberneticists argued, this symbiosis was present in history; a new age of information had replaced one of materialism. As computers and their workings grew less visceral and palpable, comparisons to brains were muted. But perhaps this was because more complex *cyborgs* (cybernetic organisms) were on the horizons of science and science fiction or because artificial intelligence appeared to be a question of software rather than hardware. But the anthropomorphic trend was a pervasive one.[10] More specifically, given the genealogy of cybernetics, it is not surprising that the human component of cyborg hybrids was frequently and vividly understood in military, and masculine, terms.

If extraterrestrial environments were both the original and the ultimate cyborg vistas, it was obvious that cybernetic research had quite practical implications for earthbound exploration as well. In one prominent version, the cyborg was the new colonizer, a being that could adapt to and eliminate geographic difference. Consider this passage from Joseph Russell's "Military Geography" contribution to the landmark 1954 text *American Geography: Inventory and Prospect*:

> Geographic studies in the field of research and development, as distinguished from intelligence, were started during the war, but have been considerably expanded during the postwar period. The studies of extreme and unfamiliar environments were undertaken to note the effects of these environments on men, equipment, and materiel. There was need for the development of new machines, new lubricants, new methods of upkeep to operate efficiently in extreme cold, or in extreme heat, or on steep slopes . . . very wet or very dry climates, very deep snow or very rugged or high terrain, which, if encountered for any extended period of time, either singly or in combination, can impair seriously the performance of machines and humans. . . . These examples point to the desirability of having ultimately an analysis of the physical environment of world regions in terms of the critical elements or combinations of elements that impede or preclude satisfactory equipment and human performance.[11]

Russell's was still a regional geography that would map areal differentiation for the purposes of military mobility, but geographic diversity was, ultimately, a factor to be overcome. Here, he suggested, was where geography could fit into a cybernetic world — as a study of natural environments that a universal man–machine had to *recognize* but could then *regulate* as much as possible.

Displaying a touch of environmental determinism, Russell remained skeptical that such extreme landscapes were of any geopolitical significance. But the need for a complete analysis, beginning with the most important regions, was required in the midst of a conflict that valued the entire globe and all of its regions as strategically relevant. When it was necessary to move from a detached worldview to a local position, Cold War explorers and those who monitored and controlled them could, he seemed to imply, remain abstracted from specific landscapes. More recent developments linking computer mapping to military hardware worn by the soldier in the field indicate that neither Russell nor the early cyborg theorists were deluded. The dream of a bloodless victory, powered by information science and detached command structures, requires that rational and technical authority order a hazardous geography.[12]

This chapter considers the *territorialization* of holistic, cybernetic worldviews in a specific strategic region: the continent of North America. My concern is not theories of cybernetics but systems and networks built using these theoretical principles for inspiration and justification. Whether these constructions were accurate extensions of theory is irrelevant. Rather, they were developed and used within the practical context of an environment that they also defined and stabilized—a continental space. I am referring to the defense laboratories of the Massachusetts Institute of Technology and the RAND Corporation and to the role at these sites of technologies such as computers and radar. By inscribing lines of division around a territorial citadel, the models produced at MIT and RAND in the early Cold War adjusted for the growing intricacy of an integrated globe. But the imagination and fabrication of North America as a vast strategic grid—a cybernetic continent—also required the mastery of the network's elusive northern frontier.

It makes little sense to divide the history of Cold War science and social science by paradigm, law, or discipline. The figures and ideas moving across these domains were linked by new modes of knowledge production, which in turn fashioned a new intellectual geography. For instance, Peter Galison has considered the attempt in atomic physics to build artificial worlds, or digital devices, which would "simulate nature in its complexity." This was initially a local process, but its advocates believed it could be replicated and coordinated across places and scales.[13] Precisely the same conviction rested at the heart of nuclear strategy. Extensive simulations of political crises run by the RAND Corporation and others were designed to bridge academic and policy circles just as their personnel rosters were merging. But neither physics nor strategy, despite significant evidence of allegiance to the state, were necessarily dependent on the continental boundaries defining the edges of this state's defensive security. Each subject could be conceived at a higher level of abstraction or employed in the service of an alternate internationalism. The production of a cybernetic continent was, instead, the task of projects that drew from physics, strategy, and much else, but did so at the specific behest of Cold War imperatives. As a result, these initiatives sought to extend the terrain of simulation from laboratories into a wider world, but they were also premised on the geopolitical division of that world.

I am indebted here to studies of "systems" that do not read technology and scientific practices through the words and deeds of great inventors, the vocabulary of determinism, or partitioned social, economic, and other lenses.[14] The precarious aspects of the continental network discussed here should be apparent. But unlike much work in science studies, this chapter does not adopt ethnographic tools, nor is the object of study a particular artifact. I am more interested in the historical relationship between technology and territory, meaning that I want to avoid the frequently "flattened spatial imaginary" of science studies.[15] The environment at the heart of this chapter is a composite, but its existence cannot be reduced to a set of shared characteristics or lines of association coalescing into an intellectual grid almost as arrogantly stable as the one under consideration.

Men, Measurement, and Machines

During the Second World War, as Norbert Wiener was working on targeting mechanisms, the Office of Strategic Services (OSS) established an elaborate program of behavioral testing to weed out unsuitable recruits. Assisted by consultants Clyde Kluckhohn, Alexander Leighton, and the social psychologist Kurt Lewin, OSS staff scrutinized the actions of over five thousand candidates in intensive three- or one-day camps in the Washington, D.C., area. The summary report of this *Assessment of Men*, unpublished until 1948, is dense and heavily mathematical, the product of "months of statistical calculation" aided by IBM. It describes a rigorous schedule of examinations, interviews, group tasks, questionnaires, and physical activities, including a "map memory" exercise, all intended to shed light on general variables such as motivation, emotional stability, leadership, and initiative. A further series of "special qualifications," including physical ability, film-observation and reportage, resistance to interrogation, and propaganda skills were also considered. The aim was to move past distinct tests to an "organismic" understanding of the entire personality, which, once crafted, could then be dissected into appropriate segments. What drove the results beyond mere psychoanalysis, the report concluded, was work in anthropology and sociology that had "furnished evidence of the determining influence of different cultural forms, ideological and behavioral."[16]

Conducted under the aegis of the OSS's Psychology Division (directed by University of California professor Robert C. Tyron), the assessment was intended to be a precautionary measure before successful candidates were sent on to other tasks. What guaranteed a successful spy was by no means certain. This was the rationale for the intensive, multifaceted analysis of large numbers of candidates. While some psychologists who worked on the assessment project later regretted their actions or were troubled by the project's repeated failure to validate data, others carried selection procedures to the much wider testing group of the public at large, convinced "that they had a valuable contribution to make toward viable human relations."[17] Their efforts were aided by the cocktail of patriotic service, practical value, and scientific opportunity that shaped Second World War and Cold War social science. Such convergence had considerable consequences.[18]

The OSS assessment was only a dramatic example of a much broader interest in military psychology and human engineering spurred by the Second World War. Yale's Robert Yerkes, known for his comparative research with primates and his direction of First World War mental testing, sought greater authority for the applied aspects of psychology that covered every type of military work, from training and morale to equipment design and punishment.[19] Employing the term *engineering* was simply an appeal to the authority of science, an acceptable means of crossing from mechanisms to humans as subjects of study. Applied psychology could be used "at every stage in the life cycle" to facilitate "adjustment," the aim of which was "matching human capacities to the technologies of modern warfare." This emphasis on psychology as a science was not received with complete enthusiasm by the larger scientific community. But the singularly destructive Second World War aided promotional campaigns for the social sciences. An age of danger required wise control and guidance. That the psychological dimensions of conflict were expanded during the Cold War was simply viewed as proof of the continued importance of human engineering.[20]

Cyborg visions of bodily control and regulation were closely aligned with attempts to *shape* populations and individuals, whether foreign or domestic. And because an accessible, substantial group of subjects existed in the ranks of the American military — whose successes, it was assumed, could be improved with the correct conditioning — social scientists were intrigued by the prospects of research in this particular "laboratory."[21] It was the mix of practical merit and theoretical potential that was so exciting. Some of the same scholars who worked with the OSS thus also prepared *Psychology for the Fighting Man* (1943), a pocketsize tome whose circulation by the end of the war totaled over five hundred thousand. Rewritten from the manuscripts of "experts" in "popular form without sacrifice of its scientific accuracy," the book, like the survival manuals discussed in chapter 2, adopted a soothing, perfunctory tone while it described gruesome events.[22] It also offered "a unifying conceptual framework" that balanced laboratory results with battlefield advice.[23]

Psychology for the Fighting Man moved smoothly from group behavior to the individual brain and from sites of combat to the inner spaces of the body. An early chapter, for instance, was titled "Sight as a Weapon" and lauded the eye as "one of the most important military instruments that the armed forces possess." The combination of eyes, ears, brain, and muscles formed the "indispensable tools of war," and military equipment was only an extension of bodily ability.[24] By establishing this hierarchy and prioritizing bodies — which were one with the mind in the rhetoric of human engineering — psychologists were claiming ground for the legitimacy of their insights. But they were also solidifying a concept of "man–machine units," later expanded in what was called *human factors* research, such that the truly effective soldier was a combination of technological design and bodily training.[25] Other aspects of the man–machine unit were concurrently under consideration in several locations, but on one university campus in particular.

Thought and Action at MIT

The connection between academic research and national security was difficult to relinquish as military money continued to flood into universities during the early Cold War. MIT was the premier recipient of this funding, and its flagship postwar project was the Lincoln Laboratory. In a 1954 speech, MIT president James Killian described the laboratory as a "major effort in the field of safety engineering."[26] The Lincoln Laboratory also fitted with the oft-repeated claim that MIT held a unique position in both the "world of action" and the "world of thought."[27] But the phrase *safety engineering* additionally reflected the belief that security was a function of science and that the defense of North America could be engineered systemically.

By the late 1950s, MIT's campus and the institute's outlying properties were dotted with interdisciplinary laboratories of electronics, nuclear science and engineering, instrumentation, and other booming fields. These sites were descendents of both the Manhattan Project and MIT's own Radiation Laboratory, the location where accounts of radar's tangled history often come to rest. The Lincoln Laboratory, which still occupies a sprawling, defoliated property in the Boston suburb of Lexington, owes its existence to both radar and the bomb. Its origin story, recounted in equal measure by historians of science and scholars of strategy, unfolds in glossy fashion. A symbolic trigger—the detonation of a Soviet atomic device in 1949—is combined with a slightly cantankerous visionary, George Valley, an MIT physicist who had resisted the lure of Los Alamos but came to his senses in a risk-laden world of two nuclear powers. Valley's concerns were both patriotic and personal; he realized that the location of his new home in Lexington, with its striking view of the Boston skyline, also offered little blast protection from that direction.[28] After discovering the inadequate condition of American air defenses, Valley took up the problem with contacts in the Air Force, who, following a series of MIT studies, greenlighted the Lincoln Laboratory, which in turn produced two remarkable Cold War feats of engineering: the Distant Early Warning (DEW) Line and the SAGE (Semi-Automatic Ground Environment) computer system.

SAGE and the DEW Line were the two most prominent and promising components of the continental defense program of the early Cold War, the center and periphery of an "emerging shield" projected and maintained by "the talents of man with the best aptitude of machines," which transformed North America into a vast cybernetic region.[29] Such technological eyes, ears, brain, and, ultimately, fists—to use the parlance of the period—were coordinated to generate televisionlike composite "air pictures" that were "drawn like maps."[30] These graphic displays combined the recording and calculation abilities of computers with the "perceptive and display talents of radar."[31] Once fully installed, SAGE divided a continental space into a grid of distinct areas. Each area possessed a central direction center, a massive concrete block shut off from the surrounding landscape. Sightings detected on individual radar scopes covering a smaller area of each sector were then plotted—first manually, and in later versions,

An aerial view of the MIT Lincoln Laboratory, Lexington, Massachusetts, circa 1956. The blockhouse in the foreground housed the AN/FSQ-7 (XD-1) SAGE computer. Photograph courtesy of The MITRE Corporation. Copyright The MITRE Corporation. All rights reserved.

automatically—on a larger, transparent grid that could be monitored by commanding officers. As a hypothetical battle developed, SAGE's fluid representational capabilities provided the appropriate "basis for the necessary human judgment."[32] According to Paul Edwards, a "SAGE center was an archetypal closed-world space: enclosed and insulated, containing a world represented abstractly on a screen, rendered manageable, coherent, and rational through digital calculation and control."[33] The echoes of period geography are apparent here, from the abstractions of vectors to the license permitted by perspectival detachment. But in the case of SAGE, these were also strategic designs, fixing in place a boundary dividing a secure heartland from an insecure exterior.

Capturing and fixing a continental space in a microworld, whether within one of thirty SAGE direction centers, the Lincoln Laboratory, or the similar environments built by the RAND Corporation for behavioral

A typical SAGE console, December 1958. National Archives and Records Administration Still Picture Division, RG 342, Series 342B, Box 839, Folder 342-B-03-003-14.

studies, was an act of *construction*. As Bruno Latour states, "Since scientific facts are made inside laboratories, in order to make them circulate you need to build costly networks inside which they can maintain their fragile efficacy. *If this means transforming society into a vast laboratory, then do it*."[34] Latour's tongue-in-cheek directive is not only applicable to cybernetics but is more generally useful for geographical studies of scientific practices, "meaningful situations or configurations of the world."[35] It is certainly the case that the creation of controlled microworlds "provides models and strategies for reconstructing the world around us," for making that crucial step between the "mastery of locally situated phenomena" and the standardization of this achievement such that it can be replicated in various contexts.[36]

But a laboratory does not have to be a physical enclosure, and what counts as science is not just laboratory work. Moreover, the authority vested in scientific facts at MIT was secondary to a larger relationship between strategy and science unfolding in the "channels of information that exist[ed] behind and beyond the lab environs."[37] The scientific spaces constructed at MIT—from the Lincoln Laboratory to the continental defense grid—were also strategic and were made possible by and designed to suit discourses of air power, military balance, and domestic security. In this sense, SAGE might have contributed to or even epitomized a technological closed world. However, boundary-producing performances of foreign policy were equally and reciprocally responsible for the artifact named SAGE—its development and its particular configuration of geography. To assert the need for an improved defense network, scientists could not gesture to the truths of a technological system that was very much an ad hoc and speculative project. Instead, they cited a strategic "reality" whose scientific status was patently precarious.

From the Laboratory to the Skies

After George Valley's Air Defense Systems Evaluation Committee (ADSEC) had presented its scathing survey of the nation's air defenses to the Air Force, Louis Ridenour, the Air Force's chief scientist, asked James Killian to set up a laboratory devoted to studying the problems identified by Valley and his colleagues. Killian and other MIT administrators were reluctant to take on what promised to be another controversial project that would divert resources and labor from the institute's educational base, but they consented under the condition that an Air Force–funded study group would first "conduct an intensive and comprehensive analysis of the overall problem of air defense."[38] They were also encouraged by the "scientific approach" adopted by ADSEC, which, not coincidentally, identified many of the air defense network's deficiencies as human and thus correctable through automation and computerization.[39]

The study group, named Project Charles, was one of several summer studies held at or associated with MIT during the late 1940s and early
1950s.[40] These were intensive, interdisciplinary, and influential gatherings
of academics and military employees addressing various facets of scientific
war, from undersea combat and civil defense to propaganda technologies and
nuclear aircraft. Killian boasted that they encouraged "unorthodox, un-
inhibited thinking," but summer studies were not limited to brainstorming
sessions.[41] They were also crude war games insofar as they began with presumptions concerning alliances, enemies, and geographical orders of safe
and dangerous places. Summer studies were additionally dubbed "systems
studies"—a more accurate description of their philosophical intent and
scope given the popularity of the word *systems* in postwar management,
engineering, and other fields that eliminated distinctions between science
and social science.[42] Known as a source of technological invention and innovation, during the early Cold War MIT also became a site of geopolitical
authority and the location for an unprecedented blurring of strategy and
science.

Beginning in February 1951 (and thus stretching the definition of a
summer study to its limit), the twenty-eight members of Project Charles,
supported by sixteen consultants, were exposed to two distinct but overlapping air defense representations: those proposed at Pentagon briefings
and three successful trial interceptions of live aircraft, which used the
Whirlwind computer that later became the heart of SAGE.[43] Charles was
also explicitly situated within a novel strategic context. The final report's
preface begins with this statement:

> For the first time in history, as a consequence of the atomic explosion in
> the Soviet Union, the United States is confronted with a really serious
> threat of a devastating attack by a foreign power. This new danger has
> necessitated major changes in the scale and methods for the defense of
> this country.

A consideration of complex technological problems involved in aircraft
detection and warning was thus dependent on assertions of American
vulnerability—a product of "geographical concentration of industry and
population"—and unpredictable Soviet behavior. American strategists and
scientists were in the awkward position of estimating the importance of *domestic* targets, as a "lack of knowledge of what we have to defend against"

was turned into an insecurity problem that saw the enemy holding "the initiative."[44] Despite this language of deep uncertainty, the Project Charles report, particularly in its appendices, is filled with the confident mathematization of combat conditions. Geopolitical danger still required technical solutions.

The participating "scientific statesmen," as President Truman called them, understood summer studies in the terms of systems engineering.[45] Such initiatives promised to integrate the requirements of laboratory research with new technologies into an "operating system" that would absorb new components while still running smoothly. Systems engineering therefore traveled beyond its predecessor—operations research—to include actual construction work, coordinating "the overall system requirement with local conditions." The field, the site of these local conditions, was thus incorporated into a systemic frame as one more set of variables. But the system was also seen as an idea with an end product. By tracing progress to completion, engineers could gain legitimacy in both industrial management and military command.[46] Systems engineering brought these realms closer to academic research and academic theorizing, and was thus the most direct exemplar of the military–industrial–academic complex.[47]

Many of the members and conclusions of Project Charles migrated into Project Lincoln, which became the Lincoln Laboratory in 1952. Before it could move into its private Lexington facilities, Project Lincoln's use of a campus lounge, with an armed guard posted at the door, elicited complaints from other MIT staff and students, a reaction that foreshadowed more substantial protests against MIT's military and intelligence ties in the 1960s. Even members of Project Charles and Lincoln were concerned with the utility of such high-profile congresses. Chief among them was the consistently caustic George Valley, who remarked in an April 1952 letter to James Killian, "Looking back . . . I remember chiefly the luncheons, which were fattening."[48] This sentiment did not prevent Valley from accepting an important role as assistant director of the Lincoln Laboratory.

Killian and his staff decided to disconnect the Laboratory's multimillion dollar expenditures from the institute budget and did the same with two other military-dependent projects, the Operations Evaluation Group and the Instrumentation Laboratory. All three classified programs were, in 1955, bureaucratically segregated under a new Division of Defense

Laboratories. Less than two years later, this was replaced by a Division of Sponsored Research, which also included MIT's Industrial Cooperation section, the office that managed and organized external contracts such as summer studies. Of course, these attempts at separation were far from complete. Several university employees, including Killian and Valley, held top-secret military and Atomic Energy Commission "Q" clearances.[49] In 1959, recalling an extraordinary period of military-supported research at MIT, the Lincoln Laboratory's director, Carl Overhage, suggested to institute president Julius Stratton that the university "may become the first of the world's great academic institutions to take a much more positive view of these 'non-academic' operations."[50]

Retired Navy Admiral Edward Cochrane (MIT's vice president for industrial and government relations), George Valley, Air Force Major General Raymond Maude, and Colonel Dorr Newton at a press conference introducing the SAGE system for continental air defense, January 1956. Photograph courtesy of The MITRE Corporation. Copyright The MITRE Corporation. All rights reserved.

The Project Charles participants were very aware that they were part of a significant debate over air defense that included, among others, members of the Air Force — particularly its powerful Strategic Air Command (SAC) — as well as the popular press. Controversy reached a feverous pitch with the conclusion of a follow-up to Project Charles, now dubbed simply the 1952 Summer Study. In addition to the divisive presence of physicist Robert Oppenheimer, what sparked criticism was the study's advocacy of a comprehensive radar chain across the northern reaches of the continent, connected to an electronic command and control system run by computers such as Whirlwind.[51]

After being leaked to the media, this proposal resulted in denunciations of a costly Maginot Line. Prominent figures associated with continental defense research at MIT weighed in to defend the deterrent of early warning as part of a balanced military stance, a natural response to the "realities of the atomic age."[52] A draft of James Killian's off-the-record remarks to the American Society of Editorial Writers in October 1953 includes the suggestion that the Lincoln Laboratory had been mistakenly "credited with making recommendations involving expenditures of astronomical sums of money."[53] In the most prominent published rebuttal, Killian and Lincoln Laboratory director Albert Hill argued in the *Atlantic* that it was "creating a dangerously false issue to charge either the military mind or the scientific mind with having a warped or distorted point of view."[54] Praising Killian and Hill's measured intervention, Vannevar Bush contrasted it with a concurrent article in *Fortune,* which, he believed, "seems to me to give [the Soviet Union] everything they might need in order to estimate our whole situation in some detail."[55] Six months earlier, the Summer Study's leader, MIT physicist and electrical engineer Jerrold Zacharias, was already registering his discontent with the consistently "hysterical" media treatment of continental defense. What was needed, he stated, was "a cold blooded analysis that sends a chill down your spine" — an analysis, in other words, that adopted the discourse of strategic science.[56] Indeed, Lloyd Berkner recalled Zacharias as one of "the great strategists of our day... who have shown that today's national strategy depends no longer on geography but on the appropriate applications of science to the whole spectrum of our national affairs."[57] Berkner did not mention his own crucial role in the same fusion of geography, science, and strategy, not least as a key member of Zacharias's Summer Study.

Just as Project Charles, as I show in the next chapter, incorporated social scientists to explore and explain the economics of vulnerability and dispersal, the 1952 Summer Study was an interdisciplinary venture that included the geographer and OSS veteran Kirk Stone as well as scholars versed in communications and transportation economics. Stone and two other participants provided an appendix, "Geographic Studies," which surveyed Arctic terrain using aerial photographs, maps, and personal recollections to discover where environmental factors "were the most favorable, or, as the case often was, the least unfavorable" for the construction of small radar stations.[58] The presence of Stone and others who were removed from the technical detail of radar technology was critical and signified that the study's organizers were considering Arctic landscapes *in advance*. Stone continued to consult for the Lincoln group after the construction of the laboratory, providing lists of contacts who might assist with the northern aspects of radar work.[59] Environmental details were not the only concern. As Zacharias observed, "One does not take a course preliminary to a Ph.D. in physics on the human problems of the Arctic."[60] Such interdisciplinary challenges were not only compatible with, but also critical to, the synthetic military research of the Lincoln Laboratory, which despite its scientific orientation was not all that distinct from MIT's concurrent strategic initiatives. Albert Hill noted in 1951 that the laboratory's first charter "provided for work in the field of scientific reconnaissance and intelligence . . . by mutual consent of MIT and the Air Force inasmuch as it was complementary to an existing program at MIT for [the] CIA."[61]

MIT also hosted a 1955 follow-up to Project Charles, an Office of Naval Research (ONR)–funded study titled Lamp Light, which featured several Charles alumni and employees of the Lincoln Laboratory. Unsatisfied with the scope of the air defense network under construction, the Lamp Light group was determined to extend it outward to the Pacific and Atlantic oceans. But Lamp Light was also a more rigorous exercise than Charles, featuring additional geographic tests for each proposed radar route:

> The simplest method of visualizing a system's effectiveness against an attack, and a good method of estimating its quality, is to take the position of an enemy and plan an operation against it, and then study the possible actions of the defense against the attack as it progresses step by step across a map.[62]

Specifically, Lamp Light participants assumed that certain cities would be targets, used the CIA's National Intelligence Estimates (some certainly inflated) to formulate an evaluation of Soviet strength, and then made calculations to ascertain how many enemy planes would have to breach North American defenses to cause fifteen million deaths. Given that map tests were being conducted for proposed systems, and given the frequency with which strategists had bandied about such numbers by 1955, it is safe to assume that the figure of fifteen million was considered *acceptable*. These tests were treated as games featuring one team commanding offensive units and the other positioning defensive forces. While the visual benefits of such map-based exercises were cited repeatedly in various summer study reports, larger, more complex simulations required the comfort of more abstract mathematical analyses.[63]

At the same time that the Lincoln Laboratory was under construction, the RAND Corporation was wrapping up a radar research project of its own. Owing to their mutual dependence on the Air Force, both groups were kept apprised of developments on the opposite coast. A November 1951 meeting in Santa Monica, California, between participants found "agreement . . . in wide areas."[64] The hallmark of these early air defense discussions, one MIT researcher wrote, "was a map or globe covered with arrays of overlapping circles" showing the need for complete, continuous radar coverage over and in front of strategic spaces.[65] At RAND, as in Project Lamp Light, intelligence estimates of the Soviet Union were used to establish the degree of threat, and mathematical models were supplemented by map exercises: "The laying out of bomber raids and air battle regions on maps, and the 'fighting through' of the campaigns, has served to call attention to important weaknesses as well as to provide a means of predicting the outcome."[66] This much, at least, was shared between the groups in Cambridge and Santa Monica. Some RAND theorists remained opposed to radar lines, going so far as to suggest that the DEW Line was proposed only because MIT scientists "could find no other use" for the expensive computer technology developed at the institute with government money.[67] These criticisms were supplemented by the more valid concern that the DEW Line's "usefulness may be greatly reduced by repeated warnings set off by Soviet spoof raids" — a flaw that had not been corrected in the design stages.[68] Others at RAND, however, saw early warning lines as practical because they allowed American bombers more time to get off the ground and avoid destruction while immobile at their respective bases. RAND employees "were often more politically conservative than those in academia."[69] But their influence on the culture of strategy during the 1950s was profound.

Simulations, Systems, and Strategy

As the foundations for the Lincoln Laboratory and its "all-out technological attack on some of the new problems of air defense" were being laid, a RAND consultant named John L. Kennedy was also taking an interest in these conundrums, but from a different perspective.[70] In August 1950, having drawn attention to the importance of "human factors affecting man–machine behavior in a man–machine system," Kennedy suggested that RAND set up a department of psychology. His request led to the creation of a Systems Research Laboratory (SRL) in May 1951. Five months later, two more psychologists and a mathematician had been added. The SRL's first project was the construction of a "fairly complete physical model" of a Tacoma, Washington, Air Defense Direction Center.[71] Computers had made feasible what an early SAGE manual called the "synthetic air defense situations" — or simulations — that were useful not only for training personnel but for more general research into individual and organizational behavior under duress in a contained environment.[72] This, in turn, would lead to a sharpening of the man–machine relationship at the heart of SAGE, which was designed so that electronic components "carry out those tasks which men do most poorly in air defense."[73] One way to address and alleviate the proliferation of uncertainty after the Second World War was to concentrate on technological control and eliminate humans from positions of command as much as possible.[74]

Nowhere was the combination of strategy and science, including the human sciences, more pervasive during the early Cold War than at Santa Monica's RAND Corporation. As the well-connected Sovietologist Philip Mosely put it, "RAND pioneered in developing an even broader range of interdisciplinary cooperation, thus learning by trial and error to blend and harness the joint efforts of, for example, strategists and political scientists, physicists and economists, in analyzing problems of tremendous import and baffling complexity."[75] Like the study groups established at MIT, RAND

Systems Research Laboratory, RAND Corporation, n.d. (early 1950s). Copyright The RAND Corporation; reproduced with permission.

was a practical outgrowth of the emergent interest in the scientific aspects of air power and nuclear war. As I noted in chapter 1, General Henry "Hap" Arnold advocated for an institution where civilians could work full time on military analysis for the Air Force.[76] RAND's reputation as a haven for brilliant and eccentric intellectuals who coolly played with the variables of combat is well known. But RAND was also a locus of significant debate and dissention not only with respect to its sponsor, the Air Force, but also internally. Put simply, the key issue of contention was whether the emphasis should be placed on fighting—and winning—a nuclear war or on preventing it by adjusting to the jittery but cautious temper of an atomic world. Regardless of where one stood, escaping the overwhelming prominence of "the bomb" in American strategy was impossible; the shrinkage and redirection of military budgeting after the Second World War favored the "air-atomic offensive."[77]

Though far from philosophically uniform, then, RAND employees were still overwhelmingly committed to what Warren Weaver at an early organizing conference called "the rational life" — the presumption that the world was fully knowable and equally manageable.[78] This meeting, attended by many leading academics, was held in New York in September 1947 and was essentially a recruiting opportunity for the corporation's new (and separate) Social Science and Economics divisions. But the shared values presumed by Weaver's statement also had a more practical and ominous aspect: the creation of a "science of war" was RAND's entry in the competition of interdisciplinarity.[79] Its various games, models, and simulations were all devoted to constructing combat environments and filling them with approximate variables ideally approaching verisimilitude.[80] At the 1947 conference, discussions were not limited to RAND's direct responsibilities but rather addressed the "identification, measurement, and control" of war's "factors" and how these could be aligned to ensure victory.[81] As an influential figure in the Rockefeller Foundation's natural sciences sector, Weaver encouraged the study of "organized complexity" and the importation of physical science into biology, and he had also directed the Second World War gun control work of Norbert Wiener under the auspices of the Office of Scientific Research and Development.[82] Philip Mirowski claims that while Weaver might have been less flashy than some of his contemporaries, he was the "anonymous entity behind the lines who left his mark on most of the nascent cyborg sciences."[83]

The search for the "deep elements," or the "strategic sense," of an atomic world drew the Yale scholar Bernard Brodie and many other social scientists to RAND at midcentury.[84] Brodie became a leading member of one set of corporation analysts who employed a familiar brand of historically dependent realist international relations theory. Another group of strategists borrowed from RAND's economists and physical scientists to present heavily quantitative studies. Both sects, however, rarely budged from a "missionary attitude" toward the Cold War: the typical "RAND style" of dispassionate scholarship attempted to impose order, effectiveness, and doctrine on what appeared to be unpredictable and irrational actions.[85] But it was in the second group that the excitement accompanying new techniques, such as operations research and game theory, proliferated. Unlike interpretative Sovietology, these techniques appeared to produce tangible results. Operations research, a product of World War II planning rooms, used interdisciplinary teams to determine the most efficient use of technologies, while game theory, notable for its conservatism and pessimism, attempted to rationalize the uncertainty of strategy by predicting an opponent's actions and selecting the appropriate response. Game theory at its most arcane involved intensive calculation, but RAND's Math Department also designed "scratch-pad" war games played by virtually everyone at the corporation. By the late 1950s, the language and techniques of game theory had fully infiltrated strategic analysis at RAND, spawning more complex heirs along the way.[86] One result was a geography of risk that emphasized limited wars in the testing grounds of third world regions and typically featured nonnuclear weapons.

Various types of simulations were also paramount, obviously, in the study of full-scale nuclear war. This was an event that could not be completely field tested, although the effects of atomic explosions on buildings and proximate troops certainly were, as I show in the next chapter. Nuclear weapons represented a direct challenge to military conventions, lending prominence to civilian strategists who conceived of the bomb as an unconventional tool of war. New expertise rested in the avant-garde at places such as RAND, where theorists were resolved to minimize the muddle of "instinct, bias, and personality" characteristic of earlier military authority and replace it with detached quantification. Since, as RAND analysts pointed out, there was no one qualified to fight a nuclear war, the best way to sift through the list of hypothetical scenarios was by scientifically gaming them. This was one of the major sources of friction between the Santa Monica modernists and their sponsors in the armed services. And yet, as Sharon Ghamari-Tabrizi has carefully argued, the distinctions were hardly so clear; RAND employees, like other Cold War strategists, also employed a language of "intuition, insight, discretion and artistry," and they were fully aware of this hypocrisy. The creativity nurtured at RAND enabled the investigation of all forms of strategic dilemma and solution, leading directly to the Strangelovian scenarios so commonly associated with the corporation and with certain RAND employees.[87]

In the more elaborate simulation games, or "diplomatic exercises," players adjourned to a special room in the RAND basement and divided into Red and Blue teams (for instance), acting out roles under the watchful eyes of umpires or managers. These ventures received scorn from many of the corporation's mathematicians and economists, who in turn were criticized for abstract calculations completely obscuring history and chance.[88] Both factions found audiences for at least some of their work in the military, which was slowly adopting operations research and systems theory, and, of course, had long staged war games. But the RAND exercises also simulated state behavior, carrying the scholarly field of international relations into the realm of active learning, and predicted "future possibilities and prospects" for American foreign policy. These games occupied the practical end of a spectrum of scenarios that encompassed more theoretical, quantitative examples of political sociology, some of which lacked *any* geographical determinants. Interestingly, all camps claimed the high ground of realism (if admittedly partial), but for divergent reasons. Nearly every drama, however, deemphasized outright victory, which made them peculiarly suitable to the Cold War stalemate. Thus, while often spectral, the spatial outlines of strategy were still crucial to the ubiquitous great-power confrontation at the heart of so many RAND games. It was important that these political exercises could, if necessary, be "global in scope, simulating not only the global political environment but also the detailed interaction of governments on a global scale."[89] Once this stable typology had been sharpened in the RAND factory, it spread to various academic settings, including MIT's Center for International Studies.

The situatedness of RAND games and gaming proposals was signified by the assignment of players to certain roles based on their area expertise. Foreign teams were directed to act realistically—that is, as they believed the governments of their states or blocs would under the set circumstances. The American group, however, was less constrained. Referees could also disqualify a move on the grounds of implausibility, an authority radically advanced when computers assumed the role of umpire. Based on a template devised in 1954, over the course of 1955 and 1956, four major games were played at RAND's headquarters; the first two were only a few days in duration, but the third and fourth each lasted about a month and were played half-time and full-time, respectively. The final game, in April 1956, included three "senior Foreign Service Officers from the Department of State." As with other projects dependent on regional awareness, game designers recognized that advances in social scientific theory could heighten the ability to forecast changes and consequences but also saw that applying this theory in appropriate ways required finely tuned abilities and assumptions—senses, skills, and speculations that might be made more acute around a map board.[90]

Political games created at RAND were intended to aid the training of military and diplomatic professionals. Students were exposed to conditions replicating demands and tensions that they were likely to encounter beyond the walls of gaming facilities. An equivalent set of concerns was driving another RAND project, one using similar cartographic technologies but addressing another facet of the strategic landscape. At the time of its development and implementation, SAGE was frequently described as the first "large scale man machine system of its kind": geographically dispersed, made up of complex equipment (including the display facilities), featuring "real-time data flow" and requiring extensive numbers of personnel with a range of technical skills. The specialization of the tasks assigned to these personnel, who were forced to deal with significant "psychological isolation," resulted in a "fractionation of the decision making process." Moreover, the heightened role of computers in the system resulted in "feeling of loss of control over the environment" among human participants.[91] All of these features meant that working SAGE direction centers were not only difficult to construct and sustain but also fascinating from the perspective of social science; they were laboratories in which the connection to an exterior space was negotiated through machines. This meant that the symbiosis of humans and machines could be studied as a controlled example of systemic *degradation*—the difference between optimal and operational performance.

John L. Kennedy's "natural habitat" was human engineering and physiological psychology, so his work at RAND was not really "straying," as he wrote in 1952, but rather expanding these interests as a result of contact with the mathematicians, physicists, and behavioral scientists who roamed the halls of the corporation's facility. All of these intellectuals, Kennedy believed, shared an interest in the "complexity of real human affairs" and the methods—"preferably scientific and objective"—required to predict the behavior not of individuals or groups but of systems, which included technological devices as necessary components. Long before the term entered the parlance of poststructuralism, Kennedy was referring to these systems as "assemblages" featuring decidedly irregular interactions of parts. Even applied mathematics, he believed, would not succeed at grasping these relations unless it (and all other methodologies) began with the premise of a "total system." New branches of mathematics, such as systems analysis — which added the far less fixed variables of "wisdom" and "common sense" to quantitative models — were, for Kennedy, more successful at this task, in part because they relied heavily on computers.[92] As a result, analysis moved closer in appearance to its subject of study. That said, SRL leaders freely admitted that mistakes and modifications were necessary because their investigations were ambitious and the "terrain [was] not well mapped."[93] Objectivity was flexible, and scientific research was an adventure subject to unexpected events and advanced by hunches and sudden revelations. But as long as all of this guesswork, contingency, and organic growth occurred within the laboratory (and the adjacent but integral observation platform), chaos could be averted.

The impressive size of 1950s computing also meant that a laboratory space was required for synthetic research. In Kennedy's SRL, first built in the back of a Santa Monica billiard hall, IBM technology controlled input by presenting a set of "complex 'real' problems to the system." This weapon system was a "low-order abstraction" of a strategic conundrum involving humans, machines, and a communications network. In the process of its construction, the "nature" of the natural sciences that Kennedy hoped to translate onto human bodies and interactions was made technical. As simulations proceeded, the laboratory itself became a kind of machine, "grinding out the interactions." Such raw compilation would, it was hoped, eventually lead to more generalizable mathematical models of these interactions. Hypotheses could be sharpened and returned to the SRL, the site of the operational test, for further testing and study.[94]

SRL exercises were spectacles, complete with rehearsal, script, and "ensemble work." This was a cybernetic form of education, for both participants and observers, which presented learning as a type of feedback, delivered in a familiar language of scientific justification.[95] But as with Project Charles and Lamp Light, the abstractions fed into the SRL system carried the traces of strategic uncertainty. The laboratory's experiments were premised on a situation similar to that faced by, or anticipated for, actual SAGE direction center personnel and relied on speculative information from intelligence estimates.

Junior college students and military personnel were used as crews in SRL tests that gradually increased the difficulty of tasks, lengthened the duration of experiments, and provided real-time results for observers such as Kennedy. RAND's J. R. Goldstein wrote offhandedly that "what had been intended as research on patterns of behavior of men working with machines under conditions of stress had now apparently produced a valuable method for training Air Force personnel."[96] The replacement of the students from the first simulation, Casey (1952), with Air Force officers and airmen for the next three—Cowboy, Cobra, and Cogwheel (1953–1954)—was therefore not surprising and proved momentous. Military culture had been imported into the laboratory, including the use of Air Force communication styles and methods of address. Within the Air Force teams of test subjects, the "excitement was obvious," leading to "restless nights and bad dreams."[97] Of course, the laboratory environment was also essential to the training of actual direction center workers. This task, dubbed the System Training Program, rapidly expanded in scale and scope. In 1951, the year of its construction, a handful of researchers were based at the SRL. By 1956, the number had increased to 850, and the previous year, a new Systems Development Division had been created to house them. Its employees had "moved out of the laboratory and into the field."[98] By 1958, an entirely distinct organization, the Systems Development Corporation (SDC), had been carved off of RAND and soon dwarfed it. The SDC addressed the practical mechanisms of human engineering that escaped the scattered approach of high-flying RAND analysts.

By the time of the SDC's incorporation, the training program was using an entire air division rather than a single "indoctrination direction center" as a unit, while actual SAGE centers were integrated into the continental zone of the North American Air Defense Command (NORAD).[99] But this expansion had already occurred, at least virtually, at RAND. In the Casey simulation, Kennedy and his colleagues acted as representatives from early-warning stations, "phoning" information to the direction center crew. But they quickly realized that this scheme was insufficient and inaccurate and added more credible stations operated by one person, followed by even more early-warning reporting facilities with larger crews. By the time of the Cobra experiment, additional centers were added in order for one direction center to communicate with the "rest of the world."[100] In the

Air Defense Training Laboratory, Systems Development Corporation, 1960s. Copyright The RAND Corporation; reproduced with permission.

quest for accurate results, simulations were repeatedly broadened, moving closer and closer to what was being copied.

Kennedy hoped that the SRL could serve as a transitional model that would bridge the scholarly gap between laboratory and field studies of organizations. RAND's own direction center was compared to equivalent "information-processing centers" such as stock exchanges and weather bureaus, but a military format was ultimately chosen because the stress-inducing inputs, particularly from radar, could be accurately replicated. Even so, RAND's stake in the defense industry was downplayed. If the corporation's Air Force ties were public knowledge, and if experimental crews were explicitly motivated by the language of security, the argument that SRL research could equally serve other organizational environments was a means of diminishing particularity. This was a claim reinforced by the steady faith in autonomous science at RAND (and in the wider sphere of military-funded cybernetics work)—a belief that the Air Force could be turned into a domain of rigorous analysis, using a language at first accessible only to scientists but later translated into a dialect suited to the military. Inside the SRL, a culture could be cultivated and observed and then classified according to both military typologies and alternative models. What guaranteed a successful transition to field conditions, however, was the perceived accuracy of the laboratory space. This was integral not just because the SRL was soon appropriated for the training of actual direction center personnel but also because the practice of simulation, once firmly bounded, was mobile and could "go on location," perhaps even to radar sites built in landscapes that did not resemble downtown Santa Monica.[101] Yet what the RAND researchers failed to acknowledge was that the field, in both general and specific terms, had been a part of their simulations from the beginning.

Frontier Engineering

In January 1949, Isaiah Bowman—the distinguished geographer, science advisor, and president emeritus of Johns Hopkins University—delivered the opening address at the fifteenth annual meeting of the American Society of Photogrammetry. He spoke on what was surely a popular and pertinent topic: "Geographical Objectives in the Polar Regions." While wide-ranging, Bowman's speech repeatedly returned to the importance of the scientific comprehension of polar environments, particularly the North American Arctic. As Bowman put it, "Survey, survey, and survey may be said to be the three basic requirements of present-day polar research, and we do not restrict the word to cartography." Viewing and traversing the Arctic from multiple perspectives, he added, "will give us better maps or maps where none exist," and the observations produced from this field-work would be "an inexhaustible spring of inspiration for the mathematical, physical, and biological syntheses that are the foundations of scientific system and law, that is, constantly improving generalization."[102]

Bowman was no Arctic expert, but his equation of fieldwork with the ability to generalize must have struck a chord among those in attendance who were familiar with the course of recent Arctic research. His speech arrived in the early stages of an extraordinarily intensive period of North American polar scholarship. This was a diverse but highly coordinated effort, unprecedented in scope and strategic significance. As the text accompanying a new 1949 *National Geographic* map of the Arctic put it, "the Northlands" were still "gradually revealing their secrets to man."[103] Interest in the Arctic increased exponentially during World War II, with the Japanese invasion of the Aleutian Islands, the establishment of "staging routes" east and west for transport of aircraft to Britain and the Soviet Union respectively, and massive construction projects such as that of the Alcan (later Alaska) Highway. Perhaps the most obvious indication of this shift in attention, as I suggested in chapter 1, was the proliferation of maps oriented over the North Pole, an air-age cartographic style appropriated after 1945 to demonstrate the surprising proximity of the Soviet Union. America was now, in the emphatic language of the period, "wide open at the top" and should "push out there for our defense."[104] For the Yale geographer Stephen Jones, air power, combined with atomic weaponry, had "thrown a spotlight on the Arctic regions."[105] Air Force general and RAND sponsor Hap Arnold was blunter: "If there is a Third World War the strategic center of it will be the North Pole."[106]

Just over a year after Bowman's speech, M. C. Shelesnyak of the ONR drafted a paper titled "The Arctic as a Strategic Scientific Area." A seminar series on "Problems of the Arctic," run jointly by the Arctic Institute of North America (AINA) and the Bowman School of Geography at Johns Hopkins University, was the occasion for presentation. Shelesnyak's thesis was that "the Arctic region allows for the conduct of scientific research in a manner which permits the securing of objects of a campaign (scientific research) for fuller understanding of natural and social phenomena." He was fond of military imagery in his published descriptions of the ONR's northern research initiatives, but this was a far more direct version. The Arctic, he stated, nurtured scientific research in three respects: it was a *frontier* lacking a "systematic body of scientific data"; a simple, homogeneous, and contained space ideal for *experimental design;* and a region of profound intellectual *interdependency* that did not allow for closed disciplinary forms of knowledge to survive.[107] Shelesnyak never elaborated on the implications of the title of his paper, but he did not need to do so. Not only was he using an area studies approach to define the Arctic as strategic, but he was also turning geopolitics into science, effacing the military interests of the ONR in the north while rendering the Arctic in cybernetic terms.

The nearly concurrent visions of a northern space propounded by Bowman, a political geographer, and Shelesnyak, a natural scientist, were also convergent. This was because the Arctic had become, by the end of the Second World War, a dual strategic and scientific frontier, likened, in the 1944 proposal that created the AINA, to "the undeveloped West in the middle of the last century."[108] Over the next fifteen years, fears of a Soviet attack on North America led to an alternate "assault" on Arctic landscapes by research teams, administrators, and troops, all pushing northward to occupy a vaguely known region.[109] This region was treated as a unique space whose very complexity required a synthetic approach that would then yield even more general outcomes: the laws demanded, as Bowman suggested, in both the sciences and the social sciences after the Second World War. The Arctic, in other words, was an ideal laboratory for interdisciplinary intellectual practices whose results would be not only regionally significant but global in implication. Precisely the same logic underwrote northern military activity: the North was understood as vital to continental defense against a gathering external threat, but, particularly for the American armed forces, it also stood as part of a larger set of hostile environments, which defined the new, world-spanning presence of the United States.

Of course, military and scientific projects of the period were not just parallel in orientation; in many cases, they were inseparable. As Shelesnyak indicated in a 1947 "discussion of the Arctic" published by the U.S. Navy,

> From the standpoint of national security, it is essential to know the intimate details of living conditions and of the natural conditions of our own territories. . . . This last frontier of exploration presents an exciting field not alone in terms of the old geographical exploration, but more in terms of the utilization of our finest and newest techniques in geophysical and biological science applied to a large and vast area of relatively unknown territory.[110]

Historians of science have repeatedly reminded us that this relationship can be found across the streams of Cold War research. In particular, as Ronald Doel notes, it is crucial to consider the many ways in which "military patrons sought to enlist scientists in efforts to control nature to further national security aims."[111] And military–scientific affiliations, the mandate of control, and the "naturalness" of nature were all singularly prominent in the Arctic during the 1940s and 1950s. But this preeminence depended on the securing of a geographic object with that same name — a performative act that took place as much in boardrooms and laboratories as it did over specific northern terrain. In this sense, the Arctic frontier was *engineered* — not just in the sense of specific landscapes and bodies as sites for technical manipulation and control but also according to more general principles of development, order, and appropriation for scientific and strategic needs.[112]

If we are to understand the North American Arctic — and, by extension, the continent — as a space for certain forms of knowledge production during the early Cold War, we must consider it as an object of knowledge itself. At the individual scales of experimental practice and moving, warring bodies, the Arctic, not surprisingly, became a difficult entity to locate, understand, and overcome. But we should not forget that debates over the significance and role of the north also raged in the schematic realm of strategy or that the increased complexity of the Cold War's many engineered systems, including the continental defense network, also resulted in a heightened sensitivity to error, as their "ever-ready" requirements made unreliability a paramount concern and a subject of dispute.[113]

The period's most characteristic northern institutions and initiatives, from the AINA and the DEW Line to military simulations and survival schools, reflected and promoted the merger of strategy and science. And they did so above and beyond nationalism — a theme of great significance to Canadian historians and literary scholars, in particular.[114] Extensive discussions of Arctic sovereignty did occur in the 1940s and 1950s, but Canadian and American interest in the Arctic during this period reveals far more continuities than disparities. These were not only political but spatial. After studying the situation, the chairman of the U.S. half of the Permanent Joint Board of Defense informed President Truman in 1951 that the existing radar networks of both countries were equally inadequate and also that "the radar coverage requirements of Canada were geographically almost identical with some of the extended early warning requirements of the United States."[115] In addition, I take seriously Michel Foucault's argument that the analysis of power should not proceed strictly from the perspective of traditional, legal sovereignty but might be more productively pursued through a strategic model premised on the practices of war and related modes of technical rationality.[116]

Writing in *Foreign Affairs* in 1953, the Canadian politician (and future prime minister) Lester B. Pearson heralded the achievements of the cartographers, policemen, missionaries, mechanics, and scientists who together had opened a new territory.[117] He and others were waging a campaign to introduce what seemed at first glance to be a distant and forbidding landscape into the popular imaginations of North American citizens — a campaign, tinged with colonial and civilizational tones, which eyed the apparent successes of Soviet enterprise and settlement in Siberia with nervous envy. But the same imaginations appealed to by Pearson were also clouded by the threat of disaster. The Arctic was, after all, only a few hours by plane from the North American heartland. As a result, an alternative vision positioned the north as an empty bulwark separating the superpowers, a vast desert that would challenge the endurance of invaders, or for that matter, a wilderness that could hide a growing enemy presence for an extended duration. In a 1953 speech, Lloyd Berkner captured this process of simultaneous distancing and connection:

> If we can economically exploit the thousands of miles between the distant warning line and our target system, we can acquire real advantage. We can track the enemy to assess his probable intentions and the composition of his forces. We can break up formations over the sea or uninhabited land wastes with atomic weapons.[118]

Rarely mentioned were the repercussions these scenarios would have for the people who had already made the north their home; one commentator went so far as to enthusiastically suggest that Arctic war promised "no dev-astated cities, no ruined civilizations, no millions of starving refugees and displaced persons."[119] What was constant in postwar discussions was a be-lief that the geographic object of defense was now actually a unified conti-nental *area* — not just selected landmarks in Canada or the United States.

Discussed during the Second World War, formed in 1945, and sup-ported by the ONR, Canada's Defence Research Board (DRB), the Carne-gie Corporation, and the national research councils of both countries, AINA boasted a list of governors that included the most prominent Arctic experts of the period, and its research program was overwhelmingly composed of government contracts.[120] In reports to sponsoring agencies — reports that typically began with or included a polar projection map, as if to remind readers once more of the Cold War's changed cartography — the AINA board explicitly appealed to the new strategic position of the Arctic in its search for funding. With American and Canadian scientific *and* defensive interests largely coinciding, as a "technical research organization," AINA was "in a position to aid these . . . by providing new information on many Arctic subjects and by acquiring new data."[121] These relationships and this pragmatic, intelligence-based view of geographic data were integral to the broader project of area studies underway in North American universi-ties, and AINA drew inspiration from regionally oriented precursors such as the Ethnogeographic Board to assemble an interdisciplinary roster of northern experts and an Arctic bibliography that could be consulted by interested parties.[122] As was so common during the Cold War, the produc-tion of scientific knowledge was invoked as a cover for classified military work. And if AINA did not initially welcome classified research, it was only because there were "so many scientific problems that do not appear to need security restrictions" and because individuals associated with the institute were free to "serve their governments on a personal basis."[123]

To AINA leaders and others with similar interests, the Arctic was at once a field for dozens of grueling, simultaneous expeditions and a "huge laboratory" — a "natural 'experimental set-up' not to be matched in tem-perate zones."[124] By the late 1950s, there was little pretense as to the insti-tute's military dependency. In 1957, a joint ONR–AINA Arctic Research Advisory Committee was created to examine "the phases of arctic research which might be of military significance." Interviewing authorities at laboratories and bases in Fort Churchill in Manitoba; Thule, Greenland,

Point Barrow, Fairbanks, and Kodiak, Alaska; and Seattle, Washington, the committee visited a remarkable list of sites that were at once specific and part of a "total environment."[125] Reporting on a similar, U.S. Army-sponsored excursion in 1959, the State Department geographer G. Etzel Pearcy described "innumerable tests of all descriptions, ranging from acclimatization of the individual to simulation of actual combat conditions. Almost every aspect of life in these centers must be studied carefully and adjustments made because of the Arctic environment." The map of Arctic militarization inscribed by such tours would have been incomprehensible fifteen years earlier. But for both Pearcy and the AINA's advocates, the academic benefits of "familiarization" were worth the "operational control" of the Army, an institution ultimately devoted to *defeating* its subject of study.[126]

The ONR–AINA Advisory Committee's report also included an appendix titled "Importance of the High Arctic to North American Defense." This statement advocated a shift in the military approach to Arctic nature. Once the "implacable laws" of the north were fully cataloged and comprehended, the liabilities of Arctic terrain and climate could be overcome, harnessed, and "transformed to the dynamic," enabling "military man to work with the cold rather than against it."[127] When successfully colonized, then, the environment became an ally in a strategic conflict that was itself, ironically, naturalized — an inevitable Cold War against an implacable foe. The language of the appendix was clearly influenced by the many military exercises that had been staged across the north since the Second World War. The 1949 *Naval Arctic Operations Handbook* had admitted that adaptation, "on a mass scale, to Arctic conditions will be a long and tedious process." But the handbook stressed that with war becoming "more global and more universal," because the Arctic was "the central area of most of the earth's land masses," its role would be crucial: *"Only by bringing life to the Arctic will the Arctic 'come to life.'"*[128] This combination of geographic centrality and presumed emptiness justified an extraordinary northern construction project.

Across the Top of the World

Exemplified by the Air Force's Arctic, Desert, Tropic Information Center (ADTIC) — launched during the Second World War, briefly discontinued, and restarted in 1947 at Air University in Alabama—the Cold War study of nontemperate climates was, as with concurrent disciplinary debates in geography, caught between the regional and the universal.[129] It borrowed from earlier colonial discourses of tropicality to create cartographies of difference, yet it concurrently sought technological solutions to the deleterious effects of alien landscapes.[130] Thus, the Arctic as a location was at once deeply strategic and profoundly simplified, reduced to a singular area whose environmental constraints, while certainly a challenge to scientific and spatial laws, could nonetheless be encompassed by them. Environment, in this case, referred to a *category*. In ADTIC publications on "survival geography," this category combined effectively with the regional outlines of area studies.[131]

The globe, according to one writer from the Canadian DRB, was a "mosaic of military regions within which the terrain elements are reasonably homogeneous or have similar diversification of environmental factors relevant to the military problem under consideration."[132] This was the strategic perspective. But the DRB and other organizations, including the Arctic Aeromedical Laboratory at Alaska's Ladd Air Force Base, were also interested in the relationship between the environment and the individual human body, a concern that led to numerous research projects on the technological maintenance of body temperature and normal moods under conditions of extreme cold. As M. C. Shelesnyak of the ONR observed, interest in the "influence of environmental conditions of relatively unknown areas upon man and his performance is widespread."[133]

If the question of human interaction with nature is fundamental to geographical scholarship, it is startling that so little attention has been paid to the military dimensions of this relationship—to wars *on* geography. Military geography is now perhaps the most moribund and certainly the least critical of all the discipline's identified subdivisions and continues to treat nature, when not obliterating it in favor of urban scenarios, as an oppositional object, or more specifically, as a force to be overcome.[134] This is hardly a novel habit. RAND political gaming proposals of the 1950s frequently added Nature as an external factor influencing and disrupting the outcome of a simulation and conceived of the environment as merely "the scene of the activity."[135] But in terms of actual military operations and the full range of nontemperate regions, it was only during the Second

World War and the subsequent Cold War that overcoming environmental constraints on warring bodies became the subject of sustained scholarly study in the laboratories and agencies of the American military. Some of this history has been discussed in previous chapters, with particular reference to the dissection of the world into regions where American forces might have to operate. But there was one strategic region that, likely because of its perceived emptiness or its location within (but on the edge of) North America, occupied a unique position on the maps of Cold War area studies: the Arctic.

The global, regional, continental, and intimately local aspects of strategic geography came together in the DEW Line, an integrated chain of radar and communications stations stretching from northwest Alaska to Baffin Island. Something of the sort had been envisioned since at least 1946, when U.S. Army Air Force planners hatched a scheme for a string of radars across the north. But the key impetus to its actual construction was the 1952 MIT Summer Study. Its purpose, Jerrold Zacharias recalled, was to design a "passive system . . . that would give us a few extra hours' warning. We were shocked that no such alert system existed."[136] The Summer Study's final report was nonetheless confident:

> Our geographical experts have examined northern Canada for sites that would be logistically accessible by means other than aircraft. These sites of the outer DEW Line would form a continuous line along which any aircraft flying at any feasible altitude above the terrain would be in the unimpeded line of sight of at least one station.[137]

It was a *physical* presence, then, that gave the DEW Line its imaginative significance as a political boundary. A series of scattered construction sites became a technological wall that was also a moral divide, marking the boundaries of security and certainty.

Fixing this boundary in place was very much an act of spatial reasoning. The Summer Study's director, Jerrold Zacharias, credited Lloyd Berkner with the DEW Line argument. According to one account,

> On being told by a senior Air Force officer that under certain circumstances and jet speeds the warning might not be more than ten minutes, Berkner exploded. "If geography can be made to work that well for the Russians," he retorted, "it can be made to work just as well for us."[138]

The Distant Early Warning (DEW) Line. From *The DEW System* (Paramus, N.J.: Federal Electric Corporation, n.d. [1958–59]), n.p.

Similar conclusions were reached in the subsequent Project Lamp Light, which was treated as an additional opportunity for consideration of a "more distant, or remote, air battle" and its merits. This led to a practical exercise in the prioritization of target sites and regions — not surprisingly, the northeastern United States finished first — but also a discussion of a "remote zone" between the line of northernmost radar coverage and the fringes of Soviet territory. Pushing this zone farther from the American heartland was the key justification for the ambitious DEW Line project. As the Lamp Light final report acknowledged, Arctic radar provided additional "time to think, to consider the situation and to decide on the best action."[139] Geography, then, would work to slow the speed of Cold War conflict, and the northern region identified by the Lamp Light group could host a relatively safe form of war or, in a different sense, it could represent an imaginative space of calm, rational thought.

The MIT Summer Study was not received well by its offensively minded sponsors in the U.S. Air Force. "From the outset," according to Zacharias, they "gave us a hard time . . . we heard that the problem was that the DEW Line was competing with the Strategic Air Command for funding."[140] On the other hand, the National Security Resources Board, an agency with a substantial investment in civil defense, appealed to the National Security Council for precisely what the Summer Study had advocated: a line that would provide maximum time for urban evacuations. According to Zacharias, "Only President Truman needed to be persuaded, which is exactly what Albert G. Hill did."[141] Seeking a compromise solution, as well as a slowdown, the Air Force contracted with Bell Laboratories and Western Electric to build installations for replicating DEW-like conditions, testing communications equipment, training personnel, and simulating attacks using Air Force bombers and the insertion of "artificial data" into the tracking system.[142] One group of such stations was on the north coast of Alaska in the vicinity of Barter Island, located near to the Navy's Arctic Research Laboratory at Point Barrow, and another was in rural Illinois. In tandem with continuing trials carried out at MIT, where Lincoln Laboratory employees were providing advice to Western Electric, these experiments navigated the political and epistemological transition from controlled laboratory enclosures and then carefully organized field trials to a quite different set of landscapes in the north.

With the arrival of a new president, Dwight Eisenhower, who immediately shifted strategic and spending priorities to favor the SAC, it took a further favorable report to the National Security Council and the testing of a Soviet hydrogen bomb to settle controversies and solidify support for the DEW Line.[143] According to David Winkler, "Eisenhower recognized that if America was to deter war through massive retaliation, it needed air defense to ensure the survival of its retaliation force."[144] The president formally approved the DEW project on February 24, 1954, and Canadian consent was subsequently secured in return for "nominal recognition of Canada's Arctic sovereignty."[145] Using maps, hydrographic charts, and aerial photographs, in 1953 Western Electric had begun low-level overflights of the Arctic to select likely sites. With the aid of a binational committee versed in operations research techniques, a route was designed that would link up with the existing Alaskan radar network, which in turn was extended into the Pacific through airborne and seaborne radar-carrying craft.[146]

In the east, the DEW Line was eventually pushed to Greenland and Iceland, and Navy radar picket ships carried the arc of surveillance to Scotland and south to the Azores. It had already been supplemented on the continent by the more southern Mid-Canada and Pinetree Lines. In all cases, routes were determined by a dizzying combination of factors: reconnaissance, studies of photographs and geographic reports, discussions with experts such as those provided by AINA, participation in northern resupply missions, and not surprisingly, extensive mathematical exercises. But alongside the dilemmas of distance and topography, site selection was forced to grapple with a more ambiguous question. Which route had the fewest gaps?

For the journalists who ventured north to file stories on this compelling venture, describing a largely electronic creation was no easy feat, and thus the radomes—the geodesic domes of fiberglass and plastic— located at many DEW stations became icons of the era. Indeed, well before Buckminster Fuller's famous creations served as symbols of alternative, communal habitation in the 1960s, similar structures were being shuttled to the Cold War's various frontiers and were put on display in the Arctic or at international exhibits and trade fairs sponsored by the U.S. Department of Commerce. Domes came to represent "industry's power to transform nature," and Fuller was well aware of their marketability in a time of geopolitical uncertainty. In the early 1950s, he aided the Air Force and Marines in designing dwellings that could be "prefabricated in America and dropped into combat zones at the first sign of Communist mischief." These creations represented yet another example of the desire to control geography through science—in this case, using "prepackaged and self-contained cultural and technological units."[147] Delighted with his success, Fuller boasted that "North Pole, South Pole, I'm all around the world."[148] But so, more importantly, were the American personnel under his domes.

The construction of the DEW Line, completed in about two years, was an extraordinary feat of *geographical engineering*, planned and sequenced in minute detail. As early as 1953, *Fortune* claimed that the con-

DEW Line site, Barrow, Alaska, December 1962. David Chesmore Photo Collection, Alaska and Polar Regions Collections, Elmer E. Rasmuson Library, University of Alaska–Fairbanks (item 2004-171-163).

cept "almost certainly incorporates more of the lessons of information theory than any other communication system yet devised."[149] By 1956, in the midst of construction, a Canadian commentator celebrated it as "a monument to the ingenuity and hardihood of the North American human being."[150] The first popular book-length study of the DEW Line, a romantic and ethnocentric piece of promotion published by Rand McNally, described the request received by Western Electric to survey a route and assess and solve logistical problems as "probably the greatest single construction order ever issued."[151]

Geographical engineering was a term coined by the Manhattan Project physicist Edward Teller, in the context of his Project Plowshare (launched in 1957), to describe the physical shaping of the earth to reflect human needs. Through the destructive power of nuclear weapons, he planned to create, among other miracles, new harbors around the world, including one in Alaska, all in the service of civilization — which of course had a specific meaning for an ardent Cold Warrior like Teller.[152]

What was novel about Plowshare and the DEW Line was that they were more ambitious in scope, but also more coordinated, than earlier mega-projects. They were situated at the apogee of a triumphal scientific modernism. Plowshare's Chariot initiative in Alaska was protested and eventually defeated, and the DEW Line was made largely obsolete almost immediately by the development of intercontinental ballistic missiles, although this did not stop the United States from upgrading it repeatedly.

The DEW Line was not a simple demonstration of the scientific mastery of nature, or for that matter, dominance over the existing human geography of the north. As James Scott notes, "Formal schemes of order are untenable without some elements of the practical knowledge that they tend to dismiss."[153] Climate was a constant concern. In addition, station sites plotted mathematically from a distance were forced elsewhere due to impossible terrain and the presence of long-standing Inuit settlements — although it is fair to say that the DEW Line's backers were not entirely accommodating with the people of the north. The DEW Line agreement itself, pushed on the United States by the Canadian government (in order to secure rights on Canadian soil), stipulated that the "Eskimos of Canada are in a primitive state of social development" and that disruption of their hunting economy and settlements be avoided when at all possible.[154] Well before the DEW legislation, the geographer and Arctic advocate Trevor Lloyd wrote that Canadians should "see that none of the contemporary military activity in the Arctic is allowed to touch the lives of the Eskimos."[155]

Of course, like the claims that science and strategy were two separate imperatives in the north, these were simplistic and futile gestures. The building and operation of the DEW Line had tremendous consequences for Inuit across the Arctic, many of whom gained employment, typically in secondary positions, on the Line.[156] All Inuit who resided close to a station, but particularly those who actually worked at stations, were exposed to the replication of "southern" culture inside prefab walls — including regular movies, religious services, and cuisine — that was yet another facet of the attempt to normalize the north. Although there was certainly a reciprocal influence on DEW employees from the south, it was clear who the respectively permanent and temporary residents were and whose lifeworlds were most significantly altered in the long term.

These consequences were strikingly apparent in a strongly worded 1957 report composed for the Canadian Department of Northern Affairs

Visitors from the south: In a staged photograph, U.S. Air Force personnel meet residents of Frobisher Bay, Canada, on March 28, 1956. National Archives and Records Administration Still Picture Division, RG 342, Series 342FH, Box 5000, Photo 4A-24501.

by J. D. Ferguson, a young sociologist on summer assignment. Flying up and down the DEW Line on a daily courier, he noted the division of native communities by gender, age, and income as a result of the DEW Line's arrival. At least 25 percent of the population of the western Arctic, he estimated, was subject to the immediate imposition — "not too strong" a word — of a "radically different kind of living pattern."[157] This pattern, according to another observer, the anthropologist Jacob Fried, was "planning in the nearly total sense," suited not to aboriginal families but to "a small isolated group of men whose only business is to tend machines." There was no need to "draw on the outside physical environment" for necessities; "they are mere employees and not settlers." In sum, a DEW station was "an example," Fried wrote, "of the complex, specialized working unit of that variety that the next centuries will see developed further in isolated regions and perhaps in outer space."[158] Station residents evoked another location, calling their quarters "submarines."[159] The DEW Line's tiny cybernetic microworlds were the epitome of Cold War military modernization — and indicative of its limits.

From the planning phase of the 1952 MIT Summer Study onward, advocates of northern radar were concerned with the behavior of personnel posted to the high Arctic, if only because human alertness was still the ultimate determinant of systemic efficacy. In his "Geographic Studies" appendix to the Summer Study Report, Kirk Stone wrote that a "unit in which every man is able to keep busy is essential to morale in the Arctic."[160] And in a 1952 statement of support for social research on isolated early-warning stations, the Department of Defense's Research and Development Board noted that the installations posed "rather unique human relations problems along the lines of motivation, team-work and sustained job satisfaction under stressful non-combat conditions." Scholars from the University of Washington conducted the most significant early inquiry on this subject, reporting to the Air Force's Human Resources Research Institute and developing models of "site efficiency" from field visits.[161]

Overall, a staff member at Alaska's Arctic Aeromedical Laboratory wrote, indoctrination for isolated work was a procedure that replaced "fantastic notions" concerning the north with "factual knowledge" of climate, culture, and terrain.[162] Such reassuring, confident dichotomies were ubiquitous in the military manuals and studies prepared during the early Cold War and were incorporated didactically into the systems of order and discipline at Arctic military installations. But ultimately the task of adjustment and normalization was a personal challenge. The perceived wildness of Arctic space and the confrontation with northern nature placed an extra burden on what Michel Foucault called *technologies of the self,* encouraging "individuals to effect by their own means, or with the help of others, a certain number of operations on their own bodies and souls, thought, conduct, and way of being."[163] Retaining normality in an "abnormal" environment, whether in the form of station design and culture or personal behavior, was essential to the stability of the continental defense network and thus of the continent itself.

The changes delivered to the Arctic by the Cold War were quite apparent to the promoters of the DEW Line. As a book published by the Western Electric Company put it, "The DEW Line men are doubly pioneers, they have opened new vistas in electronics as well as geography."[164] But the opening of these vistas required the engineering of a continental frontier in the name of both strategy and science. This process was ably summarized in a 1958 Department of Defense publication titled *The Arctic: A Hot Spot of Free World Defense*. It described the triumph over permafrost and other "problems" and accorded responsibility to the "determination and ingenuity of the fighting forces, science, and industry to build and operate such ramparts as the Thule Base and the DEW Line."[165] Here, quite literally, the fruits of the military–industrial–academic complex were on display. And those ramparts, while increasingly material, were, just as importantly, imaginative constructions, flexible enough to accommodate the boundaries of both scientific knowledge and strategic destiny.

Keep Watching

While SAGE and the DEW Line were the most spectacular components of a cybernetic continent, their contributions to what the historian Laura McEnaney has called the "militarization of everyday life" were initially minimal.[166] McEnaney's focus is civil defense, a subject also taken up in the next chapter. But during the 1950s — particularly the first half of the decade, when radar defenses were considered severely inadequate — Americans were also bombarded with radio announcements and other advertisements encouraging them to "protect your country, your town, your children" by joining the Ground Observer Corps (GOC).[167] This extraordinary if short-lived initiative, bolstered by the authoritative approval of study groups like Project Charles, created an army of civilian "skywatchers" who staffed outposts across the continent and around the clock.[168] At its peak, hundreds of thousands of volunteers occupied thousands of reconnaissance platforms across the United States.[169] One retrospective in *Air Force Magazine* claimed that the country "has not experienced anything quite like that kind of nationwide participation with the military since the GOC disbanded" in 1959.[170]

The roots of the GOC were in World War II, when over a million civilians occupied several thousand observation posts along the coasts of the United States. The initiative ended in 1944 but was restarted in 1950 and eventually grew to include over eight hundred thousand volunteers, sixteen thousand observation posts, and seventy-three "filter centers."[171]

OBSERVATION POST

LOCAL TELEPHONE SWITCHBOARD

FILTER CENTER

AIR DEFENSE FIGHTER INTERCEPT STATION

Fighter Commander takes appropriate action based upon Observer report and data from radar screen.

The Defense network. From *The Aircraft Warning Service of the United States Air Force* (Washington, D.C.: Office of the Secretary of Defense, Civil Defense Liaison, 1950), n.p.

Aided by binoculars and Air Force manuals featuring dozens of plane profiles, skywatchers were connected by phone to one of these stations, and relevant information was then sent on to the Air Defense Command (which became NORAD in 1957).[172] RAND's Systems Research Laboratory would have seemed familiar to anyone who had worked at a filter center. A confident official portrayal of the latter noted a "plotting or filter board" set up "horizontally at table height," which was "constructed to represent, in contour, the area in which its associated Observation Posts are located." A grid was superimposed over the board, allowing volunteers to accurately describe the location of aircraft. Each "plotter"—some ten to fifteen per shift—oversaw part of the board, working in tandem with four to six "filterers" who analyzed the information once it was plotted, ensuring the accuracy of a plane's track. All of this activity was watched, from an elevated platform, by "tellers"—typically three—who relayed "filtered plots" to their contacts at Ground Control Intercept stations, "where observer information is coordinated with data from the radar detectors to give the Air Defense Commander a graphic portrayal of the air activity over his area."[173] The *systemic* quality of this description, with its combination of visualization and quantification, could not have been more clearly expressed.

Associations with the Air Force lent GOC duties a romantic quality that might have blunted the boredom of such tasks, especially for teenage observers. One inductee recalled "shiny silver wing pins, a subscription to *Aircraft Flash* magazine for photos of military jets with billowing contrails, and viewings of fun Air Force movies about the earth slowly turning red, like a Sherwin-Williams paint commercial."[174] But *The Aircraft Flash*, the GOC's official magazine, also recorded the mundane events unfolding innocuously below watchtowers — "the Miami filter center picnic was a complete success, with over 900 persons enjoying a fried chicken dinner and the entertainment that followed."[175] The GOC, one Ohio participant recalled, "had a feel-good element to it."[176] Such "civic conviviality" was overshadowed by what Sharon Ghamari-Tabrizi calls "credulity offset by fatalistic forbearance." If there was one overarching justification for GOC volunteering, it was the sentiment of *defenselessness*.[177] This could be alleviated, of course, by an active contribution to continental defense, counteracting the human (or animal) condition of fear with tasks that were more suited to machines.[178]

GOC members did not spot many enemy craft; more accurately, they did not identify actual Soviet planes. As the continental defense network was constructed, however, ground observers, whether atop towers or inside radar stations, did record numerous sightings of unidentified flying objects. The Cold War flying saucer craze began in 1947 (with the first nationally reported sighting two weeks before the infamous Roswell incident) and culminated in the summer of 1952 — the year Operation Skywatch turned the GOC into a round-the-clock organization and the year the Air Force restarted its comprehensive study of UFOs under the title of Project Blue Book.[179] While directly related to a growing fascination with space exploration, UFOs were also a manifestation of a cybernetic culture obsessed with secrecy and technology. Like the outrageous scenarios of certain RAND strategists or the far-flung stations of the DEW Line, UFOs represented the outward edge of scientific and strategic discourses that strained to pull them back in, replacing reports of Roswell flying disks with stories of mundane weather balloons.

Many of those "contacted" by extraterrestrials worked on the fringes of the aerospace industry, where researchers continued the tradition of the Manhattan Project by creating interdisciplinary groups to collect knowledge on a particular problem but also developing objects — under cover of classification—from this knowledge. As Phil Patton points out, the encyclopedic, associational quality of such efforts created a realm that was simultaneously a research frontier and a zone of instability or even conspiracy.[180] This was a profoundly geographic predicament. In constructing Cold War "things" in the shape of atomic bombs, computers, or radar networks, *spaces* were also produced. These spaces included defense laboratories, the militarized landscapes of the Arctic, and ultimately the continent itself—all environments invented anew and drawn together by Cold War strategy, and all environments wherein certainty was sought but never quite achieved.

5 ANXIOUS URBANISM
Strategies for the Atomic City

> The atomic bomb has raised, in fact, the question of the survival of
> urban culture itself.
>
> —WINFIELD W. RIEFLER, preface to
> *The Problem of Reducing Vulnerability to Atomic Bombs*

Even as it diverged from George Kennan's original formulation, the term
containment could be relied on to clearly differentiate America — or a larger
sphere of allied comfort — from a threatening exterior realm. Of course, as
the case of the fragile continental defense network shows, the two spheres
were not so easily separated. In a discussion of colonialism and nationalism, Homi Bhabha has argued that "paranoid projections 'outwards' return to haunt and split the place from which they are made."[1]

In the case of the Cold War, then, it is not surprising that the doubling
back of paranoid political projections can be found in the return of containment to haunt a domestic space. Andrew Ross makes the appropriate
distinction:

> The first [conception of containment] speaks to a threat *outside* of the
> social body, a threat that therefore has to be excluded, or isolated in quar-
> antine, and kept at bay from the domestic body. The second meaning of
> containment, which speaks to the domestic *contents* of the social body,
> concerns a threat internal to the host which must then be neutralized by
> being fully absorbed.

Ross's use of the language of immunology is a deliberate reference to what
he calls "the Cold War culture of germophobia," nicely epitomized by
Kennan's description of world communism, in his 1946 Long Telegram,
as a "malignant parasite" threatening the reproductive body-spaces of the

American state and its allies.[2] Containment, therefore, was both a recommended foreign policy and a national narrative. This symbiotic role was reinforced by Kennan's closing words in "The Sources of Soviet Conduct," where he enthusiastically placed the burden of Cold War moral responsibility on a lethargic American population.[3]

For Kennan, however, some elements of American society seemed more likely than others to fail his "test of national quality."[4] The Long Telegram featured a lengthy list of groups most susceptible to communist infiltration, including "labor unions, youth leagues, women's organizations, racial societies, religious societies, social organizations, cultural groups, liberal magazines, publishing houses, etc."[5] These comments foreshadowed the imposition of Cold War surveillance programs and the proliferation of hysteric discourses that reached to the very psyche to resolve doubt over who was American and who was not. The marking of certain individuals or groups as un-American suggested not only that they represented a direct internal threat to the nation but also that they might be geographically contained. Drawing various identity categories into simple binaries, the articulation of distinction within the boundaries of the United States was thus shaped in reference to specific *sites*.

The atomic bomb, according to the influential strategist Bernard Brodie, radically altered the "significance of distance between rival powers" and lifted "to the first order of importance as a factor of power the precise spatial arrangement of industry and population within each country."[6] Like Kennan's Long Telegram, while Brodie's argument was published in 1946, during a brief American atomic monopoly, it suggestively anticipated the inevitable arrival of a Soviet challenge. His comments also confirm that the "risk society" symbolically inaugurated by the bomb — dubbed "a monster of our own creation" in one *Collier's* magazine piece — was profoundly geographical.[7] The "precise spatial arrangement" described by Brodie suggested not only that certain locations were more strategic — and thus more at risk — than others but also that American society might be geographically recalibrated to reduce this risk.

Such tasks are not easily accomplished, of course, and so more modest layers of security, in the form of civil defense schemes, were placed over this topography of turmoil. Geographies of panic were also geographies of control. Anxiety was tempered by expressions of revised and improved civic order, often *in advance* of an atomic disaster. In a 1951 speech titled

Fortress Main Street, Federal Civil Defense Administration (FCDA) head Millard Caldwell, a former segregationist governor of Florida, described his agency's "national insurance policy," which was renewable "year after year [for] protection," since Americans should be willing to stay "on the alert for 50 years, if need be."[8] Caldwell's prediction was oddly accurate, but it was the civil defense of his era, in particular, that was key to the popularization and domestication of nuclear technology and Cold War strategy. Civil defense represented a "social contract" with the bomb, a relationship necessitating the repetitive contemplation and simulation of urban and national ruination but also an agreement in which the vivid prospect of apocalypse could be wielded as a geopolitical instrument.[9]

This chapter explores the anxious American urbanism stimulated by the Cold War and its defining geopolitical, scientific, and cultural symbol, the atomic bomb. I have taken my title from W. H. Auden's Pulitzer Prize–winning "baroque eclogue" *The Age of Anxiety* (1947), set in a New York City bar during wartime. The poem fixed the United States as the inheritor of European modernism and its troubling contradictions; on another occasion, Auden referred to America as a "fully alienated land."[10] Such claims were not unique, nor were they entirely abstract. As I demonstrated in chapter 1, musings on American cultural destiny were invariably tied to strategy. But the resulting cultural Cold War was not just waged overseas. It also found form in domestic landscapes, where a long-standing ambivalence toward urban spaces was coupled with the specificities of American cities and society.

One of the great ironies of the post–World War II United States, as Robert Beauregard has observed, is that the country's well-documented prosperity did not extend to many of its urban centers. The portent of further conflict, this time with much more dramatic domestic impacts, was simply one of numerous factors urging an unprecedented abandonment of central cities by manufacturers, corporations, and populations dominated by the white middle class. Beauregard argues that atomic fear was ultimately a minor factor in this process of decentralization.[11] To be sure, no comprehensive state-led campaigns for urban restructuring were mobilized solely in the name of Cold War safety. As numerous writers were quick to note, such initiatives would not have suited a time-hardened mythology of American freedom and individualism, especially during a period of increasingly virulent anticommunism.

The aim of this chapter is not to directly contest Beauregard's claim, although examples such as highway construction and industrial dispersal, as I show, certainly complicate his conclusion. It is, instead, to illuminate the multiple ways in which the American city became Cold War terrain, a site for the localization of strategy, where the response (or lack of response) to Kennan's challenge would actually occur. As in other chapters, the singular phrasing is crucial. Although I refer to specific cities, some of which were clearly more significant than others, I am ultimately concerned with the generic idea of American urban space as a *target*—for atomic weapons, certainly, but also for the sorts of strategic investigation and explanation detailed throughout this book. From politicians, journalists, and civil defense advocates to strategists, social scientists, and even physicists, the city became a militarized "laboratory of conduct," yet another strategic space of the Cold War.[12] Like Jennifer Light, I am interested in the generation of strategic urban knowledge and the creation of an influential Cold War community of experts on urban life.[13] But I am also interested in explicitly linking the landscapes that concerned these experts to the broader geography of the Cold War. Civil defense was, after all, also a national discourse twined tightly with the continental defense network, the study of Soviet behavior, and the global horizons of geopolitical tension.

There were, of course, immediate precedents for both the destruction of cities during wartime, particularly from above, and defense programs established in response to the prospect of such destruction.[14] These were crucial to the Cold War's scenarios of urban disaster. But because such scenarios were hypothetical—that is, they were constructed in advance of catastrophe—the anxious urbanism of the Cold War was at once both prolonged and pragmatic. If it seemed inevitable, it was also manageable, although at significant cost.

Atomic Cities

The atomic bomb's importance was not understood until two were dropped on the Japanese cities of Hiroshima and Nagasaki, acts that Americans received with both jubilation and horror. The volume of subsequent philosophical commentary was extraordinary; according to David Lilienthal, the first chairman of the Atomic Energy Commission (AEC), the explosions had revealed "the ultimate fact . . . the final secret of Nature."[15] Similarly, for General Leslie Groves, the military head of the Manhattan Project, the bomb and its necessary infrastructure of laboratories, factories, and towns was, writes Peter Hales, "the apotheosis of modernity and its unspoken ends: progress, practicality, efficiency, the production of things by which, then, power might be accumulated and held."[16] The atomic age that was just beginning as the Manhattan Engineer District shut down on January 1, 1947, had its origins in a series of interconnected clandestine spaces, most notably Oak Ridge, Tennessee; Hanford, Washington; and Los Alamos, New Mexico. As the bureaucrat and lawyer Herbert Marks put it, the Manhattan Project "was a separate state, with its own airplanes and its own factories and its thousands of secrets. It had a peculiar sovereignty, one that could bring about the end, peacefully or violently, of all other sovereignties."[17]

With the close of the district, its mysterious locations and their short scientific histories were opened, at least partially, to external interest. But there was a reciprocal process underway as well: "The systems of behavior and belief that guided the actors and participants of the District spread from the sites and spaces as the fences came down. . . . Atomic spaces interpenetrated, perhaps even became, American spaces." Los Alamos, in particular, was a new and influential type of space in terms of both scale and secrecy.[18] This interpenetration occurred in at least three ways. First, surreptitious sites of military science multiplied across the United States, particularly in the West. Second, the towns of the Manhattan Project were quickly built, carved into cul-de-sacs and neighborhoods of uniform box housing, and then converted from temporary to permanent "scientific-industrial" settlements, as the Second World War ended but the production of atomic weapons did not. The number of these settlements swelled with the proliferation of Cold War research funding, but the project's "model cities" also shared many visual and physical features with a much larger list of postwar developments, starting with the paradigmatic suburb, Levittown, New York.[19] Finally, like the scholars of the Ethnogeographic Board and the Office of Strategic Services, Manhattan Project employees dispersed to dozens of academic and industrial locations after the war. Some maintained close ties with the military and the new AEC, while others were more circumspect. Under the AEC, which was established in 1946 to transfer control of atomic science to a civilian agency, the United States built a network of national laboratories that would continue the landmark "big science" structure of research initiated by the teams at Los Alamos, Oak Ridge, and Hanford.[20]

At the beginning of his superb book *Atomic Spaces,* Peter Hales includes a 1945 photograph of an officer seated in front of a world map titled "Geography of the Manhattan Project."[21] The map, which demarcates no political boundaries, is pocked with white pins; most are situated within the North American landmass, but others are scattered across the globe. It suggests that the geography of the atomic age was inescapably, immediately global. Utopian plans for new communities and the centralized design of a well-funded national laboratory system were matched by a vision of an atomic world with America at its center. The bomb was just beginning to play a key role in the creation of a geography that was paradoxically based on both secrecy and spectacle.

As they were built and operated, a modernist model of uniformity was imposed over the diversity of the Manhattan Project's scientific spaces. Participants from a wide array of educational backgrounds, ethnicities, classes, and regions, including a large number of émigré researchers, were homogenized by several policies: patriotic doctrines of obligation; regulations; campaigns of silence, secrecy, and security; the invention of new spoken, written, symbolic, and visual languages; and the local segregation and ordering of space. This was all particularly wrenching for the cosmopolitan scientists who had envisioned America as "a place of 'everywhere communities' that 'floated over time and space'. . . linked in part by the same technological and cybernetic systems that had held the scientific community together."[22] Many scientists had to be persuaded forcefully by the charismatic project leader J. Robert Oppenheimer to join him at Los Alamos, where they were isolated in what Emilio Segrè called a "beautiful and savage country" and faced an array of physical and intellectual restrictions.[23] The trade-off, of course, was an unprecedented and astonishing amalgamation of talent, a model that proved hugely important for future Cold War research, even as the fragile political unity of scientists, held together by excitement and expediency at Los Alamos, frayed further and further, first internationally and then even within the American scientific community.

The bomb's influence on *social* science in 1945 was signaled by the immediate creation, within the Social Science Research Council (SSRC), of a Committee on the Social Aspects of Atomic Energy.[24] As the implications of atomic technology were recognized in nearly every corner of intellectual inquiry, this committee was disbanded (in 1947), and its concerns were taken up by a multitude of other panels, but not before it sponsored early and important studies on public opinion and urban vulnerability in the atomic age. Unlike many of the other scientific achievements of the Second World War, after Hiroshima and Nagasaki, the bomb was never perceived as just a tool; its awesome capabilities were as much an incentive for studies in "social nature" as they were an indication that "physical nature" had been mastered.[25] It was quickly acknowledged, for instance, that the sheer destructiveness of the bomb lent credence to alternative methods of combat, particularly those targeting the "inner landscape" of *both* "national and international psyches," an approach that perfectly encapsulated the domestic civil defense programs of the early Cold War.[26] While the use of atomic weapons represented an unparalleled challenge to the virtuous reputation of modern science, another powerful discourse, particularly in the United States, sought to reassert the value of expertise, intervention, and enlightenment.[27]

Even before the creation of the SSRC committee, connections between social science and atomic weapons had been established by the United States Strategic Bombing Survey (USSBS), an initiative that actually began in November 1944, and considered for a first case study German landscapes subject to "conventional" destruction. The work of the European Survey was designed to aid the efficiency of ongoing fire-bombing campaigns over Japan, and once two atomic bombs had been dropped on August 6 and August 9, 1945, President Harry Truman authorized the formation of the Pacific Survey, which moved less than a month later to specifically assess the damage in Hiroshima and Nagasaki. But the USSBS was much more than a vehicle for the analysis of city form. German and Japanese landscapes were turned into laboratories for the collection of evidence and the testing of theories, places where the principles and methods of cutting-edge science and social science could be applied to a definite geography.[28] The results were then carried back to more familiar environs, modified to suit the physical and human geography of American cities, and turned into predictive models.

The USSBS included a Morale Division headed by the psychologists Rensis Likert and Angus Campbell, who used probability sampling in Germany to clarify the relationship between bombing and behavior. Others sent to Germany included *Age of Anxiety* author W. H. Auden, the economist John Kenneth Galbraith, and the young political scientist Gabriel Almond, who was charged with finding German documents relating to the air war and with interrogations of interned police or Gestapo "regarding problems of internal order" that had resulted as Allied bombing intensified. Much of this information was shared with members of the Office of Strategic Services. Almond was dismayed when important papers he had unearthed were treated with only "clinical detachment" by the survey researchers; this, he recalled, was his first exposure to the "mechanization" of social science, whose "fanaticism and reductionism" continued to dog him as he moved into the field of comparative politics in the 1950s.[29] Neither the abstractions of statistics and the experimental method nor the cultural relativism that posited regions as incomparable appealed to him. As I argued in chapter 3, his Cold War work (and that of others) at Yale and Princeton in modernization, development, and political culture was precisely an attempt to remain between these two poles, an approach justified by appeals to strategic practicality.

The Japanese morale team included Alexander Leighton, the psychiatrist and anthropologist who had also led the Foreign Morale Analysis Division of the Office of War Information. Several academics from this division were appointed to the USSBS. Leighton, who advised wartime social scientists to seek the "essential oneness of the material in all studies of human behavior and relationships," and who wrote a book on the social conditions in a wartime relocation camp for Japanese-Americans, used his USSBS experience as a springboard to a general discussion of applied social science in his 1949 study *Human Relations in a Changing World*.[30] The audacity of this task bears considering: an exceptional Japanese field experience—as indicated by Leighton's first-person, travelogue-format introduction and conclusion to the book—was transformed into a trial for social science, an extreme but nonetheless manageable challenge to the scientific method. Leighton's fear was that this method would not be properly employed or that it could not rise to the test of understanding a devastated Hiroshima. However, his approach also bridged the divide between a unique Japanese site and anxious *American* urban landscapes.

The perspectives and projects on atomic cities were transferred from Japan to the United States as the bomb's testing moved in a similar direction, albeit via the neocolonial landscape of Bikini Atoll in the Marshall

Islands. Just as the South Pacific's longstanding role as a laboratory for military science was perpetuated by Cold War atomic testing, the dual frontiers of science and the geographic West were prolonged in American deserts, "empty" spaces that were "never vacant enough."[31] Indeed, the use of this region for explosive experiments was intriguingly accompanied by two forms of urbanization. Las Vegas, another "atomic boom town," grew rapidly just south of the AEC's Nevada Test Site.[32] And a series of bombing exercises, begun during World War II in Utah's Dugway Proving Ground and culminating at the Nevada Test Site a decade later, repeatedly destroyed urban structures built for the purposes of simulation.

At Dugway, the German-Jewish architect Eric Mendelsohn, a prominent modernist, created a miniature Berlin suburb in exacting detail. The complex, built with Utah prison labor, was "firebombed and completely reconstructed at least three times between May and September of 1943." An equivalent Japanese village was also erected and destroyed.[33] "Among the points that had to be determined," according to the official history of the Chemical Warfare Service, "was the degree of penetration of bombs, and the time–temperature factor for igniting the typical Japanese target."[34]

The postwar incarnation of urban construction in the desert, a "survival city" west of Dugway in Nevada's Yucca Flats, received far more publicity. In the suitably titled Operation Doorstep, conducted on March 17, 1953, the AEC, Department of Defense, and FCDA collaborated to stage a small atomic blast in the vicinity of a "simulated suburb": two emblematic American homes, complete with mannequins, where "little had been left to the imagination."[35] The residences were also equipped with basement shelters, while eight outdoor shelters and a "variety of typical passenger cars" dotted the Nevada moonscape.[36] Doorstep was followed on May 5, 1955, by Operation Cue, which included a set of civil defense field exercises, from rescue to feeding services, in the wake of the explosion.[37]

Both operations produced durable images of destruction and "defined who and what was endangered by the atomic age: families, homes, consumer commodities."[38] In an echo of the Ground Observer Corps' banal community building, canned and refrigerated food was extracted from the rubble of Operation Cue and served to participants as an affirmation of the "survival" ethos.[39] But administrators in attendance at the symbolic Nevada rehearsals staged for national media audiences perceived a broader

Mannequins in a damaged "living room," Operation Doorstep, Nevada test site, March 17, 1953. Photograph courtesy of the National Nuclear Security Administration/Nevada Site Office.

purpose: a contribution to what the official Doorstep summary called "general conclusions" for the public, "which will apply in the majority of cases under the principle of the 'calculated risk'... basic to all realistic civil defense planning."[40] A pamphlet issued to residents of the area near the Nevada Test Site assured them that they were in "a very real sense active participants in the Nation's atomic test program," that each test was only conducted due to "national need," and that proper safety concerns were being addressed.[41] Whether in the terminology of cutting-edge social science or in the lurid photographs of Nevada tests published in popular magazines, the role of civil defense as a national program of awareness and response was becoming solidified. America was the target of a Cold War enemy, and as such, all of its citizens had a duty to enroll as citizens of the atomic age — an "act of governance, technoscientific practice, and democratic participation" that normalized the imagination of disaster, turning it into an opportunity for nation-building.[42] Civil defense administrators, meanwhile, were determined to manage this process carefully.

Beyond Cities

The most prominent geographic division in Cold War America was that between city and suburb, a contrast powerfully expressed by George Kennan himself. Kennan's 1950 train journey from Washington, D.C., to Mexico City convinced him that the American metropolis was a place of corruption and iniquity. As his train passed through an anonymous urban landscape during a "sinister and pitiless" dawn, Kennan noted the "desolation of factories and cinder-yards" and the "mute slabs" of skyscrapers. Later, he walked past "grotesque decay," blighted, "indecent skeletons," including "blocks of saloons and rooming houses, with seedy-looking men slouching in front of the windows of the closed stores, leaning against the walls, in the sunshine, waiting."[43] Such language was strikingly similar to that used by W. R. Burnett in his classic 1949 noir novel *The Asphalt Jungle,* made into an equally memorable film one year later by John Huston. Both the film and the novel envision "acres of hard cement" and the hard individuals — mostly men — who inhabit them. The city is a "monstrous, sprawling immensity" that demands death as the price of escape.[44] In their 1955 book *American Skyline,* the prominent urban critics Christopher Tunnard and Henry Hope Reed described a "certain Midwestern city": "The place is empty at night, ghostly and silent like a painting by Edward Hopper, when the cars have all gone home to the suburbs and the only activity that gives it any life is absent."[45] While extreme, then, Kennan's sentiments were not significantly different from those of a wide range of commentators who concluded that postwar cities were declining sites of "social and technological alienation . . . ringed by expanding centerless suburbs."[46]

For Kennan, the antithesis of the degraded city was the small, independent farm, but by 1950, this image, like his affection for nineteenth-century diplomatic history, was an anachronism replaced by the largely homogenous high-modernist pastoralism of the postwar suburbs. These suburban "citadels" were linked to visions of geopolitical primacy: they were the quintessential sites of American life and the spaces where history was being actively rewritten. Almost a decade later, the individuals and lifestyles emphasized at the 1959 American National Exhibition in Moscow were suburban, and cities had been reduced to architectural elements. Suburbs embodied order, safety, and a deeply gendered consumerism that "became as solid a pillar of the United States version of cold war culture as did its remasculinized military."[47] It was no coincidence, then, that in advance of Soviet Premier Nikita Khrushchev's reciprocal visit in 1959, President Dwight Eisenhower suggested a trip to Levittown, whose builder, William Levitt, had remarked upon completion of his creation twelve years earlier that "no man who owns his own house and lot can be a Communist . . . he has too much to do."[48]

Kennan's diagnosis of urban vice echoed a familiar, much older anti-city refrain, but it also acquired additional potency with the invention of the atomic bomb.[49] The clearest explication of this development came, in September 1949, from none other than the young Baptist evangelist Billy Graham, sermonizing two days after President Truman publicly announced the first Soviet atomic test:

> Do you know the area that is marked out for the enemy's first bomb? New York! Secondly, Chicago; and thirdly, the city of Los Angeles! We don't know how soon, but we do know this, that right now the grace of God can still save a poor lost sinner.[50]

For some families, salvation meant moving (as Washington, D.C., realtors advertised) "beyond the radiation zone" to the suburban developments that were, according to the sociologist William Whyte, becoming "the norm of American aspiration." In the introduction to the 1958 collection *The Exploding Metropolis*, a "book by people who like cities," Whyte wrote that the American city was "becoming a place of extremes—a place for the very poor, or the very rich, or the slightly odd."[51]

As Whyte's statement attests, central cities were not wholly abandoned, and those who remained behind, in addition to new arrivals who chose to or were forced to settle close to traditional downtowns, were responsible for what in hindsight appear to be some of the defining cultural achievements of the 1940s and 1950s. Abstract expressionist art, bebop, African American "protest" literature, and the Beat movement were all urban productions—and, in a different register, so was film noir. At midcentury, however, few of these movements occupied the peaks of American culture. The most successful, abstract expressionism, had been substantially decontextualized.[52] And the diverse films belatedly classified as noir included many that were "delusional journeys into panic and conservative white flight."[53] Just as the city-mysteries of Poe, Balzac, and other sensationalist authors of a century earlier had "registered the dreaded rise of the metropolis, film noir registered its decline, accomplishing a demonization and an estrangement from its landscape in advance of its actual abandonment."[54] It was precisely this quest for distinction that led one *Saturday Evening Post* writer to compare the "human tides . . . flowing out of the cities" to the "dark tides" that replaced them. "Decay and race," Robert Beauregard notes, "were thrown together in a discursive unity."[55] Thus, while the divisions between inner cities and suburbs were not always clear in practice, the power of a more simplistic imaginative geography of difference was still significant.

Noir's aesthetic conventions were also geographic; its films and fiction wandered across dark, rain-soaked streets and into seedy bars, diners, and lodgings.[56] In her 1947 tour of the United States, Simone de Beauvoir, an unabashed noir aficionado, described the same landscapes in nonfictional terms, with a cinematic reference inserted for good measure:

> American cities are too big. At night their dimensions proliferate; they become jungles where it's easy to lose your way. The second evening we wanted to see *The Killers* . . . which was playing in an outlying district. We set out on foot in the evening, thinking that after a short walk we'd catch a tram, bus, or taxi. Suddenly, we were on a dark road lined with tracks, unmoving trains, and hangars, crossed now and then by other deserted streets. We were in the heart of town yet in a desert. It began to rain violently, and in the wind and rain, we felt as forlorn as on a treeless plain — no shelter, no cars in sight.[57]

The antitheses, in all respects, of the booming suburbs, these noir worlds linger as a striking reversal of the Cold War cultural hierarchy of places. While stories such as Edgar Allan Poe's "The Man in the Crowd" (1840) presented cities as confusing terrain demanding obsessive exploration, noir's urban landscapes are forlorn, quiet places that have "lost that air of pleasurable excitement and possibility" and are instead "traumatized and resigned," following the Depression and then the Second World War. Even the crucial public realm of prewar hardboiled authors had given way to an unfocused landscape of creeping blight and moral decline.[58]

As a result, one dominant understanding of postwar American cities was a decidedly ruinous one. Cities became museumlike places best traversed and exited by the streaking automobiles that open films like *Black Angel* (1946) and *Criss Cross* (1949). These machines signal the importance of postwar highway schemes, the rise of a democratized car culture, and new understandings of dispersed urban spatiality. The personal mobility and rootlessness heralded by many noir films and the shifting, uncertain identities of their characters were literally facilitated by expanding, endless grids of alleys, arteries, and highways that were not spaces of community but rather networks of flight and anonymous transience — of both individuals and their money.[59] The supposedly public nature of these corridors is tarnished by deviant behavior, whether sexual or criminal, or panic, in the manner of a postdisaster scenario.

Well suited, then, to the obsessive investigations of noir protagonists were the shadowy, subversive inner cities that fitted smoothly into the detectivelike rhetoric of postwar anticommunism. For Senator Joseph McCarthy, a typical public housing project was "a breeding ground for Communists."[60] New arrivals to the country were of particular concern to McCarthy and others. Not only were the political sentiments of immigrants in question, but their habit of settling in cities, according to the respected New York Times military correspondent Hanson Baldwin, would increase urban vulnerability immeasurably, particularly because many of them were "depressed and ill." Unrest initiated in such "focal points of infection," Baldwin argued, would be difficult to contain; "hordes of the foreign-born, speaking no English, strangers in their own cities" constituted "a danger to themselves and to all their city neighbors."[61] In a 1946 article titled "Sociology and the Atom," William Ogburn had already posed a solution to this problem: when a designated "slum" in a city was cleared, its footprint should be left barren.[62] Peter Conrad has noted that after Hiroshima, the American city became "the choicest place for the destruction of the new bombs because, like those bombs, [it was] the product of energy in destructive excess."[63]

Civil defense is now associated, often satirically, with educational films such as the legendary Duck and Cover (1951) and its timid star, Bert the Turtle.[64] But its visibility as a component of Cold War culture is unimpeachable and is at least a partial counter to claims that civil defense agencies held little political and economic power. The geographic locus of civil defense bulletins, videos, radio alerts, and other media forms was, of course, the suburban home, where resourceful nuclear families were removed from the dense diversity and confusion of cities and were able to smoothly and rapidly follow the suggestions in such publications as the FCDA's *Home Protection Exercises* (1953).[65] Although this information frequently depicted women as industrious, it did so by encouraging mothers "to imagine themselves as warriors in training" and as part of a Cold War "civic garrison" of constrained Cold War liberalism that encompassed both public and private spaces — if only certain types of each.[66] The comforting bases of the family and Fortress Main Street were paralleled by a national framework of "distributed preparedness" that linked proper domestic behavior to the health of an American body.[67] These various geographic scales were united by similar ideals of safety, sovereignty, and fortification, universal constructions insensitive to the complexities of American life. The shelter and evacuation programs of the Truman and Eisenhower administrations, for instance, were predicated on a middle-class ideal of home and automobile ownership, which encompassed no more than 60 percent of the American population during the 1950s.[68]

Although the suburban home and its occupants may have represented a civil defense ideal, the FCDA's mandate was, ultimately, the education of a national body composed of individual minds. The agency was very active in classrooms, urging a "greater emphasis . . . on teaching a keen awareness of national dangers and the necessary precautions against them." Elementary school students should understand, among other civil defense-related subjects, "principal target areas and plans for their defense" as well as "effects of nuclear attack or major natural disaster upon our large metropolitan areas."[69] *Atomic Attack: A Manual for Survival*, published by the Council on Atomic Implications, urged families to coach their children to respond appropriately:

> Can Junior fall instantly, face down, elbow out, forehead on elbow, eyes shut? Have him try it tonight as he gets into bed. [In the event of an attack] Junior will feel the wind go by, the dirt and pebbles blown with hurricane force against his head. . . . A few cuts on the arms and legs aren't important. His playmates, standing upright, will be blown over like matchsticks. Some may get concussion, some broken bones.[70]

A 1951 article in the *Journal of Social Hygiene* by Charles Walter Clarke, a Harvard physician, chose to describe the consequences of a first strike in different terms, albeit with the same focus on adolescent individuality. Without appropriate vigilance, families would become separated and lost from each other in confusion. Supports of normal family and community life would be broken down ... there would develop among many people, especially youths ... the reckless psychological state often seen following great disasters ... moral standards would relax and promiscuity would increase.[71]

But a lack of preparation and awareness was not the only potential cause of social disorganization, and suburban promiscuity did not represent the only threat to national order. There were other peoples and places that inspired greater concern from those preoccupied with Cold War urbanism.

Managing Panic

The anxieties evoked by such complex cultural forms as film noir, of course, cannot be reduced solely to the atomic bomb. The postwar climate was responsible for "feeding, not breeding" the fear, violence, and misogyny already present in noir progenitors such as hardboiled fiction, tabloid street photography, and European existentialism.[72] Yet both Jean-Paul Sartre's oft-quoted description of Manhattan as "the Great American Desert" and Albert Camus' noir vision of New York as "a prodigious funeral pyre at midnight" seemed to take on additional valence after Hiroshima and Nagasaki, when the fallen American city became a common media image, and even more so after the first Soviet atomic test in 1949.[73] Journalists, science fiction authors, religious leaders, and scholars all rapidly "transmuted the devastation of Hiroshima into visions of American cities in smoldering ruins," inscribing concentric circles of destruction over various urban topographies.[74] "The clustered buildings and congested areas of our great cities," Hanson Baldwin wrote, "are natural 'area' targets of immense vulnerability for all the mass killers of the age," while the atomic scientist and hydrogen bomb proponent Edward Teller described them as "death-traps."[75] Even more explicitly, a radial urban model was likened to "the traditional target in rifle practice."[76]

Virtually all of these imaginative damage maps were centered precisely on the urban core — an extraordinary assumption, given the admitted inaccuracy of such bombing exercises, but also a strategic decision that created zonal models with profound structural and moral repercussions. The FCDA admitted that civil defense planning relied too heavily on an assumption of "symmetrical behavior of a nuclear burst" but argued that

Sixth-grade students and their teacher act out a scene from the Federal Civil Defense
Administration film *Duck and Cover* at Public School 152 in Queens, New York City, on
November 21, 1951. Courtesy of the Associated Press/Dan Grossi.

using concentric circles was "most useful and, in fact, is the most practical basis available for planning."[77] Whether cities were *primary* objects of
Soviet interest was not the issue; not only would such discussions potentially reduce interest in civil defense, but the simple fact was that there was
no set understanding of when an attack would come and where it would
occur.

No place or body was entirely immune from danger and thus from
fear. In the summer of 1950, *Time* described a "Chamber of Commerce
faith in their city's importance as a target" among city planners.[78] This
(un)certainty resulted in contours of risk whose gradients, delimited by
an overlapping concatenation of multiple "indicators," were actually shifting constantly, threatening to spill into adjoining districts.[79] Frightening,
unfamiliar, and profoundly disruptive, the bomb was an uncanny object,

but only properly so when given a geography, a place of impact. From this location, displacement would spread, ruining conventions of homeliness and planned order—the mythologies of both domestic security and rational cities.[80] Such ambiguity encouraged a revival of calls for the spatial independence of new communities from urban centers, which were familiar demands bolstered by new "evidence" from atomic scientists or, in a different register, by the noir division of urban spaces into safe and hazardous sectors.

The relationship between noir cities and the atomic age was also buoyed by the popular media. Reporting on a 1949 AEC study of the bomb's potential effects on the city of Washington, *Time*, borrowing from a contemporaneous movie, dubbed the nation's capital a "naked city," passively awaiting the arrival of a Russian bomb.[81] The phrasing was apt, since Jules Dassin's 1948 film *Naked City* depicts New York as a social scientific laboratory where "surveillance and interdiction [are] natural, organic functions, a form of social self-immunization."[82] This logic was precisely what lay at the heart of efforts to predict and manage panic through civil defense initiatives.

In the best-selling 1946 collection *One World or None*, Philip Morrison, a Manhattan Project physicist who had just visited a devastated Japan at the request of the War Department, transferred what he had witnessed to a more recognizable landscape:

> The streets and buildings of Hiroshima are unfamiliar to Americans. Even from pictures of the damage realization is abstract and remote. A clearer and truer understanding can be gained from thinking of the bomb as falling on a city, among buildings and people, which Americans know well.
>
> . . . The device detonated about half a mile in the air, just above the corner of Third Avenue and East 20th Street, near Gramercy Park. Evidently there had been no special target chosen, just Manhattan and its people . . . the streets were filled with the dead and dying.[83]

What made such scenarios so alarming to American readers was not necessarily the gruesome description of the bomb's victims—since this is what Morrison, John Hersey, and others had reported (however partially) from Japan—but rather the location of the destruction, in the middle of a crowded city that was the cultural capital of "the final undamaged citadel of western civilization."[84] Indeed, as the essayist and novelist E. B. White observed in *Here Is New York* (1949), for the first time, American cities were directly threatened by war, particularly the Empire City, as it possessed "a certain clear priority." His otherwise exuberant urban homage closed by anticipating the "cold shadow" of planes overhead.[85]

Perhaps the most dramatic representations of atomic disaster were presented in periodicals such as *Life, Collier's, Newsweek, Reader's Digest,* and *Time*—crucial contributors to the popular geopolitics of the early Cold War. Chilling scenarios unfolded in the pages of these magazines, in some cases well before the United States had lost the atomic monopoly. Borrowing liberally from the pronouncements of Army Air Forces General Hap Arnold, the November 19, 1945, issue of *Life* included a detailed description of a "36-hour war" beginning with the atomic bombardment of Washington, D.C., followed by a "shower of enemy rockets" on twelve other major cities, and an airborne invasion. Despite "apocalyptic destruction" including forty million deaths, the United States wins the rapid conflict through overwhelming firepower, and the last illustration depicts American technicians examining rubble in front of the still-standing and highly symbolic stone lions of the New York Public Library. Not surprisingly, few casualties are depicted in the accompanying illustrations, except a blonde woman sprawled obscenely beside a masked, cyborglike enemy soldier repairing a telephone line.[86]

Life's dramatization was one-upped by the August 5, 1950, issue of *Collier's,* titled "Hiroshima, U.S.A.," which featured a cover image of an atomic bomb detonating over midtown Manhattan. Inside, accompanied by the inhuman illustrations of Chesley Bonestell—known for his "views gazing down from a great height upon a city lit by a nuclear fireball"—associate editor John Lear fictionalized the incident.[87] Whereas *Life's* scenario was predicated upon an anonymous enemy, by 1950 this identity was no longer in question. An accompanying note from the magazine's editor made clear that Lear's account

> may seem highly imaginative. Actually, little of it is invention. Incidents are related in circumstances identical with or extremely close to those which really happened elsewhere in World War II.... Death and injury were computed by correlating Census Bureau figures on population of particular sections of New York with Atomic Energy Commission and U.S. Strategic Bombing Survey data on the two A-bombs that fell on Japan. Every place and name used is real.

[Lear] interviewed officials of the National Security Resources Board, the Atomic Energy Commission, the Defense Department; experts on nuclear physics, engineering, construction, fire and police methods, traffic, and atomic medicine.[88]

A final example, both more general and more extensive, appeared in the October 27, 1951, issue of *Collier's*, titled "Preview of the War We Do Not Want." An impressive list of literary, military, and political authorities, from Arthur Koestler to Edward R. Murrow, contributed to the detailed composition of "Operation Eggnog"—planned "to demonstrate that if The War We Do Not Want is forced upon us, we will win." While the U.S.-led United Nations force begins by avoiding centers of population, concentrating on "legitimate military targets only," American cities are directly bombed, leading to a retaliatory mission to Moscow witnessed by Murrow and, ultimately, to the occupation of the Soviet Union. Again, the story featured illustrations by Bonestell and geometric maps of Chicago and Detroit "under the bomb."[89]

These dramatizations and others like them were characterized by a similar "imagination of disaster," constructing urban uncertainty and justifying unprecedented forms of everyday American militarization.[90] In addition to the use of abstract visual representations, they relied on the selective introduction of *expertise*, particularly in the form of scientific wisdom. Using a curious mixture of graphic and sanitized language, magazines and the experts they consulted fueled nuclear fear while simultaneously rationalizing and containing it—a strategy that was central to Cold War civil defense efforts. But containment, as I have noted, was geographically sensitive. It was not just an "abstract ideology," but became "lived experience" in segregationist processes, for instance.[91]

"City people," Richard Gerstell wrote in *How to Survive an Atomic Bomb* (1950), "are the ones who have to guard most against panic." He went on to argue that "if we let prejudice of any kind enter the picture, the result can only make added trouble."[92] However, a report on morale submitted to the National Security Resources Board (NSRB) less than a month before President Truman transferred civil defense responsibilities to the new FCDA was blunter.[93] Predicting that "social disorganization" would follow an atomic attack, the authors were particularly excited by the potential for tensions in complex cities such as New York, Chicago, and Detroit:

"It is awesome to reflect on what would happen in one of these cities if colored people and white people were forced into close association in shelters, in homes, and even evacuation reception centers."[94] Seeking solutions to such predicaments, *Collier's* sent a reporter to Britain's Home Office Civil Defence School, where "A-bomb problems [were] analyzed in realistic detail on a contour map" and where "the model for mob management was India."[95] Despite moves toward equality during Second World War mobilization, the social or economic aspects of Cold War shelter and evacuation policies were rarely addressed directly by government officials. Organizations representing labor and African American constituencies lobbied vigorously, and with some success, to diversify at least the symbolic register of FCDA output. Yet this did not stop certain negative associations and buried fears from rising to the surface in more reactionary commentaries.[96]

The importance of both panic and control to civil defense was captured in an extraordinary 1953 *Collier's* article written by FCDA head Val Peterson. Citing various historical disasters and Orson Welles's infamous 1938 broadcast of *The War of the Worlds*, Peterson argued that Americans were the most "panic-prone" people on earth. War, he noted, was now a pervasive phenomenon: "Every city is a potential battleground, every citizen a target."[97] But in a continuous state of Cold War, constantly maintaining composure was paramount. To determine whether readers were panic-proof, Peterson's article included a quiz based on psychological studies carried out by the RAND Corporation, the Institute for Social Research at the University of Michigan, and other bastions of social scientific rationality. These surveys were based, in turn, on the extensive testing procedures performed on World War II soldiers—a lineage indicating the deep and subtle militarization of everyday life during the early Cold War. In addition, according to Peterson, women were more likely than men to panic. Even inside the home, the mood required to participate effectively in the struggle against the Soviet Union was one of masculine levelheadedness, precisely the approach advocated by strategists. Through the combination of self-maintenance and state-sponsored direction, an atomic attack could be not only survived but also given order. It is all the more interesting, then, that civil defense relied on both singular visions of a disaster and an indefinite challenge requiring perpetual vigilance.

When the potency of Cold War weaponry increased exponentially

Federal Civil Defense Administration poster, 1950s. National Archives and Records Administration Still Picture Division, RG 304, Series 304-P-6, FCDA 1A-6.

with the arrival of the hydrogen bomb (particularly the "practical" fusion weapon detonated in the 1954 Castle Bravo test at Bikini Atoll), the
options for survival appeared limited to shelters deep underground or
massive evacuation initiatives. Cities, according to Val Peterson, were "finished."[98] While a public shelter system was considered excessively expensive, evacuation posed alternate problems. Without clearly defined lines
of flight from cities, Peterson thundered in the pages of *Newsweek*, "we'll
have uncontrolled mobs moving about our countryside."[99] Such comments
echoed the racial covenants that had been placed on new suburban housing by both the Federal Housing Administration (until they were legally
struck down in 1950) and individual developers like William Levitt (which
lasted much longer). The postdisaster infiltration of one community into
another was, according to a prominent RAND Corporation study, a key
cause of demoralization.[100] In a November 1950 speech to the American Public Welfare Association, James Wadsworth, the acting director of
the NSRB's Civil Defense Office, decried the "take-to-the-hills" attitude
of urban "escapists," reminding his audience that "there can be no mass
stampede from our critical target areas," not only because it would be
impossible but also because it would mean "surrender of our production
centers." Civilians, Wadsworth concluded, "must 'fight our cities'" as a
Navy captain would hang onto his ship.[101] But while the FCDA and other
arms of the federal government officially opposed mass flight, the policy
did not prevent civil defense officials from participating in various evacuation simulations at the local scale.

The successes of the FCDA's campaigns may have ultimately been
mixed, but the pervasiveness of civil defense motifs in the popular media
is testament to the fecundity of civil defense iconography and, more generally, to the deepening nationalization and normalization of militarization
during the early Cold War. As the most substantial domestic propaganda
initiative (at the time) in American history, civil defense was shifting "responsibility from nuclear war from the state to its citizens by making public panic the enemy, not nuclear war itself."[102] These trends were strikingly
apparent in the FCDA's plan for an Alert America convoy, developed in
late 1951 and launched in early 1952 in Washington. A train of vehicles
traveled 36,000 miles across the United States, visiting selected "target
cities" and offering attendees a view of "dramatic visualizations" portray-
ing the extent of atomic danger:

Through photographs, movies, three-dimensional mock-ups, and scientific action-dioramas they depict the possible uses of atomic energy in both peace and war. Visitors to the exhibits see the damage that could be done to American communities by atomic bombs, nerve gas, and germ warfare. Visitors experience a vivid dramatization of a mock A-bomb attack on their own cities. They learn what they can do through civil defense to protect themselves and the freedoms they cherish.[103]

The arrival of a convoy was often supplemented by "parades, military exercises, air raid siren tests, mutual aid exercises, mobile support demonstrations, and mock raids by the Air National Guard."[104] Such spectacles would have been difficult to miss.

Alert America was followed by Operation Alert, a yearly national exercise that ran from 1954 to 1961 and featured simultaneous civil defense drills in a number of American target cities. In the first year, as a "theoretical fire storm raged," Dwight Eisenhower retired to a secret shelter, lingering underground for 25 minutes.[105] The following year, the president and thousands of federal employees left Washington for three days at various undisclosed locations. Inside an Army tent off a "mountain road lined with troops," Eisenhower found a large map of the country, dotted with blue and red pins marking air and surface "bursts" respectively. In total, sixty-one cities were "hit," causing some eight million deaths, although evacuations staged on paper saved over a million lives.[106] As Andrew Grossman recounts, in New York City, no "person or moving vehicle was on the streets," and the few who ignored air-raid sirens were quickly arrested.[107] By 1957, several additional elements of "realism," from fallout to "additional combat incidents," were incorporated into Operation Alert, and produced a stunning 54,500,000 deaths.[108] The expanded scope of simulation was dependent not only on the "reality" of ever-greater destruction but also on the equivalent premise of *imminent* thermonuclear war, reinforced by the "theater" of concurrent mushroom clouds over Nevada and Pacific atolls.[109]

Disaster Strikes City X

Academics and civil defense advocates were particularly concerned with the predicament of panic. The inaccurate and extremely popular government publication *Survival under Atomic Attack* (1950) noted in a list of tips that "a single rumor might touch off a panic that could cost your life."[110] In a more rigorous vein, disaster studies, virtually nonexistent before the Second World War, became an important interdisciplinary subject for numerous postwar research agencies, including (in addition to those listed below) the RAND Corporation and the National Opinion Research Center at the University of Chicago. Drawing inspiration from such precedents as the USSBS, and using such recent intellectual innovations as game theory and behavioral modeling, disaster scholars pushed for consistent "conceptual schema [and] . . . general theoretical categories and constructs" with the intention of hauling their subject firmly into the domain of the social sciences.[111] Lloyd Berkner, who seemed to have a stake in nearly every form of Cold War scholarship, argued that the intellectual "interest in disasters stems not so much from their spectacular aspects, as from the fact of our growing realization that even the worst events can be brought under some measure of control."[112] This was both a theoretical and practical belief: academic studies of disaster could also be translated into policy, namely as "an operational model for the 'protection' and surveillance of the emotional well-being of the American public."[113]

The National Research Council (NRC) and its partner organization, the National Academy of Sciences, made one of the most substantial bureaucratic investments in civil defense and disaster science, bringing many prominent researchers together on advisory committees within its Division of Anthropology and Psychology to identify suitable approaches and marginalize those deemed inappropriate. As with other interdisciplinary intermediaries, from the Ethnogeographic Board to the Arctic Institute of North America, the NRC produced a roster of worthy scholars, in this case those who treated cities as behavioral laboratories. As one joint report noted, social science could be brought to bear on all aspects of what was called "passive" defense, from the management of military facilities (including radar stations) to social and economic recovery following an atomic attack. High-profile research topics included the "psychology of threat" — group and individual problem solving under conditions of "danger, confusion, isolation, and deprivation" — and survival studies in austere environments. Many of the fashionable tools and techniques of Cold War social science, from simulation exercises to systems analysis, were cited as applicable to these inquiries.[114] The intention was to identify and rationalize risk by studying "the American people in as many disaster situations as possible" and extrapolating the impact of floods and tornadoes, for instance, to account for the unknown dimensions of sudden atomic attack.[115] This comparison was sensible but disturbing in the way it naturalized a profoundly human event.

That the same methods, themes, and sponsors were equally characteristic of other types of strategic studies was no coincidence; all were common to the dilemmas of life in atomic America, behind an unstable shield of security. In the midst of its program in the behavioral sciences, the Ford Foundation also funded the NRC's Committee on Disaster Studies after initial support from the Army, Navy, and Air Force medical services had expired. In a 1956 letter to a Ford Foundation representative, Glen Finch, the executive secretary of the NRC's Division of Anthropology and Psychology, outlined the major pursuits of the committee. NRC researchers, he wrote, were interested in the "patterns of social interaction and communication before, during, and after disaster," from the dysfunction of panic to the functional development of a "therapeutic community" through solidarity. If it was not already clear, Finch went on to describe the advantages of disaster study for the behavioral sciences: not only did the "concrete" events examined have a "starting point," but disasters provided a rationale for the infiltration of communities via participant observation and interviews. Finally, and most crucially, disasters were both unique and generalizable. They could not be "duplicated in the laboratory," and they incited abnormal human activity, but they were concurrently understood as aberrations from an understandable norm and were thus relatable to that standard using the same vocabulary.[116]

The FCDA briefed the NRC's Committee on Disaster Studies at the National Civil Defense Training Center in January 1953. FCDA staff presented papers on psychological warfare, medical care, and other challenges, while committee members were shown a demonstration of an air raid warning system, a mock "Rescue Street," and participated in a map exercise on "City X."[117] The following year, the committee produced a statement on the "problem of panic," which set out to delineate the term in order to grasp it objectively: "It is desirable to confine it to highly emotional behavior which is excited by the presence of an immediate severe threat, and *which results in increasing the danger for the self and for others rather than reducing it.*"[118] While hardly precise, this definition indicated that the central indicator of chaos was the viral spread of irrationality across a collective. It is not surprising, then, that one of the committee's key members, Irving Janis, later coined the term *groupthink*.[119] Threat, Steven Withey of the University of Michigan's Institute for Social Research wrote in 1957, "is a pervasive phenomenon. It is an element of much of human experience and observation."[120]

Scientific analysis of American destruction was typically applied to the hypothetical City X unless it was necessary to "emphasize certain of the bomb's effects," in which case Washington, D.C., or New York were typically substituted.[121] The FCDA matched this generic approach with publications like *Battleground U.S.A.* (1957), which outlined the civil defense plans for a "metropolitan target area" whose principal city was Battleground, an inland port in the state of "E."[122] Such imagined urban landscapes were nonetheless dependent on particular visions of spatial order and priority. There was little doubt, for instance, as to which part of City X would suffer the most grievous wounds, or, put differently, which part was most susceptible to infection.

In addition to its glossy leaflets, films, and exhibits designed for the public, the FCDA published more rarefied documents. Perhaps the most intriguing one was a 1953 manual ostensibly produced for municipal organizations titled *Civil Defense Urban Analysis*. This book shared much with concurrent attempts that mobilized the tools and language of scientific authority to compile and consider pertinent data on the Cold War's strategic spaces. In the case of cities, the FCDA recommended an initial collection of information and the presentation of these statistics cartographically. These maps might then be used to identify the area of maximum human and physical damage and to simulate an attack resulting in an accurate quantification of destruction. Scenarios such as this one were the foundation of civil defense planning; operational schemes and suitable services could be established in response. An urban analysis, then, was a practical procedure, and not just a reference tool for occasional consultation.[123]

To determine the "assumed aiming point," the FCDA urged city officials to select maps of "industrial plants and population distribution" and place over them acetate transparencies inscribed with concentric circles. Shifting this overlay "experimentally" over the various charts, points could be selected and then transferred to a base representation, preferably titled "Target Analysis Map." A line could be drawn between the two locations, and the midway position became the aiming point; this would enable one to calculate the size of a bomb required to destroy the areas around both sites. Similar procedures could be conducted for damage and casualties, or for all of the individual functions of a response unit, resulting in a series of specific maps and one master grid of the "overall defense pattern." The aiming point, however, was particularly important, the manual stated, because it was a "logical center for the pattern of civil defense ground organization of the community as a whole." Poor targeting or a related error, of course, could undo all of this plotting, but "in practically all cases" damage could still be addressed easily by wisely locating a management hub.[124] These remarkably distant instructions were accompanied by fitting cartographic examples: maps of blast effects that were *nothing but contours,* showing no urban detail underneath.

"Civil Defense Urban Analysis" was also the title of a paper presented by Milton Towner, an FCDA Education and Training Officer, during the briefing arranged for the NRC's Committee on Disaster Studies in January 1953. Towner described the various diagrams needed for a successful plan, including "isarithmic" maps: "By connecting points of the same numerical value we can draw contour lines, similar to those used by the geographer and in so doing we can know how many casualties to expect in case the center of damage appears at any point on or near any contour line."[125] His invocation of geography was appropriate, since these FCDA analyses bore a strong resemblance to the techniques of the spatial scientists who were just beginning to occupy positions of prominence in American geography departments. And yet, even more than the geography associated with the quantitative revolution, the cartography of civil defense analyses was not completely conceptual, but first specifically national and then generically urban.

Other military, academic, and philanthropic agencies were similarly hard at work on disaster research during the first decade of the Cold War. In his work funded and sponsored by the Ford Foundation, the Air Force, and Columbia University's Bureau of Applied Social Research, Fred Iklé argued that speculating on the social effects of bomb destruction was problematic because "rational planning is 'switched off' at the point of the real nuclear attack." After the explosion, Iklé postulated, the opposite occurred: "There is nothing but chaos, doom for all humanity, panic, or suicide—and immediate defeat or immediate victory." His dichotomy between rational and irrational time also had a spatial equivalent. Gestur-

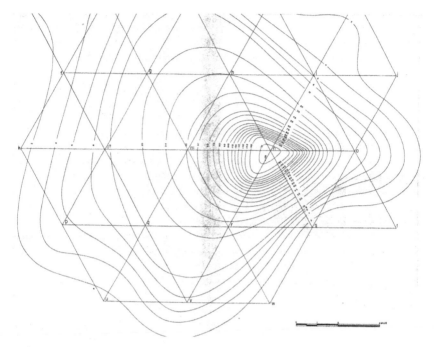

"Daytime with Warning." From Federal Civil Defense Administration, *Civil Defense Urban Analysis* (Washington, D.C.: U.S. Government Printing Office, July 1953), n.p.

ing vaguely toward both the Chicago school of sociology and subsequent Parsonian theory, Iklé summoned a functional-ecological model of urban life, arguing that a disaster would upset quantitatively observable human relations, "leaving tangible effects in the form of readjustments and measurable discrepancies." His city was an abstraction suited to equilibrium; it would readjust "to destruction somewhat as a living organism responds to injury."[126] And readjustment, as FCDA leaders repeatedly emphasized, was a matter of coordination.

Bureau of Applied Social Research (BASR) scholars have been widely credited with promoting postwar quantitative sociology, but they were also active in the integration of social science and military intelligence. Iklé's work on cities and disaster was the public face of a much larger Urban Analysis Project, or Project AFIRM, which closely linked the bureau's urbanists to the Air Force's Human Resources Research Institute (HRRI).[127] In 1950, for instance, several BASR members began collaborating with academics from the University of Chicago on an Urban Targets Research Project; the results were compiled as HRRI reports. While Chicago investigators under Philip Hauser studied the "sociological and psychological components of intra-urban target analysis," combining the spatial and temporal patterns of their home city to form a "framework for target selection," BASR researchers led by Kingsley Davis considered "inter-urban patterns of target complexes."[128] The data accumulated and models prepared for these studies were valuable for defensive planning, of course, but their appeal was both broader and more flexible—nothing less than the improvement and centralization of information on cities at a global scale. According to the BASR contribution to a 1951 HRRI report, the selection of data for inclusion in the Urban Resources Index would "facilitate systematic comparative analyses for strategic scientific purposes."[129] This meant that the creation of such an index was perfectly designed to suit Cold War operations, since these analyses could hypothetically survey any city on Earth, as a battleground or a site for strategic bombing. After all, the BASR–HRRI collaboration was launched during a period when the American atomic arsenal was limited to a relatively small number of crude "city-busters," before the hydrogen bomb–influenced Massive Retaliation policy associated with the Eisenhower administration. Research on urban disaster conducted to predict and limit chaos in American cities, therefore, was matched by a midcentury interest in *Soviet* urban targets.[130]

As I noted in chapter 3, Kingsley Davis went on to achieve significant fame as a demographer, coining such terms as "population explosion" and maintaining a lifelong interest in global urbanization patterns. That this curiosity was nurtured by the Urban Resources Index project went unmentioned in the numerous obituaries that appeared after his death in 1997. But in a March 1952 progress report on the HRRI-funded research at Columbia, he clearly delimited the orientation of the Index and the seamless connection between military and academic objectives:

> The idea of the Index is this: To gather and transcribe into a systematically arranged file the quantitative data for the cities of the world with over 100,000 inhabitants. By a logical and readily accessible arrangement of the file, quick reference can be had to any particular city or to any particular aspect of all cities or a limited class of cities. . . . The selection of data to be included in the files has been made from two points of view: (1) information important for military intelligence, and (2) information

important for economic, political, sociological, and social psychological purposes. *In most cases, these two points of view result in the selection
of identical types of data,* but we will necessarily be aided by detailed
instructions from the Human Resources Research Institute and the operational agencies in the determination of military relevance.[131]

The similarity with Yale's Human Relations Area Files, another classification project supported by the HRRI, is striking. At Columbia, cities were
effectively substituted for regional cultures as the organizational scale, but
it was the identification of both as (related) strategic spaces that allowed
the Air Force to so neatly align the methods and benefits of military intelligence and social science.

Evacuation and Simulation

As the United States and Soviet Union began to experiment with more
powerful hydrogen (or thermonuclear) weapons, it was not surprising that
Fred Iklé added evacuation planning to his research on the "social effects"
of bombing.[132] But the topic of flight from cities suggested a more visceral
form of simulation. This challenge was taken up by an NRC group for the
"operational analysis and field study of disaster problems," which staged a
series of evacuation tests in 1954.[133] With the construction of Arctic radar
fences, it was hoped that advance-warning times would be significantly
increased and that evacuation, especially in smaller cities, would become
distinctly feasible. Three locations participated: Mobile, Alabama, and
Washington State's Spokane and Bremerton. Operation Rideout, which involved a widely advertised dispersal of Bremerton's population by car and
ferry, featured the input of the University of Washington geographer William Garrison, a leader of the quantitative revolution.[134] In a memo to an
NRC Disaster Studies Committee member, Garrison mentioned that his
"observation team" had recommended the use of aerial photography and
a "Time-Space Car Tracking Plan" to Washington civil defense leaders.
His suggestions must have impressed those officials, since he and a team
of graduate students were soon enlisted to conduct research for the state's
1956 *Survival Plan*.[135] It is not surprising that spatial science was compatible with civil defense. Like the contour maps in *Civil Defense Urban Analysis*, the lines of flight and the grids of emergency response found in evacuation schemes represented a rational ideal but also the refined *unreality* of

abstraction. More specifically, Garrison's invocation of areal photography was telling: it offered the perspective and distance necessary for scientific forms of civil defense planning, quantitative urban geography, and, of course, targeting.[136]

The FCDA chose Spokane because it displayed a compact structure and a proximity to "open country," which combined to create a "suitable laboratory" for the study of response to disasters. To lend this exercise, Operation Walkout, both greater drama and greater realism,

> National Guardsmen were posted at street corners; emergency civil defense and military vehicles moved on the streets; anti-aircraft and machine gunners fired their weapons from the roof tops of several buildings; jet fighter planes and bombers flew over the area.... At 10 a.m., to simulate an attack, a bomber dropped leaflets over the city, saying "This might have been an H-bomb." The bomber missed the target area, and the pamphlets fell on an outlying residential district near one of the theoretical evacuation zones.[137]

The results of Operation Scat, a "drive out" evacuation of a Mobile neighborhood, were equally fascinating. There, researchers encountered demographic complexity and inequalities not apparent in Spokane. According to an anonymous report, most of the evacuees "were Negro," without private transportation, and demonstrated an "outstanding... conformity to the demands of the (white) authority." Due to poor communication, however, a rumor spread "that a real atomic bomb was going to be dropped on Mobile," and some voiced the fear that this was occurring to avoid school desegregation. The legitimacy of the report is less important than the indication that the results of actual simulations were quite distinct from the clean maps of mathematical analysis. As one of the NRC's representatives in Mobile noted frankly, almost every conceivable American urban target was populated by "lower class and lower middle class people, who in large part represent minority groups" — groups that were seen to be markedly different from "community leaders" and who were "not reached by the usual mass communication media." Another observer gained the impression, after speaking with a white policeman, that under conditions of disaster, racial divides might be partly breached, but only to the extent that black citizens would be picked up by white car-owners "after all the whites in the area had already been evacuated."[138]

The definitive combination of Cold War strategy, science, and anxious urbanism was Project East River, completed for the FCDA in 1952 by a group of academic institutions known as Associated Universities. Another one of the numerous summer studies affiliated with the Massachusetts Institute of Technology, East River not only reasserted the value of behavioral science to the military bureaucracy, but it also reaffirmed the mantra that fear could be channeled through training, emotion management, and self-surveillance. East River's diverse and authoritative members detected precisely what was wrong with American society and what could thus doom Western civilization: an "apathetic attitude" indicative of "individuals, institutions, and nations that have perished in the past because of the inability or unwillingness to adjust to major environmental changes."[139] These changes, the ten-part East River report made clear, were at once national and local, and they were prompted by both technological progress and geopolitical circumstance. Moreover, the link to American cities was quite apparent; Part V of the report, "Reduction of Urban Vulnerability," began with the assertion that "to keep pace with weapons development, it is essential to make urban targets less remunerative."[140] One response was to join in the widespread call for urban dispersal.

Although East River was not expected to actually oversee tests, experiments, or exercises to create new data, and was instead intended as a suitable forum for synthesis of prior research and opinion, one partial exception was made: a "selected area study" that formed Appendix V-A of the report. There, project participants, after deciding that "a typical American city did not exist for our purposes," borrowed from a recent disaster review produced under the aegis of two New York hospitals; the Rockefeller Institute; City and Suburban Homes, Inc.; and the New York Life Insurance Company. These risk-related agencies had conducted detailed land use and population studies of forty-seven Manhattan blocks and then simulated the dropping of atomic bombs over this grid, varying the location and height of the bombs as well as the number and position of shelters. The results of this study, in the form of large tables, were predictable and sanitized, facilitating an easy translation from the detailed topography of New York to "many of the features found in our larger cities" or to generic illustrations of threatened cities.[141] But as William Borden noted in *There Will Be No Time* (1946), America's largest cities were "concentrated spatially."[142]

Dispersal and Decentralization

As the Cold War solidified, many commentators began to suggest that American urban populations were excessive; atomic disasters would simply affect too many people and too many industrial sites. In July 1949, *Reader's Digest* noted that "U.S. skyscraper construction makes a perfect target."[143] A *Newsweek* special report published in 1951 described the United States as "a particularly attractive target, for most of its industrial potential lies in a 'geographical parallelogram,' with Minneapolis, St. Louis, Richmond and Boston at the corners."[144] Older, densely populated and "geographically bound" cities, seemingly impossible to disperse, were considered particularly vulnerable. This was yet another reason why New York, for instance, was a far more popular location for projected nuclear attack than less dense cities like Los Angeles and Houston.[145] The most effective and comprehensive solution to this dilemma of concentrated urbanization — but also the most contentious and expensive — was a massive program of decentralization, a proposal that had been anathema to most planners as recently as the Second World War.[146]

Though some aggressive theorists salivated at the prospect of an America speckled by evenly distributed towns of equal population, most agreed that the costs of such a utopia, ironically, would be too damaging for an American war machine dedicated to matching the Soviet Union stride for stride. However, various forms of limited or "progressive" dispersal — understood in the soft language of adaptation or even inevitability — did gain significant currency, particularly in the case of new urban developments.[147] Describing this advocacy as an intellectual responsibility rather than the promotion of "hysterical" schemes, Ralph Lapp championed "intensive social studies to understand the sociological 'make-up' of cities," studies that would simply "determine how natural trends in decentralization may be stimulated."[148] Accordingly, principles such as remote bomb production and storage, placement of war contracts in small cities, creation of widely spaced satellite towns, increased highway construction, and limited inner-city rebuilding were all fodder for debate. Ten years after overseeing the development and testing of the device that spawned these debates, General Leslie Groves — retired from the military and writing from his position as vice-president of Research and Development at Remington Rand — argued that industrial dispersal was necessary to eliminate "key points of vulnerability."[149]

The most powerful early evidence prompting calls for American urban decentralization was the USSBS's reporting on the effects of atomic bombs in Hiroshima and Nagasaki. In the classic documentary *The Atomic Café* (1982), Paul Tibbets, pilot of the *Enola Gay*, the plane that dropped "Little Boy" on Hiroshima, is shown stating that the two Japanese cities were "virgin targets," offering "a classroom experiment" for subsequent "bomb blast studies, or bomb damage studies." In addition, as an August 1946 editorial in *The American City* noted with alarm, they were targets precisely because of their concentrated population and activity, not to mention Hiroshima's level and open topography, which allowed the effects of the blast to "spread out." As a result, the USSBS cautioned, given "the similar peril of American cities . . . the value of decentralization is obvious."[150]

In the United States, a nation with a higher urban-to-nonurban ratio than Cold War rivals like China and the Soviet Union, a city was, as Bernard Brodie put it, "a made-to-order target, and the degree of urbanization of a country furnishes a rough index of its relative vulnerability to the atomic bomb." Like many writers familiar with the costs of national armament programs, Brodie strongly questioned the feasibility of the most drastic urban dispersal plans, including "linear" or "cellular" cities, suggesting that such schemes would interfere with natural urban growth. However, while he accurately asserted that the military benefits of massive, forced dispersal would not be commensurate with the costs, he concluded that a limited program of industrial and infrastructural decentralization (or "compartmentalization"), as well as a general encouragement of suburbanization, would be advantageous.[151] It was these more "realistic" topics that were tackled by all but the most fanatical of the dispersal advocates. The argument that dispersal should remain secondary to international control of atomic energy — a popular position taken by the sociologist Louis Wirth and others immediately after the Second World War — faded, along with hopes for global governance, as Cold War enmity increased.[152] Nor was there much discussion about the ability of some individuals and businesses to relocate more easily than others.

The same August 1946 issue of *The American City* also featured an anonymous article titled "Planning Cities for the Atomic Age," a platform for the views of the planner Tracy Augur, a long-time advocate of hybrid town–country landscapes who had been additionally shaken by the damage visited from the air on dense European cities during the war. In this piece, and in numerous other publications, Augur consistently laid out the case for the dispersal of cities as a defensive measure against a potential atomic attack. His argument was a relatively simple one: *space* was the best military defense against the bomb, and congested, poorly organized, and centralized cities were inviting targets. Like many proponents of decentralization, Augur was aware of the tremendous financial and social costs his plans seemed to entail, but he deflected these by stressing that the appropriate planning of inevitable new construction would not incur any additional expenses. If plotted scientifically, towns of thirty thousand to fifty thousand residents would not simply girdle an existing urban area but stand as independent clusters, inspired by the British garden city model, that were separated from one another by belts of open or agricultural land. Augur's modern suggestion that older, nineteenth-century patterns of urban life and design were made obsolete by technologies such as the radio, the telephone, and the automobile was thus fused with a premodern idealism.[153] This nostalgia was premised, as another dispersal campaigner argued, on the belief that residents of "small and medium-sized communities lead a much more natural and normal life than those in large cities."[154]

Interestingly, the ideal postnuclear community in many Cold War science fiction novels and films was either a small town or a contained, purposeful settlement such as a college or monastery.[155] These visions shared with those produced by planners, scientists, and nuclear strategists a belief in *survivability*. Whatever the genre, such scenarios routinely proposed that a sufficient number of people would live through a nuclear disaster and rapidly reconstruct American society and, in most cases, that these would likely be people "who are closely in touch with the unique spirit of America, and the values of the system of 'free enterprise.'" Not one strategist or government planner, Dean MacCannell points out, "envisaged a post-attack rebuilding by people who never much benefited from American society, or quite understood what America was all about, that is, by people who lived at a disadvantage on the margins of society."[156]

Tracy Augur's suggestions would not only eliminate malingering blight, but they would additionally improve the security of Americans, finally guaranteeing "the full benefits of the atomic age." Dispersal would lead to both revitalization and safety. As he put it, a "metropolitan area that is well organized in terms of the amenities of modern urban living and the efficient conduct of modern business will also be an area of decreased vulnerability to atomic bombs and other weapons of mass destruction." For this reason, the value of planned dispersal would not end with the closure of Cold War hostilities; it possessed logic above and beyond the exigencies of national defense. But there was also a third, related motivation: for Augur, dispersal held "equal value against the type of penetration that has become so common and so effective in modern times and which depends on the fomenting of internal disorder and unrest."[157] His advocacy of urban design suited to the atomic age, fusing contemporary technology with older arguments for garden cities, thus moved swiftly and smoothly across scales, linking national rituals to the conduct and proximity of individual bodies. These, as Michel Foucault recognized, were arguments that balanced and twinned the production of mass death, delivered by a terrible new weapon, with the potential for new forms of life — in decentralized landscapes made possible and powered by the benefits of "peaceful" atomic energy.[158]

Proposals for "regional cities" predate the Second World War and thus were not just responses to the atomic bomb, but the murmurs of support for dispersal within the federal government gave Augur and colleagues such as Clarence Stein added incentive to revive and promote various decentralization and regionalization schemes. In 1948, the National Security Resources Board began to consider the transplantation of certain government operations to Washington, D.C.'s periphery and published an early report on industry and decentralization titled *National Security Factors in Industrial Location*. It appealed to businesses to relocate so that "further urban concentrations of more than 50,000 people may be avoided." (A note at the bottom of the last page added that the suggestions sketched in the pamphlet were not intended to encourage relocation for the sake of lower wages, diminished working conditions, or the shredding of union contracts.)[159] A subsequent memo from the AEC to the NSRB reprinted in the *Bulletin of the Atomic Scientists* described the dispersal of Washington's government agencies to the suburbs, "with distances of perhaps two miles between targets," as a "fair measure of security."[160] Meanwhile, Augur had begun consulting for the NSRB in 1949. Congress rejected his major dispersal plan, which had endured many modifications, in April

1951, just days after President Truman had earned congressional wrath for firing Douglas MacArthur as the commander of American forces in Korea. Augur resigned shortly thereafter, yet some forty years later the location of many offices and laboratories with ties to the military corresponded closely to his call for a "dispersal arc" north and west of the District of Columbia.[161] His atomic-age garden cities had blended with, and been overtaken somewhat by, another type of decentralization: the "corporate regionalism" driving investment to the periphery of cities and to the American South and West.[162]

In the service of Cold war imperatives, planners debated the details of atomic physics, scientists became urban visionaries, and both groups became intimately familiar with strategic theory. From this perspective, Chesley Bonestell's god's-eye views from above, as well as the ubiquitous diagrams of concentric destruction, resembled the geometric lattices of spatial science in a manner extending beyond representational technique. Ironically, this coalescence of expertise led to crude atomic cities that were universalizing abstractions dependent on stereotypes and generalizations for authority. But they were powerful and prolific models nonetheless.

Project East River was complemented and endorsed by the nearly concurrent study on air defense at MIT, Project Charles. The latter's leaders were concerned with all aspects of continental defense, and since the details of such matters as locational patterns of population were beyond the purview of the average physicist or strategist, several economists — including Carl Kaysen, Paul Samuelson, Robert Solow, and James Tobin, all eventually towering figures in their discipline's postwar pantheon — were enlisted to provide an appendix on "Economic Aspects of Passive Defense." The result was an astonishing exposition of neoclassical reasoning, a cold-blooded summary that noted the advantages of urban concentration but then ascertained that this was a moribund factor in the atomic age:

> On any rational calculation, the possibility of enemy attack has radically changed, in favor of dispersal, the values to individuals and to society of alternative locations of particular installations, whether factories or houses. A man who is deciding whether his new house should be built in Manhattan or Fairfield, Connecticut, should now include an allowance for the distinct possibility that in Manhattan both his house and his family will be destroyed — increasing both the target attractiveness and the danger of fire.[163]

In urban studies, then, the city became a field of inquiry for a set of writers who argued that congested, poorly organized, and centralized cities were not only inviting targets but inefficient and unviable *systems*. Perhaps the most famous example of such work was the cybernetics pioneer Norbert Wiener's 1950 *Life* plan for radial "life belts" of transportation lines and essential services, separated from downtowns by "safety zones" where most construction would be prohibited.[164] This spatial distinction was essential. As the Detroit planners Donald and Astrid Monson argued in a contemporaneous article in *The American City*, without empty or agricultural interstitial areas, "the very factor which is counted on for defense is lost."[165]

Since a city, for Wiener and his colleagues, was "primarily a communications center, serving the same purpose as a nerve center in the body," the key to a livable existence was the ordered planning of informational networks. And Wiener's scheme, the magazine noted, would be useful in all circumstances. During periods of peace, quite incidentally, "it would expand and accelerate the current trend of many city dwellers toward the suburbs."[166] For early cybernetics, control was "the never-finished work of regulation which operates to bring deviations from system requirements back in line." Wiener's atomic city was thus not simply an updated version of nineteenth-century urban technical interventions. It also suggested that the governance of city life was, in addition to authoritative schemes implemented from above, a problematic of inner subjectivity and individual "participation in the networks of existence." Moreover, the cybernetic framework was a perfect example of a synoptic worldview that was not dependent on context. Understanding and designing urban systems, Wiener's vision seemed to suggest, was no different from his construction of the man–machine weapons that launched cybernetics during World War II.[167] And, in one sense, this postulation was correct.

Freedom on Four Wheels

If the decentralized city was emphatically represented as a landscape connected — or disfigured — by highways, this impression was augmented by the concern, in spectacles such as the FCDA's Alert America convey, for the "freedom" of mobility. As simulated evacuations had clearly shown, mobility was linked closely to automobile ownership, which in turn was tied to broader concerns for privacy and social order. This confluence was captured in a small pamphlet published by the FCDA in 1955. *4 Wheels to Survival* explained that the family car could "help you move away from danger," and auto owners were encouraged to keep the gas tank full, the tires firm, the trunk stocked with food and first aid supplies, and the rest of the vehicle in good shape. More interestingly, *4 Wheels to Survival* noted that tests in the Nevada desert had shown that cars could protect Americans from blast effects and provided a fairly detailed explanation of what to do with vehicles in the event of an attack. Automobiles effectively became a "small movable house" for the maintenance of the nuclear family in times of emergency, and a link to the authoritative civil defense radio broadcasts of CONELRAD (CONtrol of ELectromagnetic RADiation) that would direct traffic out of cities in orderly fashion. "Civil defense driving," the pamphlet concluded, depended on courtesy and patience: "If traffic gets stalled, don't lean on the horn. Your impatience may become someone else's panic. That could cost lives!"[168]

Like many civil defense publications of the period, *4 Wheels to Survival* is a decidedly mundane artifact, but it remains important for the ways in which American drivers and their passengers were clearly incorporated into a military geography of preconflict and postconflict mobility. The increasingly popular act of driving along highway spaces was turned into a matter of national security by normalizing risk. After the Second World War, the new interstate highways, writes Steven Goddard, "were the cathedrals of the car culture, and their social implications were staggering," changing "how Americans lived, worked, played, shopped, and even loved."[169] And the proposal to lace the nation with freeways faced no major challenges.[170] At least part of the logic advanced by the Eisenhower administration and its supporters derived from the claim that the interstates would serve the dual military purpose of aiding evacuation and facilitating the movement of troops and equipment. During the Second World War, the application of strategic language to American highways had begun in earnest, and the proposed interstate network was frequently called the "military system."[171]

Dwight Eisenhower traveled Germany's autobahns after V-E Day, a trip that helped him to "see the wisdom of broader ribbons across the land." During his first term as president, he was determined to sort out the differences among competing interest groups and "fix" American roads, finally securing support for and expanding the network proposed by the 1944 Highway

Cover of *4 Wheels to Survival* (Washington, D.C.: U.S. Government Printing Office, 1955). Courtesy of the Ohio Historical Society.

Act but never completed.[172] Ike's appointment of General Lucius Clay to head an advisory committee on the interstate system, not to mention the nomination of General Motors CEO Charles Wilson as Defense Secretary, sealed the links between the auto industry, highway construction, and defense. As Thomas Lewis notes in his history of the interstates, since nuclear

Car and home, Operation Doorstep, Nevada test site, March 17, 1953. Photograph courtesy of National Nuclear Security Administration/Nevada Site Office.

fear had transfixed the nation, huge numbers of Americans believed their city or town would be on the list of Soviet targets. "The case for defense," in other words, "could help sell highways to Congress" and the public, and so the initiative was renamed, with the momentous Federal-Aid Highway Act of 1956, to the National System of Interstate and Defense Highways. Given enough warning, citizens would theoretically "be able to pack the family car and head out of town on one of the new superhighways."[173]

The chief of transportation for the U.S. Army, however, estimated that a full-scale war would mandate the evacuation of some seventy million Americans. And as Helen Leavitt later revealed in *Superhighway-Superhoax,* interstate builders had never actually consulted with the Pentagon on such matters as the proper height of bridges so as to allow the transit of Atlas missiles.[174] But it was ultimately American drivers, or taxpayers, who had to support the massive expenditures of the interstate program, and whether highways would *actually* aid national defense was secondary. In his 1958 essay "The Highway and the City," Lewis Mumford captured this dynamic precisely, calling the linking of highway construction to defense "specious, indeed flagrantly dishonest." He went on to indict the "religion of the motorcar" and argued that the nascent "highway program will, eventually, wipe out the very area of freedom that the private motor car promised to retain." Predicting only additional congestion if motorists fled the metropolis en masse, Mumford suggested ironically that the construction of highways had already produced "the same result upon vegetation and human structures as the passage of a tornado or the blast of an atom bomb," leaving behind "a tomb of concrete roads and ramps covering the dead corpse of a city."[175] These were particularly strong words from a member of the Regional Planning Association of America, but Mumford, Augur, and others found minimal satisfaction in many of the suburban developments of the 1950s and the expansive roads built to access and connect them.[176]

As the MIT scientist and continental defense advocate Jerrold Zacharias put it in an autobiographical manuscript, "When we played war games in the air defense business, we would begin by writing off New York City."[177] He is echoed by Guy Oakes, who claims that by the end of the 1950s, the FCDA had essentially "written off the possibility of protecting urban populations" barring impossibly seamless evacuations. This conclusion was undoubtedly influenced by more destructive weapons and the "discovery" of fallout, but also by demographic changes in many metropolitan landscapes.[178] Cities such as New York had become the urban equivalent of "national sacrifice zones," the contaminated nuclear landscapes starting to spread across rural America.[179] As headlines blared that an "H-BOMB CAN WIPE OUT ANY CITY," a new urban and national geography was emerging.[180] This was "centrifugal space": the decentralized landscape of freeways and sprawl that marks, for Edward Dimendberg, the end of film noir as well as the end of "the metropolis of classical modernity, the centered city of immediately recognizable and recognized spaces." The circulation of information and automobiles had replaced the movements of pedestrians, while high-modernist freeways and the suburbs they accessed had become symbols of both progress and security.[181]

Urban Geopolitics

In this chapter, I have built on the claim that postwar America was characterized by a powerful disillusion for urban life. Central cities, for many, were degenerate spaces, repositories of extreme cultures, classes, and races, and were threatened from above and within. The language of anxious urbanism may well have been symbolic camouflage for broader fears.[182] However, this process also operated in reverse: discussions on the status of cities were appropriated and encouraged by Cold War strategy and by technology-inspired changes to the theory and practice of warfare. It was precisely the *domestic geography* of Cold War risks that led to the scientific planning schemes, some more drastic than others, designed to order and manage urban spaces while concurrently maintaining the various symbolic distinctions between central city and suburb. While the resemblance is compelling, these schemes presented imagined settlements as frequently more rational and ordered than most of the actual suburban landscapes constructed after the Second World War. For the Monsons, the suburban growth of the 1940s was "without plan and [was] largely an extension of the amorphous sprawl of the central cities." Planning this spontaneous, inevitable decentralization appeared to be a natural step forward.[183]

Support for decentralization initiatives, along with calls for evacuation in advance of attack, had waned substantially by the end of Eisenhower's presidency, for several reasons. Some influential strategists had concluded that cities would be secondary to rural military targets in the event of a nuclear strike. The development of new weaponry, particularly intercontinental ballistic missiles, had furthered the futility of evacuation despite the expensive warning lines established across the north of the continent. Civil defense, however, lingered on in various manifestations, from fallout shelter programs and urban simulations to the texts of popular culture, through the duration of the Cold War—and beyond.[184] As Lee Clarke puts it, the enduring dialectic of panic and control found in civil defense plans and the broader corpus of disaster studies allows authorities to "lay claim to mastery and thoughtfulness" even in the cases of "problems for which there are no solutions."[185]

But perhaps the most intriguing and persuasive reason for the gradual disappearance of dispersal discussions was that by the late 1950s they had become, through a subtle slippage, largely a "benign discourse over structural changes like suburban high schools and shopping malls."[186] According to Alan Wolfe, "malls did not so much replace the city as re-create it in controlled circumstances located at some remove from where people actually lived."[187] And studies such as Project East River had noted that dispersal policy was "in line with general trends" of postwar urban growth.[188]

Under conditions of nuclear deterrence, Cold War American cities, Mac-Cannell concludes, became "defense weapons": places required not only to receive an atomic bomb but to "*absorb* the hit so that damage minimally spill[ed] over to surrounding areas." Even after the sixth Operation Alert, in 1959, Chicago's African American paper, the *Defender*, was still lamenting that "They Forgot the Southside."[189] Narratives of urban decline and the various distinctions maintained and encouraged between central city and suburb were of very specific strategic value—in channeling money not spent on inner-city improvement to the national arsenal but also in consistently locating, through a potent combination of lurid drama and rational science, the locus of atomic danger in the heart of America's cities.[190] Such circular histories are a telling reminder of the peoples and places literally left behind by Cold War strategy.

CONCLUSION
Into Space

> [The] cyborg is the awful apocalyptic *telos* of the "West's" esca-
> lating dominations of abstract individuation, an ultimate self
> untied at last from all dependency, a man in space.
>
> — DONNA HARAWAY, *Simians, Cyborgs, and Women*

Popular Cold War science fiction films such as *Invaders from Mars* (1953), *Them!* (1954), and *Invasion of the Body Snatchers* (1956) figured the perils of communism through the trope of bodily replication. Although "reds" remained a distinctly alien category, the threat posed by these others could not be contained within a field of visible externality. Rather, danger emerged from *inside*, producing a problem of indeterminate identities and insecurity.[1] In *Body Snatchers*, personal vulnerability is a product of sleep — when the temptations of the unconscious prevail over self-vigilance. Set on the aptly named world of Altair IV, *Forbidden Planet* (1956) positioned the inhuman as native to the individual psyche, located within a savage Freudian id. A "monster from the id" is unleashed through the interface between human mind and machine, appearing only when the two become singular and indistinguishable.

Science fiction cinema "positively invites psychoanalytic readings" of extraterrestrials as "civilization's conflict with the primitive unconscious."[2] The Cold War study of cybernetics also challenged the foundations of human identity, forging new monsters defined against but produced by humanity. Cybernetic creations such as computers, biogenetically engineered organisms, and cyborgs continue to further "those forms of ambivalence reserved for the Other that is the measure of ourselves."[3] As the idea of "the human" has been extended and perfected, it has also been revealed as a fragile concept.

Forbidden Planet opens on United Planets Cruiser C-57-D, a military ship on a rescue mission to Altair IV, where the spacecraft Bellerophon had landed some twenty years earlier, at which time it promptly ended all communication with Earth.[4] Arriving in the planet's atmosphere, C-57-D and its all-male, all-white, all-American crew is "scanned" and contacted by a survivor, Dr. Morbius (the Bellerophon's philologist). Though Morbius urges Commander J. J. Adams not to land, this warning is ignored. Upon arrival, the crew encounters Robby, an advanced robot who escorts the commander, a lieutenant, and the ship's doctor (Doc) to Morbius's home, where they are met by the philologist and his daughter Altaira. After Morbius refuses to return to Earth with his rescuers, Adams decides to contact home for further instructions. These plans, however, are foiled by a mysterious presence that enters C-57-D at night and destroys valuable communications equipment. Suspecting Morbius, Adams and Doc return to the house to confront him and find the philologist emerging from behind a secret panel in his study. Thus exposed, Morbius reveals a vast complex of alien technology below his home. Built by the Krel, a "mighty and noble race" who approached "freedom from physical instrumentality" but mysteriously perished on the threshold of this "supreme accomplishment," the underground world includes a vast expanse of nuclear reactors, still drawing energy from the planet's core, and a "plastic educator" responsible for Morbius's substantially improved intellect.

The discovery of these wonders is halted by the concurrent death of the United Planets radioman, torn limb from limb in a manner identical to the deaths of the Bellerophon's crew. An "invisible monster" returns to C-57-D that night, materializing under blaster fire as a massive, shrieking beast and killing several more men. It vanishes only when Altaira wakes her father in his laboratory; behind him, several gauges wink out. Returning to the house once more, Adams and Doc encounter Altaira. As she speaks with the commander, Doc runs to the laboratory, where he uses the "brain-boosting machine" with the knowledge that Morbius barely survived its effects. Carried back into the house by Robby, Doc manages to gasp "monsters from the id" before he dies. When informed that the id is the "elementary basis of the subconscious mind," Adams suggests that the building of machines that responded to Krel thoughts also released the "secret devils" of the id during dream-sleep. The latest instantiation of the beast is Morbius himself—a speculation proven when Robby's programmed prohibition against harming humans prevents the robot from destroying the monster as it advances upon the home. Unsafe even behind the Krel-metal laboratory doors, Morbius confronts the monster and collapses, but not before directing Adams ("son") to throw a switch that will initiate a self-destruct sequence. Adams and Altaira leave on C-57-D, using Robby as a cybernetic navigator. From the safety of space, the remaining crew watches Altair IV explode on a televisual screen, leaving Adams to muse: "It will remind us that we are, after all, not God."

To situate *Forbidden Planet* in the United States of the early Cold War is to recognize the double and conflicting role of atomic energy as the source of power and fear. These concerns are explicitly manifest in the film when Morbius reveals to Adams that Robby is a creation made possible by early translations of the computerized Krel library. When Adams suggests that the systemic knowledge within such marvelous possessions should be returned to Earth, Morbius states that "man is unfit, as yet, to receive such knowledge" and cites his "own conscience and judgment" in the decision to dispense information at his discretion.[5] John Trushell has argued that Morbius's reluctance parallels the ethical stance taken by those scientists who had developed the atomic bomb but opposed the creation of a hydrogen superbomb. The 1949 Report of the General Advisory Committee of the Atomic Energy Commission, a body directed by Robert Oppenheimer, stated that "mankind would be far better off not to have a demonstration of the feasibility of such a weapon until the present climate of world opinion changes." The arrogant dissenter Morbius could thus be read as an allegory of Oppenheimer, whose opposition to the H-bomb ultimately led to accusations of communist associations and the withdrawal of his security clearance after a controversial "special hearing" in 1954.[6]

Advanced atomic technology is strikingly represented in *Forbidden Planet* when Morbius, Adams, and Doc descend into the vast underground Krel complex of "nine thousand two hundred thermonuclear reactors" on "seventy-eight hundred levels." Incomprehensibly large, devoid of "direct wiring," and perfectly self-maintaining after two thousand centuries, the subterranean world is impeccably inhuman, a technoscientific marvel of "limitless power" far beyond the conceptual mastery of human minds. "Man does not behold the face of the Gorgon and live," states Morbius, and yet it is only *man* who can attempt to confront this mechanized other, while Altaira, simulated in miniature by the brain booster, remains an object in the masculine field of vision. In an unsurprising assertion of Cold War strategic tropes, *Forbidden Planet* figures both science and the military as definitively masculine. Although differentially representative of the compassionate father and the assertive son, deity and offspring, Morbius and Adams struggle for the same trophy.

The earliest formal definition of a cyborg, or cybernetic organism, concerned the need for control systems that would assist bodily homeostasis during future space voyages.[7] As Manfred Clynes, the scientist who coined the term, remarked, *cyborgs* referred to "persons who can free themselves from the constraints of the environment to the extent that they wished." Clynes wishfully believed that this was only a physiological predicament, an "enlargement of function" that would not alter the *nature* of humans. But the premise of freedom from context was also the dream of postwar social scientists, from the market planners of economics to the aptly named "space cadets" of quantitative geography.[8] However, what Clynes did not mention was that the stability of "man and his mind in strange settings," as one student of the Arctic noted, required "some degree of simulation of his previous sensory environment if he is to continue to function satisfactorily." This was a lesson learned from psychologists with an interest in "space travel."[9]

Despite the status of cybernetics as a science of management, the attempted transformation of the world into common codes and machinic assemblages also contained ambiguities and contradictions, "something that always exceeds control," leading to what John Johnston calls "information multiplicity."[10] Nowhere is this viral quality more evident than in Norbert Wiener's distinction between positive and negative cybernetic machines. Like Robby the Robot, the former is positioned, as Katherine Hayles puts it, "alongside man as his brother and peer," while the latter is rigid and inflexible, depicted "through tropes of domination and engulfment," stripping humans of autonomy or agency. Key to Wiener's division is the maintenance of bodily boundaries, which define the autonomous self. According to Hayles, "The danger of cybernetics is that it can potentially annihilate the liberal subject as the locus of control," a problem that was enunciated concurrently with anxieties concerning communist penetration into individual minds — and ultimately the body of the state.[11]

In their refusal to position the observer, or scientist, inside the structure of study, Wiener and his cyberneticist colleagues continued the long

Cyborgs in the laboratory: feeding at "altitude," Wright Air Development Center, Wright Patterson Air Force Base, Ohio, April 22, 1958. National Archives and Records Administration Still Picture Division, RG 342, Series 342-B, Box 655, Folder 342-B-T-027-2.

tradition of modest witnessing in science: they posed transcendental truth claims concerning the systemic organization of the world from an invisible position.[12] But the philosophical dislocation of objective science has produced an alternate "horizon of intelligibility," which Michel de Certeau dubbed *science/fiction*. Noting that the unreal multivocality of fiction haunts the singular privilege of science, de Certeau proposed an interspace, or heterology, at the juncture of these two categories.[13] The politicized recognition of fiction in science is an acknowledgment of something repressed, not unlike Donna Haraway's juxtaposition of the "actual and figural . . . as constitutive of lived material-semiotic worlds."[14] But in both Cold War science fiction and Cold War popular science, the division between fact and speculation was particularly blurry when discussion turned to one subject: outer space.

Forbidden Planet was one of countless period films and novels that depicted "panic and relief, invasion and dispersal, collapse and recovery."[15] These dualisms were also given geographic dimensions; they are staged at the scale of bodies, laboratories, cities, nations, regions, and globes. But a historical geography of the American Cold War is incomplete without acknowledgment of a final strategic environment, treated as external but most certainly an earthly creation, which brought geopolitics, science, and culture together in wonderful and frightening ways.

With its artful combination of "high culture allusion and high camp effects," *Forbidden Planet* appeals provocatively to various strategic spaces and, more importantly, to their interdependence.[16] The low-slung residence on Altair IV, complete with an ideal robot mother figure and huge reactors pulsing in the basement, is in certain respects an archetypal home for a suburban nuclear family, or perhaps a survivalist refuge. In the manner of the continental defense network, the American crew attempts to set up a shield around C-57-D that will alert them to the presence of an invading force. The theme of hostile, foreign regions is clearly apparent in the exploratory designs of the "United Planets" force—a step outward from the United States—and the explicit rewriting of Shakespeare's *The Tempest* to suit a new imperial frontier. And, finally, the planet itself, seen exploding from a distance, projects a possible, nightmarish future for Earth, characterized by the mishandling of atomic science, the ambitious models of cybernetics, and the totalitarian utopianism of technocracy. That the vision of Altair IV from the departing spaceship is itself a god's-eye view is deeply ironic given the moralizing of Commander Adams, but the apocalypse is made possible only through the mediating presence of a screen, as if the scene were a simulation.

What enables the coalescence of Cold War contours is not a specific film but the motif of outer space. Interlocutors of American interest in the "final frontier" have documented the extensive pre-*Sputnik* fascination with matters extraterrestrial, an interest that was not just romantic but, for some, quite practical. The very first report of Project RAND, for instance, was titled *Preliminary Design of an Experimental World-Circling Spaceship*, a contribution, received with significant skepticism by the Air Force, that was ultimately quite prescient.[17] After *Sputnik*, RAND resuscitated its largely dormant interest in space exploration. A group of RAND staff produced an influential *Space Handbook* detailing the history and facets of astronautics, and Irwin Cooper of RAND's Mathematics Division wrote a short report on the cyborg "triangle" of man, machine, and space.[18] Given that scientists such as Wernher von Braun and Heinz Haber were hosting Disney's *Man in Space* television series (which began in 1955), concocting speculative scenarios of space travel with the artist Chesley Bonestell, *and* consulting with the American military on rocketry and aerospace medicine, the distance between Cooper's "man" and the heroic figures confronting a dark universe on the cover of science fiction paperbacks of the 1950s was minimal indeed. Haber himself penned an anticipatory 1956 *National Geographic* article, featuring lavish illustrations by magazine staffer William Palmstrom, outlining the new "'space' view of the world" that would be made possible by future satellites.[19]

Sputnik, MIT president James Killian wrote in an account of his White House role as science advisor to Dwight Eisenhower, was not itself a weapon that threatened American security "but the thrust that launched the satellite was another matter."[20] In this respect, the geopolitics of space was an extension of the global views described in chapter 1. Just four months after *Sputnik's* flight, a lecture on space by the RAND Social Science Division member Joseph Goldsen outlined an axiom directly derived from Halford Mackinder: "Who controls space controls the universe; who dominates the space above the air dominates the world."[21] Never one to pass up a hyperbolic proclamation, Senate Majority Leader and future president Lyndon Johnson similarly argued in the wake of the Soviet satellite that "control of space means control of the world." Johnson's equivalence was literal: the victors in the space race would be "masters of infinity," possessing the power to dominate nature. Technology, particularly in the form of instruments of sight, was essential to this authority. More important than an "ultimate weapon," Johnson claimed in a contemporaneous speech, was "the ultimate *position*—the position of total control over the earth that lies somewhere out in space."[22]

Even when phrased more inclusively, in the tones of humanity's extraterrestrial ambitions, such rhetoric was bluntly redolent of American *planetary* imperialism. In a broader sense, Johnson's "position" was equivalent to the ontological location (or professed nonlocation) from which Michel Foucault's transparent society could be made legible, not to mention the basis of the computerized world models just beginning to take shape in the late 1950s.[23] Before actual photographs taken from this vantage of total

Wernher Von Braun grins in Washington, D.C., on February 1, 1958, as he holds a globe marked with the path of the U.S. satellite *Explorer I*, launched the previous day. Von Braun and his team at the Army Ballistic Missile Agency developed the rocket that propelled *Explorer I*, the first American Earth satellite, into orbit. Courtesy of the Associated Press.

control were displayed and distributed, Johnson was imagining a new, more thorough way of grasping the globe. He was furthering an abstraction of enmity and a derealization of vision, made manifest during World Wars I and II, that militarized the world in a quest for national security, turning it into a "series of strategic coordinates and various symbolic entities within the coordinates."[24] The conquest of space, to borrow from one popular title of the period, was thus reciprocally related to the earthly control of natures and cultures.[25]

This relationship took several forms. The convergence of earth scientists during the early Cold War produced advances in geodesy, cartography, and photogrammetry that led directly to the clandestine CORONA satellites. These spacecraft were vehicles for the collection of *technical intelligence,* the abstract spatial science gaining support within American intelligence agencies. The CORONA project was designed to replace lower-flying reconnaissance ventures that were growing increasingly dangerous, but it also became a crucial aid to remote sensing and geographic information systems. The early spark was a trove of geodetic data retrieved from German storage sites — part of the race for scientific information and experts, staged as the Third Reich collapsed, known as Project Paperclip. New mapping facilities built in the United States by the military branches during the Second World War subsequently shifted to accommodate the expanding and ultimately planetary map datums required by long-range bombers and intercontinental ballistic missiles. But this shift also reflected the globalization of strategic spaces. Dwight Eisenhower's Open Skies proposal of 1955 was the public acknowledgment that global surveillance was a necessity, and after it was rejected by the Soviet Union, both states instead pursued secret observation machines (CORONA and ZENIT).[26]

By the end of the 1950s, new understandings of geodesy were but a sliver of the booming field of space science. The Soviet and American satellite programs, moreover, were nominally part of the International Geophysical Year (IGY), an exercise in global scientific practice scheduled from July 1957 to December 1958. Again, the internationalist impulses of the IGY ran afoul of Cold War strategy. IGY principles and the more general "freedom of space" were used as nonmilitary justifications for the American satellite program. The American armed forces established and maintained monitoring stations in the Arctic and other locations.[27] Eisenhower's administration fully understood that the Soviet Union was doing the same, and unlike many in Congress, the media, and the public, the president and his staff showed little alarm when *Sputnik* was orbited. The IGY was the brainchild of Lloyd Berkner, the influential broker between scientific and military communities. Berkner was obsessed with the human challenge to the secrets of an external physical environment, and his "expansive technocratic vision" fit neatly with both the IGY initiative and the environment of outer space more generally. As I noted in chapter 1, his interest in the tensions between national survival and principled internationalism in science dated to at least 1950, when he produced a report for the State Department on "Science and Foreign Relations," featuring a secret appendix discussing scientists as "intelligence-gatherers."[28]

In numerous publications and speeches before, during, and after the IGY, Berkner advertised 1957 as the year that Americans would be, or had been, "catapulted into a three-dimensional geography of the universe."[29] This was not just a matter of reaching space, since he also heralded the authority gained by spreading scientists across the surface of the earth on IGY projects. These scholars, Berkner stated, "must observe at the same time; they must make the measurements according to its agreed standards. Only then can we understand the gross events that encompass the Earth and affect us all." To scrutinize the whole of the planet, he acknowledged, certain inaccessible regions would require "high adventure."[30] The "interplanetary travel" Berkner discussed in a 1959 address to the American Geographical Society was thus a straightforward extension of such exploits.[31]

Lessons learned from explorations in Earth's hostile regions, in other words, were directly applicable to the hypothetical realm of galactic discovery. In Washington, D.C.'s National Air and Space Museum, a prominent quote from 1953 still greeted visitors to a room stuffed with cyborg flight suits, fifty years later: "The protective envelope that the well dressed traveler must don before he ventures into space presents many problems of engineering that require study and solution."[32] Little wonder, then, that theories of human engineering and psychology were drawn from Arctic field experiences to illustrate the perils of presence in outer space or that the science fiction writer Robert A. Heinlein could draft, in 1949, the outline of a "Baedeker of the Solar System."[33] But *Forbidden Planet*—not to mention the fraught history of actual space travel—should remind us that the confidence of those entering and contemplating space as a strategic environment was profoundly fragile. The same was true of the commanding views from satellites, rockets, and, later, space shuttles: these were intermediate positions, facing a whole Earth but also a universe of new regions that made one world seem comparatively diminutive and time–space compression seem less dramatic.

In March 1958, James Killian's Presidential Science Advisory Committee released a report titled "Introduction to Outer Space." Among the key contributors was Edwin Land, the brilliant founder of Polaroid and a frequent contributor to Cold War reconnaissance projects, including the design of the U-2 spy plane. The report listed four factors that gave "importance, urgency, and inevitability" to devices that could propel humans into space and assist them once there: "the compelling urge of man to explore, and to discover; the defense objective; the factor of national prestige; and the new opportunities space technology offers for scientific observation and experiment." The realm of outer space may have been novel, requiring "introduction." But not only were these overlapping rationales the stuff of *Forbidden Planet's* anticipatory vision, they were also all present at every strategic scale considered in this book. Indeed, the drive to map and classify Earth's surfaces; maintain security; enhance the status of national culture "among the peoples of the world"; turn regions into scientific laboratories; and translate laboratory constructions into urban, continental, or global models all reflect earlier ambitions, many of them colonial.[34] They were not uniquely Cold War phenomena. And yet only in the United States during the 1940s and 1950s were these factors all united under the banner of modern militarization — a condition that was not static but that nonetheless aligned certain strategic ideas and practices with specific spaces.

ACKNOWLEDGMENTS

In the fall of 1998, I wrote an essay on Cold War geopolitics for a graduate seminar in the Department of Geography at the University of British Columbia. Some of the ideas and fewer of the words from that paper are present in this text, which means that I have been reading, thinking, and writing about the subjects in this book for more than a decade.

I first sent a version of this manuscript to the University of Minnesota Press in 2004. I thank Carrie Mullen for her initial enthusiasm and Jason Weidemann for his subsequent patience and encouragement. Several reviewers of the manuscript provided invaluable guidance and criticism, and I am grateful for their labor on a task that does not earn sufficient academic recognition. John Agnew, who endured this role twice, deserves double thanks for his encouraging and wise commentaries. Staff at the Press, particularly Kristian Tvedten, Danielle Kasprzak, Emily Hamilton, Rachel Moeller, and Nancy Sauro, were consistently helpful and supportive. Carol Lallier's copy editing was relentless and excellent.

Although financial sustenance for scholarly work can appear more abstract than personal debts, the research for this book would have been impossible without the generous support of the Arctic Institute of North America, Green College (UBC), the Killam Trust, Memorial University, the Rockefeller Archive Center, the Social Sciences and Humanities Research Council of Canada (Doctoral Fellowship 752–2000–2091 and Post-Doctoral Fellowship 756–2003–0349), the University of British Columbia, and the University of Toronto.

During numerous research trips across the continent, I encountered many archivists and librarians at the locations listed in the following pages. Thank you, collectively, for your patience, advice, and encouragement. Throughout my travels, and at conferences and colloquia, I also met or corresponded with a host of impressive scholars, including a number from different disciplines who no doubt viewed me and my project with some skepticism. Those who were generous with their time and support include John Agnew, Laura Cameron, Simon Dalby, Anne Godlewska, Sharon Ghamari-Tabrizi, Hugh Gusterson, Stephen Graham, Gregg Herken, Edward Jones-Imhotep, David Kaiser, Scott Kirsch, Simon Marvin, David Neufeld, Naomi Oreskes, Susan Schulten, Joan Schwartz, Neil Smith, and Gerard Toal. Friends such as Ari Goelman, Maighna Jain, and Sarah Koch-Schulte made the journeys much more manageable.

Derek Gregory introduced me to the rich and diverse subject of historical geography, and he assisted and inspired me from the first lecture of his that I witnessed in 1994 to his final suggestions for this document. I owe him an immeasurable intellectual and personal debt. Within the geography department of the University of British Columbia, I am grateful for the constructive criticism and company of Trevor Barnes, Matthew Evenden, Cole Harris, David Ley, Geraldine Pratt, Olav Slaymaker, and Graeme Wynn. Conversations and coursework with Richard Cavell, Elizabeth Haiken, and Diane Newell helped me to think beyond disciplinary confines.

My graduate student colleagues at the University of British Columbia were a source of challenge and amusement, and I am delighted that friendships with Kate Geddie, Arn Keeling, David Nally, Richard Powell, Alex Vasudevan, and Jamie Winders survived both the pressures of that intense environment and the inevitable dispersal that followed. Life at two UBC colleges, Green and St. John's, brought me into contact with a list of extraordinary individuals too long to name in full, but I single out Tim Dewhirst, Mitchell and Stefanie Gray, Emily Lai, Maged Senbel, and Vanessa Timmer for special praise. I thought Aaron Hunter, my roommate at Green for a year, a first-rate fellow, and that was before I met Jess Forster. Their presence in Toronto is propitious, to say the least.

My old friends Patrick Gill, Jarrett Martineau, and Peter Orth still know how to make me forget the stresses of academe. So did Lisa Brocklebank, and she is present on these pages in uncountable ways.

In 2003, I moved from Vancouver to Toronto, and the University of Toronto's Munk Centre for International Studies quickly became a place of intellectual refuge and congeniality. I thank Marketa Evans, Lou Pauly, Wesley Wark, and especially Rick Halpern for supporting my unorthodox work; my terrific officemate Ian Cooper for conversations about politics, philosophy, and much else; and Phil Kelly for regularly interrupting his sabbatical to drink tea. During this period as a postdoctoral fellow I fortuitously encountered Whitney Lackenbauer, who shared my Arctic dreams and has become a singular collaborator and friend.

Between classes, I continued to revise this manuscript in the Department of Geography at Memorial University. I thank Keith Storey and the rest of the department's faculty, graduate students, and staff for welcoming me so generously, as well as the students in my cultural and political geography courses, who taught me more than a thing or two about Newfoundland and Labrador. Chris Lockett and Danine Farquharson stand out in my memories of what seemed like a very short year in an extraordinary environment.

The past few years of my professional life have been spent in the Department of Geography and Program in Planning at the University of Toronto. I am surrounded by a host of talented colleagues, in the department and beyond, who have inspired me, directly and indirectly, to finish the project. Two department chairs, Joe Desloges and Virginia Maclaren, made my return to U of T uncomplicated and enjoyable, and several of my peers deserve special recognition as friends and mentors: Alana Boland, Deborah Cowen, Gunter Gad, Emily Gilbert, Kanishka Goonewardena, Jason Hackworth, Debby Leslie, Robert Lewis, Minelle Mahtani, Scott Prudham, Katharine Rankin, Sue Ruddick, Matti Siemiatycki, and Rachel Silvey. The department is home to a cohort of brilliant and lively graduate students, and I single out Jason Burke, J. P. Catungal, Patrick Vitale, and those who braved my graduate courses in 2008. I have been rescued many times by the department's peerless staff, including Ayesha Alli, Mary-Marta Briones-Brid, Susan Calanza, Donna Jeynes, Siri Hansen, Bruce Huang, Marianne Ishibashi, and Marika Maslej.

I suspect that my positive impression of Toronto also has something to do with Vanessa Fleet, whom I quite appropriately first met in an archive in 2004. She read just about every word in this book, but that is the least of her contributions to my life.

The one consistent element across the past twelve years has been my family. Our numbers are not large, and we are divided by hemispheres and time zones, but when we end up in the same location (Sydney, 2007, for instance), the results are certainly impressive. Thank you to Doug, Steve L., Don, Pam, Neil, Stephanie, Vanessa W., Ursula, Steve S., and my four remarkable and much-missed grandparents, Helen, Terry, Alice, and James. I dedicate this book to Janice, William, Jillian, and Paul Farish: my reasons are infinite.

NOTES

Introduction

1. Mark Langer, "Disney's Atomic Fleet," *Animation World Magazine* 3, no. 1 (1998), http://www.awn.com. In his May 8, 1959, invitation letter to "Dick Nixon," Walt Disney mentioned the "fleet of eight atomic submarines taking people under water to the world of liquid space." http://davelandweb.com (accessed March 24, 2009).

2. For the speech, see Ira Chernus, *Eisenhower's Atoms for Peace* (College Station: Texas A&M University Press, 2002).

3. Langer, "Disney's Atomic Fleet"; Michael Smith, "Advertising the Atom," in Michael J. Lacey, ed., *Government and Environmental Politics: Essays on Historical Developments since World War Two* (Baltimore: Woodrow Wilson Center Press and Johns Hopkins University Press, 1991), 233–62.

4. Eric Schlosser, *Fast Food Nation: The Dark Side of the All-American Meal* (New York: Houghton Mifflin, 2001), 38. *The Dawn of Better Living* was produced in 1945.

5. Ibid., 39; J. P. Telotte, "Disney in Science Fiction Land," *Journal of Popular Film and Television* 33, no. 1 (2005): 12–21.

6. When "Missileman von Braun" appeared on the cover of *Time,* the magazine acknowledged his past but claimed that he had "only one interest: the conquest of space." "Reach for the Stars," *Time,* February 17, 1958, 20. For von Braun's relationship with Disney, see Michael J. Neufeld, *Von Braun: Dreamer of Space, Engineer of War* (New York: Knopf, 2007), 275–80.

7. Heinz Haber, *The Walt Disney Story of Our Friend the Atom* (New York: Simon and Schuster, 1956); Schlosser, *Fast Food Nation,* 39.

8. Haber, *The Walt Disney Story of Our Friend the Atom.*

9. Marc Eliot, *Walt Disney: Hollywood's Dark Prince* (New York: Carol, 1993), xviii. See also Schlosser, *Fast Food Nation,* 38.

10. T. M. P. [Thomas M. Pryor], review of *Victory through Air Power,* by Alexander P. de Seversky, directed by Perce Pearce, *New York Times,* July 19, 1943, http://www.nytimes.com.

11. Eliot, *Walt Disney,* 165–68; Alexander P. de Seversky, *Victory through*

Air Power (New York: Simon and Schuster, 1942); Caren Kaplan, "Mobility and War: The Cosmic View of US 'Air Power,'" *Environment and Planning A* 38 (2006): 395–407.

12. Richard Schickel, *The Disney Version: The Life, Times, Art and Commerce of Walt Disney*, rev. and updated ed. (New York: Simon and Schuster, 1985), 274.

13. Neal Gabler, *Walt Disney: The Triumph of the American Imagination* (New York: Vintage, 2006), 498.

14. Gladwin Hill, "Disneyland Gets Its Last Touches," *New York Times*, July 9, 1955, 32. "Within six months a million customers had entered the park, and by 1960 the annual number of visitors numbered around 5 million." Steven Watts, *Magic Kingdom: Walt Disney and the American Way of Life* (Boston: Houghton Mifflin, 1997), 387.

15. Quoted in Thomas M. Pryor, "Land of Fantasia Is Rising on Coast," *New York Times*, May 2, 1954, 86.

16. Warren Susman, with the assistance of Edward Griffin, "Did Success Spoil the United States? Dual Representations in Postwar America," in *Recasting America: Culture and Politics in the Age of Cold War*, ed. Lary May (Chicago: University of Chicago Press, 1989), 19–37.

17. Quoted in Pryor, "Land of Fantasia."

18. "Premier Annoyed by Ban on a Visit to Disneyland," *New York Times*, September 20, 1959, 1. See also Gladwin Hill, "The Never-Never Land Khrushchev Never Saw," *New York Times*, October 4, 1959, XX11.

19. The concept of imaginative geographies is associated with Edward Said; see his *Orientalism* (London: Penguin, 1978) and *Culture and Imperialism* (New York: Knopf, 1993). An important elaboration is Derek Gregory, "Imaginative Geographies," *Progress in Human Geography* 19, no. 4 (1995): 447–85. I have also drawn from Gregory's definition in *The Dictionary of Human Geography*, 4th ed., ed. R. J. Johnson et al. (Malden, Mass.: Blackwell, 2000), 372–73.

20. "Regime of truth" is from Michel Foucault, "Truth and Power," in *Power/Knowledge: Selected Interviews and Other Writings, 1972–1977*, ed. Colin Gordon (New York: Pantheon, 1980), 109–33. See also Nikolas Rose, *Governing the Soul: The Shaping of the Private Self*, 2nd ed. (London: Free Association Books, 1999), xiv.

21. David N. Livingstone, *The Geographical Tradition: Episodes in the History of a Contested Enterprise* (Oxford: Blackwell, 1993), 304.

22. The reference is to Benedict Anderson, *Imagined Communities: Reflections on the Origin and Spread of Nationalism*, 2nd ed. (London: Verso, 1991).

23. See Sally A. Marston, "The Social Construction of Scale," *Progress in Human Geography* 24, no. 2 (2000): 219–42.

24. Amy Kaplan, "Violent Belongings and the Question of Empire," *American Quarterly* 56, no. 1 (2004): 11.

25. As Henri Lefebvre put it, "Abstract space is not homogeneous; it simply has homogeneity as its goal, its orientation, its 'lens.'" *The Production of Space,* trans. Donald Nicholson-Smith (Cambridge, Mass.: Blackwell, 1991), 287.

26. Richard K. Ashley, "The Geopolitics of Geopolitical Space: Toward a Critical Social Theory of International Politics," *Alternatives* 12 (1987): 409.

27. Derek Gregory, *The Colonial Present: Afghanistan, Palestine, Iraq* (Malden, Mass.: Blackwell, 2004), 2. Gregory is following Michel Foucault, *The Order of Things: An Archaeology of the Human Sciences* (New York: Vintage, 1994 [1971]), xv.

28. See John Lewis Gaddis, *We Now Know: Rethinking Cold War History* (Oxford: Clarendon, 1997).

29. See the essays in Ellen Schrecker, ed., *Cold War Triumphalism: The Misuse of History after the Fall of Communism* (New York: New Press, 2004).

30. Robert J. Mayhew, review of D. Graham Burnett, *Masters of All They Surveyed: Exploration, Geography, and a British El Dorado* (2000), *Professional Geographer* 53, no. 3 (2001): 436.

31. Gearóid Ó Tuathail, "Introduction to Part Two: Cold War Geopolitics," in *The Geopolitics Reader,* 2nd ed., ed. Gearóid Ó Tuathail, Simon Dalby, and Paul Routledge (New York: Routledge, 2006), 59–73.

32. Jim George and David Campbell, "Patterns of Dissent and the Celebration of Difference: Critical Social Theory and International Relations," *International Studies Quarterly* 34, no. 3 (1990): 289.

33. Ó Tuathail, "Introduction to Part Two," 61.

34. See, for example, Neil Smith, "The Lost Geography of the American Century," *Scottish Geographical Journal* 115, no. 1 (1999): 1–18.

35. See Neil Smith's own retrieval: *American Empire: Roosevelt's Geographer and the Prelude to Globalization* (Berkeley: University of California Press, 2003). There are, of course, other lost geographies of the Cold War period that I do not address here.

36. Said, *Culture and Imperialism,* 7.

37. On this subject, see Ron Robin, *The Making of the Cold War Enemy: Culture and Politics in the Military-Industrial Complex* (Princeton, N.J.: Princeton University Press, 2001).

38. See, for instance, the BBC obituary of George Kennan, "American Cold War Architect Dies," March 18, 2005, http://news.bbc.co.uk.

39. Examples include Preston E. James, *All Possible Worlds: A History of Geographical Ideas* (Indianapolis: Odyssey Press, 1972) and the more recent and more contextual Livingstone, *The Geographical Tradition.*

40. The recent work of Trevor Barnes is a notable exception. See "Geography's Underworld: The Military-Industrial Complex, Mathematical Modelling, and the Quantitative Revolution," *Geoforum* 39 (2008): 3–16.

41. Edward Soja, *Postmodern Geographies: The Reassertion of Space in Critical Social Theory* (London: Verso, 1989), 37.

42. Philip Mirowski, *Machine Dreams: Economics Becomes a Cyborg Science* (Cambridge: Cambridge University Press, 2002), 379.

43. Joseph Rouse, *Engaging Science: How to Understand Its Practices Philosophically* (Ithaca, N.Y.: Cornell University Press, 1996), 238.

44. I am alluding to Ian Hacking, *Representing and Intervening: Introductory Topics in the Philosophy of Natural Science* (Cambridge: Cambridge University Press, 1983).

45. Foucault is quoted in Chris Philo, "History, Geography, and the 'Still Greater Mystery' of Historical Geography," in *Human Geography: Society, Space, and Social Science,* ed. Derek Gregory, Ron Martin, and Graham Smith (Minneapolis: University of Minnesota Press, 1994), 273, 272.

46. Richard K. Ashley and R. B. J. Walker, "Speaking the Language of Exile: Dissident Thought in International Studies," *International Studies Quarterly* 34, no. 3 (1990): 265.

47. Philo, "History, Geography, and the 'Still Greater Mystery,'" 273–77; Richard K. Ashley and R. B. J. Walker, "Reading Dissidence/Writing the Discipline: Crisis and the Question of Sovereignty in International Studies," *International Studies Quarterly* 34, no. 3 (1990): 388.

48. Christian G. Appy, "Introduction: The Struggle for the World," in *Cold War Constructions: The Political Culture of United States Imperialism, 1945–1966,* ed. Christian G. Appy (Amherst: University of Massachusetts Press, 2000), 1–8. The reference is to James Burnham, *The Struggle for the World* (New York: John Day Company, 1947).

49. Philo, "History, Geography, and the 'Still Greater Mystery,'" 261.

50. Tom Englehardt, *The End of Victory Culture: Cold War America and the Disillusioning of a Generation* (New York: Basic Books, 1995).

51. Michel Foucault, "The Eye of Power," in *Power/Knowledge,* 149 (original emphasis).

52. Michel Foucault, "Questions on Geography," in *Power/Knowledge,* 77.

53. Stuart Elden, *Mapping the Present: Heidegger, Foucault and the Project of a Spatial History* (London: Continuum, 2001), 3.

54. Yves Lacoste, *La Géographie, ça sert, d'abord, à faire la guerre* (Paris: Maspéro, 1976).

55. Foucault, "Questions on Geography," 77.

56. Ibid., 69.

57. Gearóid Ó Tuathail, "The Critical Reading/Writing of Geopolitics: Rereading/Writing Wittfogel, Bowman and Lacoste," *Progress in Human Geography* 18, no. 3 (1994): 326–27.

58. Michel Foucault, *"Society Must Be Defended": Lectures at the Collège de France, 1975–76,* trans. David Macey (New York: Picador, 2003), 15.

59. Edward Barrett, *Truth Is Our Weapon* (New York: Funk and Wagnalls, 1953); Richard D. McKinzie, "Oral History Interview with Edward W. Barrett," July 9, 1974, http://www.trumanlibrary.org.

60. A copy of this speech is at http://www.eisenhower.archives.gov.

61. Eisenhower said that when he warned in his famed speech of "public policy" becoming "the captive of a scientific-technological elite," he was referring to Wernher von Braun and Edward Teller. Neufeld, *Von Braun,* 352–53.

62. Foucault, *The Order of Things,* 344. I use *human sciences* and *social sciences* interchangeably.

63. Human Relations Area Files, *Laboratory for the Study of Man: Report 1949–1959* (New Haven, Conn.: Human Relations Area Files, 1959), 3.

64. Matthew G. Hannah, *Governmentality and the Mastery of Territory in Nineteenth-Century America* (Cambridge: Cambridge University Press, 2000).

65. See Richard Slotkin, *Regeneration through Violence: The Mythology of the American Frontier, 1600–1860* (Middletown, Conn.: Wesleyan University Press, 1973).

66. Karl T. Compton, address before Convocation at the University of Hawaii, March 18, 1947, in MC 416, Karl T. Compton Papers, Box 1, Folder 10, Institute Archives, Massachusetts Institute of Technology, Cambridge, Mass. (hereafter MIT). Compton's rhetoric matched that of his colleague Vannevar Bush, whose influential 1945 report to President Roosevelt was titled *Science: The Endless Frontier: A Report to the President by Vannevar Bush, Director of the Office of Scientific Research and Development, July 1945* (Washington, D.C.: United States Government Printing Office, 1945). A copy is available at http://www.nsf.gov.

67. Andrew Ross, "Introduction," in *Science Wars,* ed. Andrew Ross (Durham, N.C.: Duke University Press, 1996), 6; Daniel Bell, *The End of Ideology: On the Exhaustion of Political Ideas in the Fifties* (Glencoe, Ill.: Free Press, 1960).

68. See Ulrich Beck, *Risk Society: Towards a New Modernity,* trans. Mark Ritter (London: Sage, 1992).

69. Michel Foucault, "What Is Enlightenment?" in *The Essential Works of Foucault, 1954–1984,* vol. 1, *Ethics: Subjectivity and Truth,* ed. Paul Rabinow (New York: New Press, 1997), 315.

70. David Campbell, "International Engagements: The Politics of North American International Relations Theory," *Political Theory* 29, no. 3 (2001): 446.

71. Michael Sherry, *In the Shadow of War: The United States since the 1930s* (New Haven: Yale University Press, 1995), xi.

72. Candace Vogler and Patchen Markell, "Introduction: Violence, Redemption, and the Liberal Imagination," *Public Culture* 15, no. 1 (2003): 1–10.

73. Michael Sherry, "A Hidden-Hand Garrison State?" *Diplomatic History* 27, no. 1 (2003): 163–66.

74. Both quotes are from Catherine Lutz, *Homefront: A Military City and the American Twentieth Century* (Boston: Beacon Press, 2001), 3. See also Sherry, *In the Shadow of War,* xi.

75. David C. Engerman, "Bernath Lecture: American Knowledge and Global Power," *Diplomatic History* 31, no. 4 (2007): 599–622.

76. "Remarks by General Hoyt S. Vandenberg, Chief of Staff, U.S. Air Force, before the Board of Directors of the Advertising Council," January 15, 1953, Box 91, Binder 2, Hoyt S. Vandenberg Papers, Library of Congress Manuscript Division, Washington, D.C., (hereafter LOC), 1.

77. Said, *Orientalism,* 54.

78. David Campbell, *Writing Security: United States Foreign Policy and the Politics of Identity,* rev. ed. (Minneapolis: University of Minnesota Press, 1998), 61–62, original emphasis. Campbell is quoting Richard K. Ashley, "Foreign Policy as Political Performance," *International Studies Notes* 13, no. 2 (1987): 51–54.

79. Campbell, *Writing Security,* 169.

80. Stuart Hall, "When Was 'the Post-Colonial'? Thinking at the Limit," in *The Postcolonial Question: Common Skies, Divided Horizons,* ed. Iain Chambers and Linda Curti (London: Routledge, 1996), 252.

81. Michel de Certeau, *The Practice of Everyday Life,* trans. Steven Randall (Berkeley: University of California Press, 1984), 35–36 (original emphasis). See also Campbell, *Writing Security,* 214.

82. Archival sources, Philip Mirowski notes, are often "purged of military evidence." *Machine Dreams,* 160n.

83. Stuart W. Leslie, *The Cold War and American Science: The Military-Industrial-Academic Complex at MIT and Stanford* (New York: Columbia University Press, 1993).

84. William E. Burrows, *This New Ocean: The Story of the First Space Age* (New York: Random House, 1998).

1. Global Views

1. Denis Cosgrove, *Apollo's Eye: A Cartographic Genealogy of the Earth in the Western Imagination* (Baltimore: Johns Hopkins University Press, 2001), ix, x.

2. John Agnew, *Geopolitics: Revisioning World Politics* (London: Routledge, 1998), 11–12.

3. Daniel Lerner, "American Wehrpolitik and the Military Elite," *New Leader* (April 26, 1954): 21.

4. H. J. Mackinder, "The Geographical Pivot of History," *Geographical Journal* 23, no. 4 (1904): 422.

5. Quoted in Jonathan Haslam, *No Virtue Like Necessity: Realist Thought in International Relations since Machiavelli* (New Haven, Conn.: Yale University Press, 2002), 173.

6. Sir Halford J. Mackinder, "The Round World and the Winning of the Peace," *Foreign Affairs* 21, no. 4 (1943): 595–605. This essay was largely a reassessment of Mackinder's famous "heartland" formulation; he pronounced it "more valid and useful today than it was either twenty or forty years ago" (603). Its seeming timelessness prevented Mackinder from acknowledging that air power would effect "permanent changes in strategic conditions" (602).

7. Jeremi Suri, "The Cold War, Decolonization, and Global Social Awakenings: Historical Intersections," *Cold War History* 6, no. 3 (2006): 354.

8. Denis Cosgrove, "Contested Global Visions: *One-World, Whole-Earth,* and the Apollo Space Photographs," *Annals of the Association of American Geographers* 84, no. 2 (1994): 281.

9. Kenneth A. Osgood, "Hearts and Minds: The Unconventional Cold War," *Journal of Cold War Studies* 4, no. 2 (2002): 95. The three categories are discussed in Gearóid Ó Tuathail and Simon Dalby, "Introduction: Rethinking Critical Geopolitics: Towards a Critical Geopolitics," in *Rethinking Geopolitics,* ed. Ó Tuathail and Dalby (London: Routledge, 1998), 1–15.

10. Quoted in Ellen Herman, *The Romance of American Psychology: Political Culture in the Age of Experts* (Berkeley: University of California Press, 1995), 135. Eisenhower's speech was delivered at Columbia University's Bicentennial Dinner on May 31, 1954.

11. Cosgrove, "Contested Global Visions," 281.

12. Dwight D. Eisenhower, "Proclaiming Our Faith Anew," *Department of State Bulletin,* February 2, 1953, 169.

13. Amy Kaplan, "Homeland Insecurities: Reflections on Language and Space," *Radical History Review* 85 (2003): 87.

14. "It's a Smaller World," *Scientific American* (July 1956): 50. This paragraph was influenced by Bruno Latour, "Circulating Reference: Sampling the Soil in the Amazon Forest," in *Pandora's Hope: Essays on the Reality of Science Studies* (Cambridge, Mass.: Harvard University Press, 1999), 24–79.

15. See Peter Hulme's review of Cosgrove, *Apollo's Eye,* in *Annals of the Association of American Geographers* 92, no. 3 (2002): 607–9; Susan Schulten, *The Geographical Imagination in America, 1880–1950* (Chicago: University of Chicago Press, 2001), 133.

16. "Air: What's in It for the U.S.?" *Time,* February 15, 1943, http://www.time.com.

17. Consolidated Vultee Aircraft Corporation, *Maps . . . and How to Under-stand Them* (New York: Consolidated Vultee Aircraft Corporation, 1943), 10 (original emphasis).

18. Archibald MacLeish, "The Image of Victory," in Hans Weigert and Vilhjalmur Stefansson, eds., *Compass of the World: A Symposium on Political Geography* (New York: MacMillan, 1944), 7. See also Cosgrove, *Apollo's Eye,* 242–47.

19. Hans Weigert and Vilhjalmur Stefansson, "Introduction," in *Compass of the World,* x.

20. Council on Books in Wartime, U.S. Office of War Information, *A War Atlas for Americans* (New York: Simon and Schuster, 1944), n.p.

21. Susan Schulten, "Richard Edes Harrison Reorients the World," in *Journeys of the Imagination,* ed. Ronald Grim and Roni Pick (Boston: Boston Public Library, 2006), 23.

22. Schulten, "Richard Edes Harrison and the Challenge to American Cartography," *Imago Mundi* 50 (1998): 185.

23. Wilbur Zelinsky, "Richard Edes Harrison, 1901–1994," *Annals of the Association of American Geographers* 85, no. 1 (1995): 190.

24. Schulten, "Richard Edes Harrison and the Challenge to American Cartography," 187; Schulten, *The Geographical Imagination in America,* 145. The most famous statement of air age internationalism, of course, was Wendell L. Willkie, *One World* (New York: Simon and Schuster, 1943): "When I say that peace must be planned on a world basis, I mean quite literally that it must embrace the earth. Continents and oceans are plainly only parts of a whole, seen, as I have seen them, from the air" (84). The map in the back of the book diagramming Willkie's travels was a Harrison-like view from above, centered on the North Atlantic. See also Schulten, "Richard Edes Harrison Reorients the World," 25.

25. Nicholas J. Spykman, *The Geography of the Peace* (New York: Harcourt, Brace and Co., 1944), 18, 19.

26. Schulten, "Richard Edes Harrison Reorients the World," 23.

27. "Perspective Maps: Harrison Atlas Gives Fresh New Look to Old World," *Life,* February 28, 1944, 56.

28. Wallace W. Atwood, "Global or World Geography," *Journal of Geography* 43, no. 6 (1944): 202.

29. W. H. Lawrence, "Roosevelt to Warn U.S. of Danger; Asks All to Trace Talk on Maps," *New York Times,* February 21, 1942, 1, 8. In December 1941, National Geographic Society president Gilbert Grosvenor presented Roosevelt with a "special map cabinet" containing maps of varying scales "mounted on rollers (like window blinds), so the Commander in Chief, turning around in his swivel chair, could

pull down a map of any area of the world he wished to study." John B. Garver Jr., "The President's Map Cabinet," *Imago Mundi* 49 (1997): 153.

30. Examples include George Renner, *Geographic Education for the Air Age (A Guide for Teachers and Administrators)* (New York: Macmillan, 1942), 1; Dan Stiles, "Why Not Teach Geography?" *Harper's*, May 1943, 626–32.

31. Richard Edes Harrison and Robert Strausz-Hupé, "Maps, Strategy, and World Politics," *Infantry Journal* 51, no. 5 (1942): 43.

32. Arthur H. Robinson, "The President's Globe," *Imago Mundi* 49 (1997): 143, 146, 152.

33. Council on Books in Wartime, *A War Atlas for Americans,* 84.

34. Both quotes are in Schulten, "Richard Edes Harrison and the Challenge to American Cartography," 174, 175.

35. Harrison was referring to one of his maps, titled "The World Divided," published in the August 1941 issue of *Fortune* (pages 48–49) before the United States entered the Second World War. He suggested that "because the U.S. has not yet taken military action, the map is political rather than strategical" (49), but this did not sit well with his visual presentation. See also Schulten, *The Geographical Imagination in America,* 215.

36. Gilbert Grosvenor, "Maps for Victory: National Geographic Society's Charts Used in War on Land, Sea, and in the Air," *National Geographic,* May 1942, 667.

37. Schulten, "Richard Edes Harrison and the Challenge to American Cartography," 185. Harrison's disdain for academic cartography continued into the 1950s; in 1953, he wrote that "cartography, as a well rounded profession, does not exist in this country." "Cartography in Art and Advertising," *Professional Geographer* 2, no. 6 (1953): 15. Still, he was "extremely well connected in American geography." Robert McMaster and Susan McMaster, "A History of Twentieth-Century American Academic Cartography," *Cartography and Geographic Information Science* 29, no. 3 (2002): 309.

38. H. W. Weigert, "Maps Are Weapons," *Survey Graphic* 30, no. 10 (1941): 528.

39. Richard Edes Harrison, "Making Maps Tell the Truth," *Travel* 80, no. 2 (1942): 10–13, 30.

40. Walter W. Ristow, "Air Age Geography: A Critical Appraisal and Bibliography," *Journal of Geography* 43, no. 9 (1944): 333, 334. Ristow moved to the Library of Congress after the Second World War.

41. Atwood, "Global or World Geography," 202.

42. See, for example, Stefan T. Possony and Leslie Rosenzweig, "The Geography of the Air," *Annals of the American Academy of Political and Social Science* 299 (May 1955): 1–11.

43. Renner, *Geographic Education for the Air Age,* 1; George Renner, "Air

Age Geography," *Harper's*, June 1943, 38. For a critical response, see W. J. Luyten, "Those Misleading New Maps," *Harper's*, October 1943, 449.

44. Ben D. Wood, "Foreword," in Hubert A. Bauer, *Globes, Maps, and Skyways (A Text for High School Students)* (New York: MacMillan, 1942), v.

45. Renner, *Geographic Education for the Air Age*, 1.

46. Richard Edes Harrison, "The War of the Maps: A Famous Cartographer Surveys the Field," *Saturday Review of Literature*, August 7, 1943, 24.

47. Richard Edes Harrison, "The Face of One World: Five Perspectives for an Understanding of the Air Age," *Saturday Review of Literature*, July 1, 1944, 5 (original emphasis). For Harrison's disagreement with George Renner, a professor of geography at Columbia's Teacher's College, see ibid., 5, 6. Renner gained infamy with a June 1942 *Collier's* article that loftily redrew the world map along cultural and imperial lines. Reactions, from Walter Lippmann to Isaiah Bowman, were intensely negative. Karen DeBres, "George Renner and the Great Map Scandal of 1942," *Political Geography Quarterly* 5, no. 4 (1986): 385–94.

48. Council on Books in Wartime, *A War Atlas for Americans*, 1. This prominent book was prepared with the assistance of the Office of War Information (OWI) and the advice of Richard Edes Harrison; the political geographer Harold Sprout wrote the text. See also Edgar A. Mowrer and Marthe Rajchman, *Global War: An Atlas of World Strategy* (New York: William Morrow, 1942), 6.

49. Nicholas J. Spykman, "Geography and Foreign Policy II," *American Political Science Review* 32, no. 2 (1938): 236.

50. Spykman, *The Geography of the Peace*, 5.

51. Army Service Forces Manual M103–1, *Geographical Foundations of National Power, Section One* (Washington, D.C.: U.S. Government Printing Office, 1944), 1, 4. See also John Agnew and Stuart Corbridge, *Mastering Space: Hegemony, Territory and International Political Economy* (London: Routledge, 1995), 63–64.

52. Agnew and Corbridge, *Mastering Space*, 95.

53. DeBres, "George Renner," 392.

54. Isaiah Bowman, "Geography vs. Geopolitics," *Geographical Review* 32 (1942): 646–58.

55. Derwent Whittlesey, *German Strategy of World Conquest* (New York: Farrar and Rinehart, 1942).

56. Ó Tuathail, *Critical Geopolitics*, 129.

57. Ibid., 111, 112.

58. Schulten, *The Geographical Imagination in America*, 136.

59. Ó Tuathail, *Critical Geopolitics*, 117; see also 131–32. For two contemporaneous assessments of Haushofer's theories, see Robert Strausz-Hupé, "Geopolitics," *Fortune*, November 1941, 111–112, 114, 116, 199; Lt. Col. Sidman P. Poole, "Geopolitik—Science or Magic," *Journal of Geography* 43, no. 1 (1944): 1–12.

During the war, Poole was chief of the Military Intelligence Division's Geographic Branch; he was on leave from his academic position at Syracuse University.

60. Ó Tuathail, *Critical Geopolitics,* 126–27, 283 n.48, 49; see also 133.

61. Ibid., 134 (my emphasis), 139.

62. Timothy Mitchell, *Rule of Experts: Egypt, Techno-Politics, Modernity* (Berkeley: University of California Press, 2002), 118.

63. Michael H. Gorn, *Harnessing the Genie: Science and Technology Forecasting for the Air Force, 1944–1986* (Washington, D.C.: Office of Air Force History, 1988), 12.

64. H. H. Arnold, "Memorandum for Dr. Von Karman — AAF Long-range Development Program," November 7, 1944, reprinted in *Air Force Magazine,* August 1984, 71.

65. *Toward New Horizons: A Report to General of the Army H. H. Arnold Submitted on Behalf of the A. A. F. Scientific Advisory Group by Th. Von Karman — Volume 1: Science, the Key to Air Supremacy,* submitted December 15, 1945, Box 58, Carl A. Spaatz Papers, LOC, 1, 5, 7.

66. Ibid., 12.

67. John Johnston, "Machinic Vision," *Critical Inquiry* 26 (Autumn 1999): 27–48; James Der Derian, "The (S)pace of International Relations: Simulation, Surveillance, and Speed," *International Studies Quarterly* 34, no. 3 (1990): 295–310.

68. H. H. Arnold, "Science and Air Power," *Air Affairs* 1, no. 2 (1946): 184.

69. H. H. Arnold, "Air Power for Peace," *National Geographic,* February 1946, 137, 144. See also Arnold, *Global Mission* (New York: Harper, 1949).

70. Cosgrove, "Contested Global Visions," 280.

71. Stephen B. Jones, "Global Strategic Views," *Geographical Review* 45, no. 4 (1955): 498. Jones is quoting from the *U.S. Air Force Manual AFM 1–2* (April 1954).

72. Sherry, *In the Shadow of War,* 81, 86.

73. Hanson W. Baldwin, *The Price of Power* (New York: Harper and Brothers, 1947), 24, 28, 142, 302, 303 (original emphasis); Burnham, *The Struggle for the World.*

74. Robert Latham, *The Liberal Moment: Modernity, Security, and the Making of Postwar International Order* (New York: Columbia University Press, 1997), 3.

75. See, for instance, Leslie W. Hepple, "The Revival of Geopolitics," *Political Geography Quarterly* 5, no. 4 (suppl.) (1986): S21–S36. Hepple admits that "decline" and "revival" occur first and foremost "at the level of geopolitical language, and it could be argued that whilst geopolitics as a term was avoided because of its Nazi connotations, geopolitical interpretation and analysis continued, but sailed under other colours as strategic studies or even political geography" (S23).

76. Derek Gregory, *Ideology, Science and Human Geography* (London: Hutchinson, 1978), 19.

77. Jones, "Views of the Political World," 309.

78. Isaiah Bowman, "A Department of Geography," *Science*, December 24, 1943, 564.

79. Jones, "Views of the Political World," 310, 311, 326.

80. The first two quotes are from Jones, "Views of the Political World," 326; the third is from Jones, "Global Strategic Views," 497. Interestingly, these two papers were combined, with very few alterations, into "Global Strategic Views," in *Military Aspects of World Political Geography* (Maxwell Air Force Base, Ala.: Air University, 1959), 38–67.

81. Jones, "Global Strategic Views," 505, 506, 508. See also Stephen B. Jones, "The Power Inventory and National Strategy," *World Politics* 6, no. 4 (1954): 421–52.

82. Stephen B. Jones, "A Unified Field Theory of Political Geography," *Annals of the Association of American Geographers* 44, no. 2 (1954): 111–23.

83. Richard Hartshorne, "The Functional Approach in Political Geography," *Annals of the Association of American Geographers* 40, no. 2 (1950): 95, 96, 103.

84. William C. Olson and A. J. R. Groom, *International Relations Then and Now: Ongoing Trends in Interpretation* (London: HarperCollins, 1991), 98–100, 104; Harald Kleinschmidt, "Realism," in *The Nemesis of Power: A History of International Relations Theories* (London: Reaktion, 2000), 195–216. Morgenthau, it should be noted, was also a prominent opponent of the Vietnam War.

85. Nicholas J. Spykman, *America's Strategy in World Politics: The United States and the Balance of Power* (New York: Harcourt, Brace, 1942).

86. Stanley Hoffmann, "An American Social Science: International Relations," *Daedalus* 106, no. 3 (1977): 44, 45, 46–47.

87. The IIS moved to Princeton University and became the Center of International Studies in 1951. On the dispute that led to the shift, see Fred Kaplan, *The Wizards of Armageddon* (New York: Simon and Schuster, 1983), 189.

88. Spykman, *The Geography of the Peace*, 43.

89. See Frederick S. Dunn to Pendleton Herring, September 29, 1947, Series III, Subseries A, Box 379, Folder "Yale University: Institute of International Studies, 1943–1957," Carnegie Corporation of New York Archives, Columbia University Rare Book and Manuscript Library, New York (hereafter CCNY).

90. The story of Yale-as-crucible is told best in Kaplan, *The Wizards of Armageddon*, 19–22. See also Gene M. Lyons and Louis Morton, *Schools for Strategy: Education and Research in National Security Affairs* (New York: Praeger, 1965).

91. Gene M. Lyons, *The Uneasy Partnership: Social Science and the Federal*

Government in the Twentieth Century (New York: Russell Sage Foundation, 1969), 172. See also Jim George, *Discourses of Global Politics: A Critical (Re)Introduction to International Relations* (Boulder: Lynne Rienner, 1994), 12, 70–71.

92. Jon Barnett, *The Meaning of Environmental Security: Ecological Politics and Policy in the New Security Era* (London: Zed Books, 2001), 27.

93. Robert D. Dean, *Imperial Brotherhood: Gender and the Making of Cold War Foreign Policy* (Amherst: University of Massachusetts, 2001), 63, 12; Arthur M. Schlesinger, *The Vital Center: The Politics of Freedom* (Boston: Houghton Mifflin, 1949).

94. Kaplan, *Wizards of Armageddon,* 223. For examinations of these themes, see Carol Cohn, "Sex and Death in the Rational World of Defense Intellectuals," *Signs: Journal of Women in Culture and Society* 12, no. 4 (1987): 687–718; K. A. Cuordileone, "'Politics in an Age of Anxiety': Cold War Political Culture and the Crisis in American Masculinity, 1949–1960," *Journal of American History* 87, no. 2 (2000): 515–45.

95. Arnold Wolfers, "Theory of International Politics: Its Merits and Advancement," May 1954, in RG (Record Group) 3.1, Series 910, Box 8, Folder 69, Rockefeller Foundation Papers, Rockefeller Archive Center, Tarrytown, New York (hereafter RF), 4.

96. Hans Morgenthau, "The Theoretical and Practical Importance of a Theory of International Relations," ibid., 5–6. Paul Nitze, who spoke about his role on the State Department's Policy Planning Staff, joined Morgenthau, Wolfers, and others at the conference.

97. George, *Discourses of Global Politics,* 84. See also Hoffmann, "An American Social Science," 47–48.

98. Harry Truman, "The Truman Doctrine," in *The Geopolitics Reader,* ed. Gearóid Ó Tuathail, Simon Dalby, and Paul Routledge (London: Routledge, 1998): 58, 59, 60; Ó Tuathail, "Introduction: Cold War Geopolitics," in ibid., 48.

99. Gearóid Ó Tuathail and John Agnew, "Geopolitics and Discourse: Practical Geopolitical Reasoning in American Foreign Policy," *Political Geography* 11, no. 2 (1992): 190–204.

100. Kennan's two statements are reprinted in Thomas H. Etzold and John L. Gaddis, eds., *Containment: Documents on American Policy and Strategy, 1945–50* (New York: Columbia University Press, 1978): 50–63, 84–90. The source of the "X" article is *Foreign Affairs* 25, no. 4 (1947): 566–82.

101. George Kennan, "Moscow Embassy Telegram #511: The Long Telegram," in Etzold and Gaddis, *Containment,* 55.

102. George Kennan, "The Sources of Soviet Conduct," in Etzold and Gaddis, *Containment,* 86–87.

103. Frank Costigliola, "'Unceasing Pressure for Penetration': Gender, Pathology, and Emotion in George Kennan's Formation of the Cold War," *Journal of American History* 83, no. 4 (1997), 1310, 1333, 1331.

104. Ó Tuathail and Agnew, "Geopolitics and Discourse," 199.

105. Anders Stephanson, "Comment on an Aspect of Pietz's Argument," *Social Text* 19–20 (1988): 79. See also John L. Gaddis, *Strategies of Containment: A Critical Appraisal of Postwar American National Security Policy* (New York: Oxford University Press, 1982), 56. Kennan was writing missives similar to his two infamous manifestos *before and during* World War II, when his words were effectively ignored. It took a peculiar conjunction of circumstances for his ideas to reach a mass audience. Anders Stephanson, "Fourteen Notes on the Very Concept of the Cold War," in Ó Tuathail and Dalby, *Rethinking Geopolitics*, 62–85; Englehardt, *The End of Victory Culture*, 91–95.

106. Lippmann is quoted in John Lewis Gaddis, "Was the Truman Doctrine a Real Turning Point?" *Foreign Affairs* 52, no. 2 (1974): 390. Burnham's reaction is summarized in Stephanson, "Fourteen Notes," 66.

107. George, *Discourses of Global Politics*, 85 (original emphasis). See also Paul N. Edwards, *The Closed World: Computers and the Politics of Discourse in Cold War America* (Cambridge, Mass.: MIT Press, 1996), 10.

108. Stephanson, "Fourteen Notes," 81.

109. Campbell, *Writing Security*, 169.

110. "NSC-68," April 14, 1950, in Etzold and Gaddis, *Containment*, 387, 384.

111. Gaddis, "Was the Truman Doctrine a Real Turning Point?" 392, 398. See also Gaddis, *Strategies of Containment*, chapter 4; Stephanson, "Fourteen Notes," 80; Campbell, *Writing Security*, 23. The differences between Nitze and Kennan, who were friends, was starkly apparent in the debate over the construction of the more powerful hydrogen bomb in the early 1950s; Nitze supported it, and Kennan did not.

112. James Capshew, *Psychologists on the March: Science, Practice, and Professional Identity in America, 1929–1969* (Cambridge: Cambridge University Press, 1999), 2.

113. Lloyd V. Berkner, "Science as the Handmaiden of Politics," lecture at St. John's College, Annapolis, February 26, 1959, in Box 18, Lloyd V. Berkner Papers, LOC, 2, 6, 8. See also Berkner, "Science and National Strength," *Bulletin of the Atomic Scientists* 9, no. 5 (1953): 154–55, 180–81.

114. C. P. Snow, *The Two Cultures and the Scientific Revolution* (New York: Cambridge University Press, 1959).

115. Peter Novick, *That Noble Dream: The "Objectivity Question" and the American Historical Profession* (Cambridge: Cambridge University Press, 1988), 293, 296–

97; David A. Hollinger, "Science as a Weapon in *Kulturkämpfe* in the United States during and after World War II," *Isis* 86 (1995): 440–54.

116. Lloyd V. Berkner, "Science and International Politics," Kennecott Lecture, University of Arizona, February 3, 1959, in Box 17, Berkner Papers, LOC, 8. See also Berkner, "Man Attempts to Understand His Environment," *Journal of Astronautics* 3, no. 3–4 (1956): 53–58.

117. Lloyd V. Berkner, "Assault on the Secrets of the Earth," *New York Times Magazine,* January 27, 1957, 15.

118. See, for example, Robert Kargon and Stuart Leslie, "Imagined Geographies: Princeton, Stanford and the Boundaries of Useful Knowledge in Postwar America," *Minerva* 32, no. 2 (1994): 121–43.

119. Karl Compton, "Some Educational Effects and Implications of the Defense Program," *Science,* October 17, 1941, 369.

120. MIT *President's Report* 90, no. 2 (1954), MIT, 10. On MIT's leading status as a nonindustrial defense contractor, see Leslie, *The Cold War and American Science,* 14.

121. James B. Conant, "The Role of Science in Our Unique Society," *Science,* January 23, 1948, 77.

122. Quoted in John Trumpbour, "Harvard, the Cold War, and the National Security State," in *How Harvard Rules: Reason in the Service of Empire,* ed. John Trumpbour (Boston: South End Press, 1989), 52.

123. Gyan Prakash, *Another Reason: Science and the Imagination of Modern India* (Princeton, N.J.: Princeton University Press, 1999), 7.

124. Peter Galison, "Physics between War and Peace," in *Science, Technology and the Military,* ed. Everett Mendelsohn, Merritt R. Smith, and Peter Weingart (Dordrecht: Kluwer Academic, 1988), 50 (original emphasis).

125. Mirowski, *Machine Dreams,* 309.

126. William H. Whyte, *The Organization Man* (Garden City, N.Y.: Doubleday, 1956); David Kaiser, "The Postwar Suburbanization of American Physics," *American Quarterly* 56, no. 4 (2001): 851–88.

127. Hollinger, "Science as a Weapon," 446.

128. Gregg Herken, "In the Service of the State: Science and the Cold War," *Diplomatic History* 24, no. 1 (2000): 107.

129. Jessica Wang, *American Science in an Age of Anxiety: Scientists, Anticommunism, and the Cold War* (Chapel Hill: University of North Carolina Press, 1999), 1, 215.

130. S. S. Schweber, "The Mutual Embrace of Science and the Military: ONR and the Growth of Physics in the United States after World War II," in Mendelsohn, Smith, and Weingart, *Science, Technology and the Military,* 5.

131. See Wang, *American Science in an Age of Anxiety*; Kai Bird and Martin Sherwin, *American Prometheus: The Triumph and Tragedy of J. Robert Oppenheimer* (New York: Vintage, 2005).

132. For examples of the varying perspectives, see Paul Forman, "Behind Quantum Electronics: National Security as a Basis for Physical Research in the United States," *Historical Studies in the Physical and Biological Sciences* 18, no. 1 (1987): 149–229; Daniel J. Kevles, "Cold War and Hot Physics: Science, Security, and the American State, 1945–56," *Historical Studies in the Physical and Biological Sciences* 20, no. 2 (1990): 239–64. A major forum of scientific activism and discussions of science–government relations was the *Bulletin of the Atomic Scientists,* which debuted in November 1945.

133. Joseph Manzione, "'Amusing and Amazing and Practical and Military': The Legacy of Scientific Internationalism in American Foreign Policy, 1945–1963," *Diplomatic History* 24, no. 1 (2000): 23.

134. Ibid., 21. On Wiener's wartime work, see Peter Galison, "The Ontology of the Enemy: Norbert Wiener and the Cybernetic Vision," *Critical Inquiry* 21 (1994): 228–66.

135. See Norbert Wiener, "A Scientist Rebels," *Bulletin of the Atomic Scientists* 3, no. 1 (1947), 31. Even after rebelling, Wiener could not resist "making himself the intellectual spark-plug of [MIT] laboratories that existed only because they served ends that he deplored." Forman, "Behind Quantum Electronics," 227.

136. Dexter Masters and Katharine Way, eds., *One World or None: A Report to the Public on the Full Meaning of the Atomic Bomb* (New York: McGraw Hill, 1946). See also Paul Boyer, *By the Bomb's Early Light: American Thought and Culture at the Dawn of the Atomic Age* (Chapel Hill: University of North Carolina Press, 1994 [1985]), 76–81.

137. Bush, *Science: The Endless Frontier*. See also Walter A. McDougall, "The Cold War Excursion of Science," *Diplomatic History* 24, no. 1 (2000): 117–27.

138. G. Pascal Zachary, *Endless Frontier: Vannevar Bush, Engineer of the American Century* (Cambridge, Mass.: MIT Press, 1999 [1997]), 223.

139. Manzione, "'Amusing and Amazing and Practical and Military,'" 32 (original emphasis).

140. U.S. National Security Training Commission, *The Price of Liberty: A Condensed Version of the First Report to the Congress by the National Security Training Commission* (Washington, D.C.: U.S. Government Printing Office, February 1952), n.p. (original emphasis).

141. Karl T. Compton, "Address of Welcome to the Photogrammetry Society at Its Meeting at the Harvard Institute of Geographical Exploration, Thursday, September 21, 1950," *Photogrammetric Engineering* 16 (December 1950): 660.

142. James B. Conant, *On Understanding Science* (New Haven: Yale University Press, 1947), xiii. See also Steve Fuller, *Thomas Kuhn: A Philosophical History for Our Times* (Chicago: University of Chicago Press, 2000), 171.

143. Mirowski, *Machine Dreams,* 157–62.

144. The two articles are Karl T. Compton, "If the Atomic Bomb Had Not Been Used," *Atlantic Monthly,* December 1946, 54–56, and Henry L. Stimson, "The Decision to Use the Atomic Bomb," *Harper's,* February 1947, 97–107. On Conant's role, see Barton J. Bernstein, "Seizing the Contested Terrain of Early Nuclear History: Stimson, Conant, and Their Allies Explain the Decision to Use the Atomic Bomb," *Diplomatic History* 17, no. 1 (1993): 35–72.

145. Mirowski, *Machine Dreams,* 156. The full title of the course was Natural Sciences 4: On Understanding Science, which led to the book of the same name. For more on the course, see James G. Hershberg, *James B. Conant: Harvard to Hiroshima and the Making of the Nuclear Age* (Stanford, Calif.: Stanford University Press, 1993), 409–11.

146. Conant, *Understanding Science,* chap. 4; Fuller, *Thomas Kuhn,* 150–78. Like Robert Oppenheimer, Conant was strongly opposed to the development of the hydrogen bomb, although neither resigned from the Atomic Energy Commission's General Advisory Committee in protest, "because," Conant wrote to a friend, "I did not want to do anything that seemed to indicate we were not good soldiers." This was a decision he later regretted. Quoted in Bird and Sherwin, *American Prometheus,* 429.

147. Quoted in Chernus, *Eisenhower's Atoms for Peace,* xii, xvii, xviii.

148. Ibid., 94, 9, 105.

149. Spencer Weart, *Nuclear Fear: A History of Images* (Cambridge, Mass.: Harvard University Press, 1988), 424; see also 171–76.

150. Ronald E. Doel and Zuoyue Wang, "Science and Technology," in *Encyclopedia of American Foreign Policy,* vol. 3, 2nd ed., ed. Alexander Delonde, Richard Dean Burns, and Fredrik Logevall (New York: Charles Scribners' Sons, 2001), 450, 451.

151. McDougall, "The Cold War Excursion of Science," 120.

152. David A. Hollinger, "From Species to Ethnos," in *Postethnic America: Beyond Multiculturalism* (New York: Basic Books, 1995), 55, 56.

153. Edward Steichen, *The Family of Man* (New York: Museum of Modern Art, 1955), 4.

154. Eric J. Sandeen, *Picturing an Exhibition: The Family of Man and 1950s America* (Albuquerque: University of New Mexico Press, 1995), 4. *Life's* circulation in the mid-1950s was about six million, and its photographs, some of which were used by Steichen, overwhelmingly depicted the white, middle-class nuclear family (ibid., 9).

155. Henry Steele Commager, *The American Mind: An Interpretation of American Thought and Character since the 1880's* (New Haven, Conn.: Yale University Press, 1950), 411. See also Sandeen, *Picturing an Exhibition,* 7.

156. Sandeen, *Picturing an Exhibition,* 58, 74–75. See also Hollinger, "From Species to Ethnos." For one of the more incisive criticisms of the exhibition, see Roland Barthes, "The Great Family of Man," in *Mythologies,* trans. Annette Lavers (New York: Hill and Wang, 1972), 100–2.

157. Sandeen, *Picturing an Exhibition,* 72, 99, 119.

158. Quoted in Stephen J. Whitfield, *The Culture of the Cold War* (Baltimore: Johns Hopkins University Press, 1991), 72.

159. Lizabeth Cohen, *A Consumers' Republic: The Politics of Mass Consumption in Postwar America* (New York: Vintage, 2004), 8, 127. Cohen is gesturing to Walt Rostow's landmark *The Stages of Economic Growth, a Non-Communist Manifesto* (Cambridge: Cambridge University Press, 1960).

160. "Red Sales," *Time,* July 6, 1959, 14.

161. I take "propaganda" from Henry Hazlitt, "Portrait of Russia?" *Newsweek,* July 20, 1959, 81. For "representational paintings," see "Peaceful Coexistence," *Time,* July 13, 1959, 11. The rest of the quotes are from "Red Sales."

162. "Peaceful Coexistence," 11.

163. Hazlitt, "Portrait of Russia?" He was referring to the 1958 World's Fair in Brussels.

164. Quoted in "Better to See Once," *Time,* August 3, 1959, 13.

165. Whitfield, *The Culture of the Cold War,* 73–74.

166. "Better to See Once," 13–15.

167. "Encounter," *Newsweek,* August 3, 1959, 16, 17.

168. "Remarks of the Vice President of the United States, Richard Nixon on the occasion of the opening of the American National Exhibition in Moscow, Sokolniki Park, July 24, 1959," MC 420, Box 6, Folder 197, MIT, 2. See also Karal Ann Marling, *As Seen on TV: The Visual Culture of Everyday Life in the 1950s* (Cambridge, Mass.: Harvard University Press, 1994), 243; Sandeen, *Picturing an Exhibition,* 4; Whitfield, *The Culture of the Cold War,* 74.

169. Scott Saul, "Outrageous Freedom: Charles Mingus and the Invention of the Jazz Workshop," *American Quarterly* 53, no. 3 (2001): 390. Berlin's essay is "Two Concepts of Liberty" (1958), in *Four Essays on Liberty* (Oxford: Oxford University Press, 1982), 118–72.

170. On Fuller, see Alex Soojung-Kim Pang, "Dome Days: Buckminster Fuller in the Cold War," in *Cultural Babbage: Technology, Time and Invention,* ed. Francis Spufford and Jenny Uglow (London: Faber and Faber, 1996), 167–92.

171. Watts, *The Magic Kingdom,* 283.

172. Sandeen, *Picturing an Exhibition,* 128, 136–37, 155.

173. Timothy Mitchell, "The World as Exhibition," *Comparative Studies in Society and History* 31 (1989): 222.

174. Simone de Beauvoir, *America Day by Day*, trans. Carol Cosman (Berkeley: University of California Press, 1999), 13, 390.

175. Timothy Brennan, *At Home in the World: Cosmopolitanism Now* (Cambridge, Mass.: Harvard University Press, 1997), 222.

176. Serge Guilbaut, *How New York Stole the Idea of Modern Art: Abstract Expressionism, Freedom and the Cold War*, trans. Arthur Goldhammer (Chicago: University of Chicago Press, 1983), 3.

177. "Bikini," *Fortune*, December 1946, 156–61.

178. Guilbaut, *How New York Stole the Idea of Modern Art*, 96.

179. Ibid., 106, 107, 158.

180. Andrew Ross, *No Respect: Intellectuals and Popular Culture* (New York: Routledge, 1989), 60.

181. Andreas Huyssen, *After the Great Divide: Modernism, Mass Culture, Postmodernism* (Bloomington: Indiana University Press, 1986), 190, 193. Perhaps adversarial American modernism did not precisely narrow; it moved further underground. See David Cochran, *America Noir: Underground Writers and Filmmakers of the Postwar Era* (Washington, D.C.: Smithsonian Institution Press, 2000).

182. Frances Stonor Saunders, *Who Paid the Piper? The CIA and the Cultural Cold War* (London: Granta, 1999), 3. For a more nuanced reading of the CIA in Europe, see Hugh Wilford, "Playing the CIA's Tune? The *New Leader* and the Cultural Cold War," *Diplomatic History* 27, no. 1 (2003): 15–34.

183. Novick, *That Noble Dream*, 295.

184. Baldwin, *The Price of Power*, 4. See also Cosgrove, "Contested Global Visions," 281.

2. Regional Intelligence

1. Carl E. Schorske, "The New Rigorism in the Human Sciences, 1940–1960," in *American Academic Culture in Transformation: Fifty Years, Four Disciplines*, ed. Thomas Bender and Carl E. Schorske (Princeton, N.J.: Princeton University Press, 1997), 309, 310.

2. See Nigel Thrift, "Taking Aim at the Heart of the Region," in Gregory, Martin, and Smith, *Human Geography*, 200–31; Trevor Barnes and Matthew Farish, "Between Regions: Science, Militarism, and American Geography from World War to Cold War," *Annals of the Association of American Geographers* 96, no. 4 (2006): 807–26.

3. Tim Cresswell, *Place: A Short Introduction* (Malden, Mass.: Blackwell, 2004), 16.

4. Achille Mbembe, "Necropolitics," trans. Libby Meintjes, *Public Culture* 15, no. 1 (2003): 11–40.

5. Bowman, "A Department of Geography," 564.

6. "World Regions in the Social Sciences" (May 1943), prepared for the Committee on World Regions, Social Science Research Council, by Earl J. Hamilton, in RG 87 (Records of the Ethnogeographic Board), Box 21, Folder "Committee on World Regions," Smithsonian Institution Archives (hereafter SIA), 11, 20; see also p. 1.

7. *Committee on World Regions, 1943; Joint Exploratory Committee on World Area Research, 1945–1946; Committee on World Area Research, 1946–1953,* Accession 1, Series 1, Subseries 19, Box 229, Folder 1386, SSRC Papers, Rockefeller Archive Center, Tarrytown, New York (hereafter SSRC), 6. For Langer's receipt of "World Regions," see Metchild Rossler, "Geographers and Social Scientists in the Office for Strategic Services (OSS) 1941–1945," in *Geography and Professional Practice,* ed. V. Berdoulay and J. A. van Ginkel (Utrecht: Nederlandse Geografishe Studies No. 206, 1996): 77.

8. Mortimer Graves, "An Ever-Normal Granary of Area Experts," April 30, 1945, in RG 87, Box 7, Folder 5, SIA, 1. See also Martin Lewis and Kären E. Wigen, *The Myth of Continents: A Critique of Metageography* (Berkeley: University of California Press, 1997), 162–63.

9. See "Information Furnished to War Agencies," in the Card Index File, RG 87, Box 2, SIA. The inquiries arrived from the OSS Agricultural section and the Army Air Corps respectively.

10. Mike Featherstone, "Archiving Cultures," *British Journal of Sociology* 51, no. 1 (2000): 170.

11. Quoted in ibid., 162.

12. Jorge Luis Borges, "The Library of Babel," in *Labyrinths: Selected Stories and Other Writings* (New York: New Directions, 1964), 51–58.

13. Thomas Richards, *The Imperial Archive: Knowledge and the Fantasy of Empire* (London: Routledge, 1993), 143.

14. Michael J. Shapiro, *Violent Cartographies: Mapping Cultures of War* (Minneapolis: University of Minnesota Press, 1997).

15. Felix Driver, "Editorial: Field-work in Geography," *Transactions of the Institute of British Geographers* 25, no. 3 (2000): 267–68.

16. Neil Smith and Anne Godlewska, "Introduction: Critical Histories of Geography," in *Geography and Empire,* ed. Godlewska and Smith (Oxford: Blackwell, 1994), 7 (original emphasis).

17. Smith, *American Empire,* xviii (original emphasis).

18. C. G. Abbot, "Report of the Secretary," in *Annual Report of the Board*

of Regents of the Smithsonian Institution . . . from the Year Ended June 30, 1943 (Washington, D.C.: U.S. Government Printing Office, 1944), SIA, 2, 4.

19. C. G. Abbot, "Smithsonian Enterprises," *Science*, November 6, 1942, 417–19; Pamela M. Henson, "The Smithsonian Goes to War: The Increase and Diffusion of Scientific Knowledge in the Pacific," in *Science and the Pacific War: Science and Survival in the Pacific, 1939–1945*, ed. Roy Macleod (Dordrecht, Netherlands: Kluwer, 2000), 27–50.

20. *Annual Report of the Board of Regents of the Smithsonian Institution*, 470.

21. "Second Oral History Interview with Frank A. Taylor, Research Associate, Office of Museum Programs, February 6/74," Frank A. Taylor Oral History, SIA, 50.

22. Herbert W. Krieger, *Island Peoples of the Western Pacific: Micronesia and Melanesia* (Washington, D.C.: Smithsonian Institution, War Background Study #16, September 1943), 1.

23. Carl A. Guthe, "The Ethnogeographic Board," *Scientific Monthly* 57, no. 2 (August 1943): 188–91. The board was operated experimentally for six months, and in December 1942, further funding applications were made. See the material in Series III, Subseries A, Box 239, Folder "National Academy of Sciences-National Research Council — Ethnogeographic Board," CCNY.

24. *Survival on Land and Sea* (Washington, D.C.: Publications Branch, Office of Naval Intelligence, United States Navy, 1943), 1. Of the first edition, 200,000 copies were "rapidly exhausted," and so were the 550,000 copies of a second, revised edition, published in 1944. "Brief Summary of the Activities of the Ethnogeographic Board for the Period August 1, 1943 — July 31, 1944," Series III, Subseries A, Box 239, Folder "National Academy of Sciences-National Research Council — Ethnogeographic Board," CCNY, 2.

25. Frederick Simpitch, "Fit to Fight Anywhere," *National Geographic*, August 1943, 233.

26. "The Earth-wide Battle," *New York Times*, January 25, 1942, E6.

27. David N. Livingstone, "Tropical Hermeneutics: Fragments for a Historical Narrative — An Afterword," *Singapore Journal of Tropical Geography* 21, no. 1 (2000): 92–98.

28. Lincoln R. Thiesmeyer and John E. Burchard, *Combat Scientists* (Boston: Little, Brown, 1947).

29. Roy Macleod, "Combat Science: OSRD's Postscript in the Pacific," in *Science and the Pacific War*, 14.

30. *A Field Collector's Manual in Natural History* (Washington, D.C.: Smithsonian Institution, 1944), 2. See also Egbert H. Walker, "Natural History in the Armed Forces," *Scientific Monthly* 61, no. 4 (1945): 307–12.

31. Henson, "The Smithsonian Goes to War," 39.

32. *Survival on Land and Sea,* 1.

33. The editorial is in RG 87, Box 1, Folder 1, SIA.

34. *Survival on Land and Sea,* 2.

35. Ibid., 53, 54.

36. Latour, "Circulating Reference," 24–79.

37. Michael Heffernan, "Mars and Minerva: Centres of Geographical Calculation in an Age of Total War," *Erkunde* 54 (2000): 320–33.

38. "A Partial List of Groups and Organizations Working on Survival Techniques (Including the securing, organizing or preparing of manuals)," May 19, 1943, RG 87, Box 1, Folder 4, SIA.

39. "Tropicana," n.d. [1943], RG 87, Box 1, Folder 9, SIA, 1.

40. Wendell C. Bennett, "The Ethnogeographic Board," *Smithsonian Miscellaneous Collections* 107, no. 1 (April 14, 1947): 1, 2.

41. Carle E. Guthe (chairman), "The Ethnogeographic Board—June 16 to October 16, 1942—A Report to the Sponsoring Institutions," RG 87, Box 13, Folder 1, SIA, 1; "Ethnogeographic Board: First Meeting [Minutes]," August 3, 1942, RG 87, Box 27, Folder "Board Meeting—Dec. 9, 1944," SIA, 1.

42. William D. Strong, "The Ethnogeographic Board," *Science,* October 23, 1942, 381.

43. Bennett, "The Ethnogeographic Board," 3. Isaiah Bowman, one of the few geographers on the board, never attended meetings due to other obligations and "lack of interest." Ibid., 13.

44. "Anthropology during the War and After," draft report received by the NRC on March 23, 1943, RG 87, Box 17, Folder "NRC Committee," SIA, 10–11. Strong resigned on July 31, 1944, and returned to Columbia University; Arctic specialist Henry B. Collins replaced him.

45. Richards, *The Imperial Archive,* 21; Clifford Geertz, "Thick Description: Toward an Interpretive Theory of Culture," in *The Interpretation of Cultures: Selected Essays* (New York: Basic Books, 1973), 3–30.

46. Bennett, "The Ethnogeographic Board," 63. See also Guthe, "The Ethnogeographic Board," 190.

47. W. Duncan Strong, "Director's Report of Progress for the Period January 14 to August 1, 1943, the Ethnogeographic Board," Series III, Subseries A, Box 239, Folder "National Academy of Sciences–National Research Council—Ethnogeographic Board," CCNY, 3.

48. Bennett, "The Ethnogeographic Board," 86–90, 96.

49. These lists are all in RG 87, Box 4, SIA.

50. "Here's a Spot if You Know Strange Lands," *Washington Post,* October 4,

1942, R4 (also clipped in RG 87, Box 13, Folder 2, SIA). See also the form letter written by Henry B. Collins, dated June 13, 1945, in Box 154, Folder "Ethnogeographic Board — Responses to Requests," Henry Bascom Collins Papers, National Anthropological Archives, Suitland, Md. (hereafter NAA).

51. For the groups consulted by the board, a discussion of the Area Roster, and sample questionnaires, see Bennett, "The Ethnogeographic Board," 27–38, 116–24.

52. Ibid., 40. See also George P. Murdock, "The Cross-Cultural Survey," *American Sociological Review* 5, no. 3 (1940): 361–70. The idea behind the survey dated to 1928, when Murdock decided to build a bibliography of all known cultures. Rebecca Lemov, *World as Laboratory: Experiments with Mice, Mazes, and Men* (New York: Hill and Wang, 2005), 147, 151.

53. Richards, *The Imperial Archive,* 6.

54. Lemov, *World as Laboratory,* 7.

55. Clellan S. Ford, "Human Relations Area Files: 1949–1969 — A Twenty-Year Report," *Behavioral Science Notes* 5 (1970): 5.

56. The reference here is to Schorske, "The New Rigorism."

57. Lemov, *World as Laboratory,* 149.

58. Quoted in Henson, "The Smithsonian Goes to War," 32. Other questions included *Where are the best beaches and water sources? How do you eat a coconut? Which of the locals may be friendly to the Japanese?* and *When should you pat a native on the head?* Lemov, *World as Laboratory,* 159.

59. "Brief Summary of the Activities of the Ethnogeographic Board for the Period August 1, 1943 — July 31, 1944," CCNY, 1; *Function and Scope of the Human Relations Area Files, Inc.* (New Haven, Conn.: Human Relations Area Files, 1953), 4; Bennett, "The Ethnogeographic Board," 40–41; George P. Murdock et al., *Outline of Cultural Materials,* 3rd rev. ed. (New Haven, Conn.: Human Relations Area Files, 1950), xiii.

60. Murdock et al., *Outline of Cultural Materials,* xiii. The Strategic Index of Latin America was subsequently renamed the Strategic Index of the Americas.

61. Quoted in Barry M. Katz, *Foreign Intelligence: Research and Analysis in the Office of Strategic Services, 1942–1945* (Cambridge, Mass.: Harvard University Press, 1989), 18. See also Lemov, *World as Laboratory,* 158.

62. John Borneman, "American Anthropology as Foreign Policy," *American Anthropologist* 97, no. 4 (1995): 665. The HRAF project continues; see http://www.yale/edu/hraf.

63. George P. Murdock, *Social Structure* (New York: MacMillan, 1949).

64. Robin W. Winks, *Cloak and Gown: Scholars in the Secret War* (New York: William Morrow, 1987), 45.

65. *Function and Scope of the Human Relations Area Files,* 3.

66. Derek Gregory, "Power, Knowledge and Geography," in *Explorations in Critical Human Geography: Hettner-Lecture 1997* (Heidelberg: Department of Geography, University of Heidelberg, 1998), 9–40.

67. Lemov, *World as Laboratory,* 149.

68. For an example, see Military Government Handbook OPNAV 50E-1, *Marshall Islands* (Washington, D.C.: Office of the Chief of Naval Operations, August 17, 1943). In Britain, teams composed chiefly of Oxford and Cambridge geographers produced a set of similar Handbooks for the Intelligence Division of the Royal Navy. Thirty-one large, area-specific handbooks (a total of 58 volumes) were published. Hugh Clout and Cyril Gosme, "The Naval Intelligence Handbooks: A Monument in Geographical Writing," *Progress in Human Geography* 27, no. 2 (2003): 153–73.

69. Ford, "Human Relations Area Files," 9.

70. By the mid-1960s, Yale's file held 65.8 million of these cards. Lemov, *World as Laboratory,* 148.

71. Clellan Ford to John W. Gardner, January 6, 1955, Series III, Subseries A, Box 174, Folder "Human Relations Area Files," CCNY, 2. See also Murdock et al., *Outline of Cultural Materials,* xv; Eric R. Wolf and Joseph G. Jorgensen, "A Special Supplement: Anthropology on the Warpath in Thailand," *New York Review of Books,* November 19, 1970, http://www.nybooks.com.

72. Ira Bashkow, "The Dynamics of Rapport in a Colonial Situation: David Schneider's Fieldwork on the Islands of Yap," in *Colonial Situations: Essays on the Contextualization of Ethnographic Knowledge,* ed. George W. Stocking Jr. (Madison: University of Wisconsin Press, 1991), 180–85; Lemov, *World as Laboratory,* 179. In the waning months of the war, Murdock "actually served as a police officer in military affairs, not just civil affairs, at Okinawa," where his "anthropological and Institute experience" was, in his words, "most useful." Lemov, *World as Laboratory,* 158, 160.

73. Ronald Rainger, "Science at the Crossroads: The Navy, Bikini Atoll, and American Oceanography in the 1940s," *Historical Studies in the Physical and Biological Sciences* 30, no. 2 (2000): 349–71; Lemov, *World as Laboratory,* 170–87.

74. George P. Murdock, "New Light on the Peoples of Micronesia," *Science,* October 22, 1948, 423.

75. Quoted in Bashkow, "The Dynamics of Rapport," 182.

76. Murdock, "New Light," 424. Yale's John Embree challenged his colleague's colonialist voice: "A Note on Ethnocentrism in Anthropology," *American Anthropologist* 52, no. 3 (1950): 431.

77. Schuyler C. Wallace, "The Naval School of Military Government and Administration," *Annals of the American Academy of Political and Social Science*

231 (January 1944): 32. See also Sydney Connor, "The Navy's Entry into Military Government," *Annals of the American Academy of Political and Social Science* 267 (January 1950): 8–18. According to Murdock, more officers serving in the Trust Territory passed through the School of Naval Administration at Stanford University; "New Light," 423.

78. Murdock et al., *Outline of Cultural Materials,* xxi, xxii, xxiii.

79. "History of the Beginnings of Human Relations Area Files," October 1949, Series III, Subseries A, Box 174, Folder "Human Relations Area Files," CCNY, 1.

80. Lemov, *World as Laboratory,* 173.

81. Strong, "Director's Report of Progress," CCNY. See also Bennett, "The Ethnogeographic Board," 77–78. The anthropologist Elizabeth Bacon also worked on the survey of university area courses, but she did not receive authorial credit for any publications. "Brief Summary of the Activities of the Ethnogeographic Board for the Period August 1, 1943 — July 31, 1944," CCNY, 7.

82. Bennett, "The Ethnogeographic Board," 110.

83. Carl E. Guthe, "The Future of the Ethnogeographic Board," September 8, 1944, in Box 153, Folder "Ethnogeographic Board — John E. Graf's File, 1944–1946," Collins Papers, NAA, 2.

84. Lewis and Wigen, *The Myth of Continents,* 163.

85. Ibid., 166–67.

86. For the other components of the ASTP, see U.S. Army Service Forces, Army Specialized Training Division, *Essential Facts about the Army Specialized Training Program* (Washington, D.C.: U.S. Government Printing Office, 1943), 5.

87. Wayland J. Hayes and Werner J. Cahnman, "Foreign Area Study (ASTP) as an Educational Experiment in the Social Sciences," *Social Forces* 23, no. 2 (1944): 160. See also Bennett, "The Ethnogeographic Board," 107; William N. Fenton, *Area Studies in American Universities* (Washington, D.C.: American Council on Education, 1947), preface; Robert J. Matthew, *Language and Area Studies in the Armed Services: Their Future Significance* (Washington, D.C.: American Council on Education, 1947), 4.

88. Quoted in Louis E. Keefer, *Scholars in Foxholes: The Story of the Army Specialized Training Program in World War II* (Jefferson, N.C.: McFarland, 1988), 38.

89. Hayes and Cahnman, "Foreign Area Study," 163.

90. William Nelson Fenton, "Integration of Geography and Anthropology in Army Area Study Curricula," *American Association of University Professors Bulletin* 32, no. 4 (1946). For the topics covered in the geography, history, and "institutions" and "cultures" curriculum categories, see Matthew, *Language and Area Studies,* 79–81.

91. Fenton, *Area Studies,* 81.

280 NOTES TO CHAPTER 2

92. Svend Riemer, "Individual and National Psychology: A Problem in the Army Area Study," *Social Forces* 22, no. 3 (1944): 256.

93. Fenton, "Integration," 698.

94. Hayes and Cahnman, "Foreign Area Study," 161.

95. Fenton, "Integration," 702.

96. *The Army Specialized Training Program to June 1944: ARMY Specialized Training Bulletin No. 8* (Washington, D.C.: Headquarters, Army Service Forces, June 1944), 1.

97. Army Service Forces Manual M-101, *Atlas of World Maps, Army Specialized Training Program* (Washington, D.C.: Headquarters, Army Service Forces, November 1943), 1. In April 1944, the Army having "reached its authorized strength," the ASTP's Basic Phase was terminated and replaced by a program for reserves. *The Army Specialized Training Program to June 1944*, 1, 4–5.

98. Army Service Forces Manual M103–1, *Geographical Foundations of National Power, Section One*, vii, viii. One year later, Harold and Margaret Sprout released *Foundations of National Power: Readings on World Politics and American Security* (Princeton, N.J.: Princeton University Press, 1945), a collection intended to extend the same concerns into the postwar period. A second, fully revised edition was published in 1951.

99. Fenton, "Integration," 703, 704.

100. Matthew, *Language and Area Studies*, xii.

101. Hayes and Cahnman, "Foreign Area Study," 162. See also Charles S. Hyneman, "The Wartime Area and Language Courses," *American Association of University Professors Bulletin* 31, no. 3 (1945): 438–39.

102. Army Service Forces, Army Specialized Training Division, *Fifty Questions and Answers on Army Specialized Training Program* (Washington, D.C.: U.S. Government Printing Office, 1943), 3.

103. Herman, *The Romance of American Psychology*, 305.

104. Federico Neiburg and Marcio Goldman, "Anthropology and Politics in Studies of National Character," trans. Peter Gow, *Cultural Anthropology* 13, no. 1 (1998): 69, 70.

105. On the Foreign Morale Analysis Division, established in the spring of 1944, see Alexander H. Leighton, *Human Relations in a Changing World: Observations on the Use of the Social Sciences* (New York: Dutton, 1949). Kluckhohn left the Division in the summer of 1945 "to become director of an OWI over-all planning and coordinating unit for psychological warfare to Japan, [which] provided a strategic link between the research and planning programs" (297).

106. Ruth Benedict, *The Chrysanthemum and the Sword: Patterns of Japanese Culture* (New York: Houghton Mifflin, 1946); Margaret Mead, *And Keep Your Powder Dry: An Anthropologist Looks at America* (New York: Morrow, 1942).

107. Christopher Shannon, "A World Made Safe for Differences: Ruth Benedict's *The Chrysanthemum and the Sword*," *American Quarterly* 47, no. 4 (1995): 662. See also Virginia Yans-McLaughlin, "Science, Democracy, and Ethics: Mobilizing Culture and Personality for World War II," in *Malinowski, Rivers, Benedict and Others: Essays on Culture and Personality*, ed. George W. Stocking Jr. (Madison: University of Wisconsin Press, 1986), 207; Judith S. Modell, *Ruth Benedict: Patterns of a Life* (Philadelphia: University of Pennsylvania Press, 1983), 268, 276. On Boas and World War One, see David Price, "Lessons from Second World War Anthropology: Peripheral, Persuasive and Ignored Contributions," *Anthropology Today* 18, no. 3 (2002): 14.

108. Carleton Mabee, "Margaret Mead and Behavioral Scientists in World War II: Problems in Responsibility, Truth, and Effectiveness," *Journal of the History of the Behavioral Sciences* 23, no. 1 (1987): 10. Some of the interviews were conducted with residents of United States internment camps, ibid., 6–7; Orin Starn, "Engineering Internment: Anthropologists and the War Relocation Authority," *American Ethnologist* 13, no. 4 (1986): 700–20.

109. John F. Embree, "Applied Anthropology and Its Relation to Anthropology," *American Anthropologist* 47, no. 4 (1945): 636. Embree's harsh criticism of national character methods is partially contradicted by his 1943 booklet *The Japanese*, published as Number 7 in the Smithsonian's *War Background Studies* series, where he traces Japanese insecurity and rage to practices of child-rearing. John W. Dower, *War without Mercy: Race and Power in the Pacific War* (New York: Pantheon, 1986), 128.

110. Price, "Lessons from Second World War Anthropology," 18; Yans-McLaughlin, "Science, Democracy, and Ethics," 196.

111. On Kluckhohn, see Dower, *War without Mercy*, 138–39. See also Price, "Lessons from Second World War Anthropology," 19; Mabee, "Margaret Mead," 8–9.

112. John W. Dower, *Embracing Defeat: Japan in the Wake of World War II* (New York: Norton, 2000), 219, 220. See also Dower, *War without Mercy*, 118–46. At a December 1944 conference on Japanese Character Structure in New York, attended by over forty prominent social scientists, Talcott Parsons and Margaret Mead both described Japanese conformity as analogous to adolescent behavior in the United States. Ibid., 131–32.

113. Shannon, "A World Made Safe for Differences," 675.

114. Pendleton Herring, "The Social Sciences in Modern Society," *Items* 1, no. 1 (1947): 6.

115. See Dower, *Embracing Defeat*.

116. Leonard W. Doob, "The Utilization of Social Scientists in the Overseas Branch of the Office of War Information," *American Political Science Review* 41,

no. 4 (1947): 652. Doob, a Yale psychologist, was Ruth Benedict's supervisor for a time at the OWI.

117. Neiburg and Goldman. "Anthropology and Politics," 65. See also Modell, *Ruth Benedict,* 269, 277.

118. Benedict, *The Chrysanthemum and the Sword,* 20–21, 28; Modell, *Ruth Benedict,* 278–79, 281; Shannon, "A World Made Safe for Differences," 667; Neiburg and Goldman, "Anthropology and Politics," 65–66.

119. Modell, *Ruth Benedict,* 292. Notwithstanding "a consistent record of Mead's service contributing to America's Cold War military status quo, the FBI investigated her as a possible communist or subversive largely because she articulated and advocated anthropological views regarding the equality of all peoples." David Price, *Threatening Anthropology: McCarthyism and the FBI's Surveillance of Activist Anthropologists* (Durham, N.C.: Duke University Press, 2004), 255.

120. Margaret Mead and Rhoda Métraux, eds., *The Study of Culture at a Distance* (Chicago: University of Chicago Press, 1953). See also Wolf and Jorgensen, "A Special Supplement."

121. Mead and Métraux, eds., *The Study of Culture at a Distance,* 397.

122. This section builds on Barnes and Farish, "Between Regions," 815.

123. JANIS 75, Chapter I, *Joint Army-Navy Intelligence Study of Korea: Brief* (Washington, D.C.: Joint Intelligence Study Publishing Board, April 1945), 3. A full copy is in Box 1 of the Leonard S. Wilson Collection, Georgetown University Special Collections (hereafter LWGU). Wilson, a Carleton College geography professor, was a lieutenant in the U.S. Naval Reserve and Deputy Chief of the Map Division (where he directed the Map Information Section) of the OSS's Research and Analysis Branch during World War II.

124. On oceanography, see Rainger, "Science at the Crossroads." On Sverdrup's complicated relationship with the military, see Naomi Oreskes and Ronald Rainger, "Science and Security before the Atomic Bomb: The Loyalty Case of Harald U. Sverdrup," *Studies in the History and Philosophy of Modern Physics* 31, no. 3 (2000): 309–69.

125. Richard Hartshorne et al., "Lessons from the War-time Experience for Improving Graduate Training for Geographic Research," *Annals of the Association of American Geographers* 36 (1946): 198.

126. Joseph A. Russell, "Military Geography," in *American Geography: Inventory and Prospect,* ed. Preston E. James and Clarence F. Jones (Syracuse, N.Y.: Syracuse University Press, 1954), 491.

127. This capsule history is drawn from the Web site for the *World Factbook,* the annual summary and update produced to complement Intelligence Studies (http://www.cia.gov/library/publications/the-world-factbook). See also "Directive on Joint Army and Navy Intelligence Studies (JANIS)," July 1, 1943, RG 226 (Records of

the Office of Strategic Services), Entry 001, Box 1, Folder 9, National Archives and Records Administration, College Park, Md. (hereafter NARA); M. Crane, "The National Intelligence Surveys," n.d., RG 59 (Records of the Department of State), Entry 1595, Box 9, Lot 69D 267, NN3-93-102, NARA.

128. Quoted in Barnes and Farish, "Between Regions," 815.

129. Chauncy D. Harris, "Edward Louis Ullman, 1912–1976," *Annals of the Association of American Geographers* 67, no. 4 (1977): 595–600.

130. Edward L. Ullman, "Notes on Organization of Topographical Intelligence," December 13, 1944, RG 226, Entry 001, Box 1, Folder 18, NARA. One history notes the rampant discord in 1943 between regionalists and functionalists. Bradley F. Smith, *The Shadow Warriors: O.S.S. and the Origins of the C.I.A.* (New York: Basic Books, 1983), 364.

131. Ullman, "Notes on Organization of Topographical Intelligence" (original emphasis).

132. President Franklin Roosevelt established the Office of the Co-ordinator of Information (COI) in July 1941; in June 1942 it was reconfigured as the OSS.

133. Trevor J. Barnes, "Geographical Intelligence: American Geographers and Research and Analysis in the Office of Strategic Services, 1941–1945," *Journal of Historical Geography* 32 (2006): 150. Marcuse, Neumann, and other émigrés joined R&A in 1943, after extensive security vetting. Barry M. Katz, "The Criticism of Arms: The Frankfurt School Goes to War," *Journal of Modern History* 59, no. 3 (1987): 439–78.

134. Quoted in both Smith, *The Shadow Warriors,* 362, and Katz, *Foreign Intelligence,* 17. See also Barry Katz, *Herbert Marcuse and the Art of Liberation: An Intellectual Biography* (London: Verso, 1982), 112.

135. "Memorandum on the Functions of the Research and Analysis Branch," October 30, 1942, RG 226, Entry 145, Box 2, Folder 24, NARA.

136. Quoted in the CIA's own history, *The Office of Strategic Services: America's First Intelligence Agency* (2000), http://www.cia.gov. See also Barnes, "Geographical Intelligence," 152.

137. On the R&A's diversification and its limits, see Smith, *The Shadow Warriors,* 378–79.

138. Ibid., 362. See also Novick, *That Noble Dream,* 281.

139. Quoted in Barnes, "Geographical Intelligence," 153.

140. William L. Langer, "Scholarship and the Intelligence Problem," *Proceedings of the American Philosophical Society* 92, no. 1 (1948): 44.

141. Quoted in Smith, *The Shadow Warriors,* 363.

142. Barnes, "Geographical Intelligence," 154.

143. Quoted in Winks, *Cloak and Gown,* 115. See also Sigmund Diamond, *Compromised Campus: The Collaboration of Universities with the Intelligence Community, 1945–1955* (New York: Oxford University Press, 1992), 10, 53.

144. Smith, *The Shadow Warriors*, 387–88. The direct link between the OSS and subsequent area studies research must be supplemented with an acknowledgment of the equivalent postwar emphasis placed on *training*, a mandate derived more from programs like the ASTP. David C. Engerman, "The Ironies of the Iron Curtain: The Cold War and the Rise of Russian Studies in the United States," *Cahiers du Monde russe* 45, no. 3–4 (2004): 468.

145. Smith, *The Shadow Warriors*, 372, 382.

146. On Robinson and the Map Division, see Arthur H. Robinson, "Geography and Cartography Then and Now," *Annals of the Association of American Geographers* 69, no. 1 (1979): 97–102; Chauncy D. Harris, "Geographers in the U.S. Government in Washington, DC, during World War II," *Professional Geographer* 49, no. 2 (1997): 245–56.

147. Barnes, "Geographical Intelligence," 153, 162.

148. Meredith F. Burrill, "Reorganization of the United States Board on Geographical Names," *Geographical Review* 35, no. 4 (1945): 647.

149. Leonard S. Wilson, "Lessons from the Experience of the Map Information Section, OSS," *Geographical Review* 39, no. 2 (1949): 307, 310. See also Wilson, "Assignment in Britain," October 20, 1944, in Box 3, LWGU. For a description of the Map Division and the other components of R&A, see "Research and Analysis Branch—Organization," August 1, 1943, RG 226, Entry 145, Box 2, Folder 24, NARA.

150. Leonard S. Wilson, "Library Filing, Classification, and Cataloging of Maps, with Special Reference to Wartime Experience," *Annals of the Association of American Geographers* 38, no. 1 (March 1948): 8.

151. Carleton S. Coon, *A North Africa Story: The Anthropologist as OSS Agent, 1941–1943* (Ipswich, Mass.: Gambit, 1980), 3.

152. One example is Captain Stockbridge H. Barker, *So You're Going Overseas!* (Washington, D.C.: Infantry Journal, 1944).

153. Special Services Division, Army Service Forces, United States Army, in cooperation with the Office of Strategic Services, *A Pocket Guide to Alaska* (Washington, D.C.: War and Navy Departments, 1943), 3.

154. Samuel A. Graham and Earl C. O'Roke, *On Your Own: How to Take Care of Yourself in Wild Country—A Manual for Field and Service Men* (Minneapolis: University of Minnesota Press, 1943), 148.

155. Yans-McLaughlin, "Science, Democracy, and Ethics," 203.

156. Quoted in David Price, "Gregory Bateson and the OSS: World War II and Bateson's Assessment of Applied Anthropology," *Human Organization* 57, no. 4 (1998): 382. See also Yans-McLaughlin, "Science, Democracy, and Ethics," 202–3; Smith, *The Shadow Warriors*, 391.

157. Memo, Gregory Bateson to General Donovan, "Influence of Atomic

Bomb on Indirect Methods of Warfare," August 18, 1945, RG 263, Entry 15, Box 2, Folder 35, NARA (original emphasis).

158. See William J. Donovan, "Strategic Services in 'Cold War,'" *Naval War College Review* 6, no. 1 (1953): 31–42. In Graham Greene's 1955 novel *The Quiet American,* one character asks another if Alden Pyle, the title character, is a member of the OSS. The reply is telling: "The initial letters are not very important. I think now they are different."

159. Ray S. Cline, *Secrets, Spies and Scholars: Blueprint of the Essential CIA* (Washington, D.C.: Acropolis Books, 1976), 47. See also Rhodri Jeffreys-Jones, *Cloak and Dollar: A History of American Secret Intelligence* (New Haven, Conn.: Yale University Press, 2002), 117–18.

160. Jeffreys-Jones, *Cloak and Dollar,* 142.

161. Katz, *Foreign Intelligence,* xii.

162. Jeffreys-Jones, *Cloak and Dollar,* 142. See also Smith, *The Shadow Warriors,* 365. The presence of OSS components in the CIA followed a failed transplant (the Interim Research Intelligence Service) into the State Department, although it was there that the "Donovan legacy" was kept alive. Ibid., 409.

163. Hartshorne's blustery opinion and Langer's unhelpful reply are in RG 226, Entry 001, Box 4, Folder 1, NARA. See also Smith, *The Shadow Warriors,* 364; Barnes, "Geographical Intelligence," 156.

164. Barry M. Katz, "The OSS and the Development of the Research and Analysis Branch," in *The Secrets War,* 46, 47. See also Jeffreys-Jones, *Cloak and Dollar,* 146.

165. Edward L. Ullman, "The Future of JANIS," March 1946 [probably March 2], RG 263 (Records of the Central Intelligence Agency), Entry 17, Box 5, Folder 65, NARA.

166. The SSRC Problems and Policy Minutes are in the bound volume *Committee on World Regions, 1943,* 60–61.

167. See Andrew Kirby, "What Did You Do in the War, Daddy?" in Godlewska and Smith, *Geography and Empire,* 300–15.

168. Preston E. James, "The Service of Geography in Government," in *The Clark Graduate School of Geography: Our First Twenty-Five Years,* ed. Wallace W. Atwood (Worcester, Mass.: Clark University, 1946), 51.

169. Kirby, "What Did You Do in the War, Daddy?" 304.

170. Derwent Whittlesey, "War, Peace, and Geography—An Editorial Foreword," *Annals of the Association of American Geographers* 31, no. 2 (1941): 79.

171. Eugene Van Cleef, "Training for Geographic Research: A Symposium," *Annals of the Association of American Geographers* 34, no. 4 (1944): 181.

172. George F. Deasy, "Training, Professional Work, and Military Experience of Geographers, 1942–1947," *Professional Geographer* 6 (December 1947): 1–14.

173. Quoted in Schulten, *The Geographical Imagination in America*, 210.

174. Stephen B. Jones, "Field Geography and Postwar Political Problems," *Geographical Review* 33 (1943): 446.

175. Kirk H. Stone, "Geography's Wartime Service," *Annals of the Association of American Geographers* 69, no. 1 (1979): 89.

176. Kirby, "What Did you Do in the War, Daddy?" 311.

177. Edward A. Ackerman, "Geographic Training, Wartime Research, and Professional Objectives," *Annals of the Association of American Geographers* 35, no. 4 (1945): 121, 127, 122, 129. See also Hartshorne et al., "Lessons," 204; James, *All Possible Worlds*, 451. As Ackerman noted, those publications identified as "systematic" before the Second World War had been composed by "amateurs," or were in physical geography, and those that did not fit either of these two categories invariably focused on only the United States. "Geographic Training," 124–25.

178. Barnes, "Geographical Intelligence," 158.

179. Leonard S. Wilson, "Geographic Training for the Postwar World: A Proposal," *Geographical Review* 38 (1948): 577.

180. Ackerman, "Geographic Training," 127, 139. For a related distinction between "geographer cartographers" and "carto-technicians," see Erwin Raisz, "Introduction," *Professional Geographer* 2, no. 6 (1950): 10. There was a gendered implication to Ackerman's discussion. F. Webster McBryde, who worked in the Geographic Branch of the War Department's Military Intelligence unit, recalled an interview with the editor of *Mademoiselle* for an article on geography's wartime role, "especially in aspects of most interest and suitability to women, such as cartography and drafting." "Origin of the American Society for Professional Geographers: Take-Over and Cover-Up by AAG Number 1," in *The American Society for Professional Geographers: Papers Presented on the Occasion of the Fiftieth Anniversary of Its Founding*, ed. E. Willard Miller (Washington, D.C.: Association of American Geographers, April 1993), 10.

181. Harris, "Geographers in the U.S. Government," 249.

182. Ackerman, "Geographic Training," 134, 129. See also Wilson, "Geographic Training."

183. Charles W. J. Withers, "Constructing 'the Geographical Archive,'" *Area* 34, no. 3 (2002): 304.

184. Matthew, *Language and Area Studies*, 172, 173.

185. "Social Science Considerations in the Planning of Regional Specialization in Higher Education and Research," March 10, 1944, RG 3.2, Series 900, Box 31, Folder 165, RF, 2. Rockefeller was the only foundation active in "non-Western studies" before World War II, but the funding and intellectual approach were deemed inadequate during the war. George M. Beckmann, "The Role of the

Foundations," *Annals of the American Academy of Political and Social Science* 365 (November 1964): 12–22.

186. Matthew, *Language and Area Studies,* 150.

3. Illuminating the Terrain

1. Lewis and Wigen, *The Myth of Continents,* ix–x. On the concept of a third world and its Cold War associations, see Odd Arne Westad, *The Global Cold War: Third World Interventions and the Making of Our Times* (Cambridge: Cambridge University Press, 2007), 2–3.

2. John C. Flanagan et al., *Current Trends: Psychology in the World Emergency* (Pittsburgh, Pa.: University of Pittsburgh Press, 1952).

3. Carl E. Pletsch, "The Three Worlds, or the Division of Social Scientific Labor, Circa 1950–1975," *Comparative Studies in Society and History* 23, no. 4 (1981): 582. See also Novick, *That Noble Dream,* 293.

4. James Bamford, *Body of Secrets: Anatomy of the Ultra-Secret National Security Agency* (New York: Anchor, 2002), 37.

5. Donald Young, "Organization for Research in the Social Sciences in the United States," *UNESCO International Social Science Bulletin* 1, no. 3–4 (1949): 105, 99–100. Young was director of the Russell Sage Foundation.

6. Henry A. Kissinger, "Military Policy and Defense of the 'Grey Areas,'" *Foreign Affairs* 33, no. 3 (1955): 416–28.

7. Bill Readings, *The University in Ruins* (Cambridge, Mass.: Harvard University Press, 1996), 45. See also Vincente L. Rafael, "The Cultures of Area Studies in the United States," *Social Text* 41 (1994): 94–95; Engerman, "Bernath Lecture," 607.

8. Ann Laura Stoler with David Bond, "Refractions Off Empire: Untimely Comparisons in Harsh Times," *Radical History Review* 95 (2006): 96.

9. Bruce Cumings, *Parallax Visions: Making Sense of American–East Asian Relations at the End of the Century* (Durham, N.C.: Duke University Press, 1999), 173–204.

10. Mary Poovey, "For What It's Worth . . ." *Critical Inquiry* 30, no. 2 (2004): 431.

11. Sherman Kent, *Strategic Intelligence for American World Policy* (Princeton, N.J.: Princeton University Press, 1949), 116, 119, 121.

12. Christopher Simpson, "U.S. Mass Communication Research, Counterinsurgency, and Scientific 'Reality,'" in *Ruthless Criticism: New Perspectives in U.S. Communication History,* ed. William S. Solomon and Robert W. McChesney (Minneapolis: University of Minnesota Press, 1993), 325.

13. William T. R. Fox, "Civil–Military Relations Research: The SSSC Committee and Its Research Survey," *World Politics* 6, no. 2 (1954): 279.

14. Roger L. Geiger, "American Foundations and Academic Social Science, 1945–1960," *Minerva* 26, no. 3 (1988): 315.

15. Simpson, "U.S. Mass Communication Research," 330.

16. Robin, *The Making of the Cold War Enemy,* 5.

17. Mark Solovey, "Riding Natural Scientists' Coattails onto the Endless Frontier: The SSRC and the Quest for Scientific Legitimacy," *Journal of the History of the Behavioral Sciences* 40, no. 4 (2004): 393–422.

18. Lloyd V. Berkner, "Can the Social Sciences Be Made Exact?" *Proceedings of the Institute of Radio Engineers* 48, no. 8 (1960): 1377.

19. Jean Converse, *Survey Research in the United States: Roots and Emergence, 1890–1960* (Berkeley: University of California Press, 1987), 1. See also Robin, *The Making of the Cold War Enemy,* 6–7.

20. Trevor J. Barnes and Matthew Hannah, "The Place of Numbers: Histories, Geographies, and Theories of Quantification," *Environment and Planning D: Society and Space* 19 (2001): 379.

21. Foucault, *The Order of Things,* 344.

22. Schorske, "The New Rigorism in the Human Sciences, 1940–1960," 309–29. See also Lyons, *The Uneasy Partnership,* 80.

23. Derek Gregory, *Geographical Imaginations* (Cambridge, Mass.: Blackwell, 1994), 55.

24. Robin, *The Making of the Cold War Enemy,* 7.

25. Leland C. Devinney, "Guideposts to RF's Research Program in the Social Sciences," June, 4, 1948, in "Future Program in the Light of Reduced Budget," RG 3.1, Series 910, Box 3, Folder 18, RF.

26. Both quotes are from Capshew, *Psychologists on the March,* 181.

27. Donald Young and Paul Webbink, "Current Problems of Council Concern in Research Organization," *Items* 1, no. 3 (1947): 1. *Items* was SSRC's in-house journal.

28. Quoted in Ellen Condliffe Lagemann, *The Politics of Knowledge: The Carnegie Corporation, Philanthropy, and Public Policy* (Middletown, Conn.: Wesleyan University Press, 1989), 142.

29. James Allen Smith, *The Idea Brokers: Think Tanks and the Rise of the New Policy Elite* (New York: Free Press, 1991), 18.

30. Russell Sage Foundation, *Effective Use of Social Science Research in the Federal Services* (New York: Russell Sage Foundation, 1950), 7, 11, 21, 42. The authors were Pendleton Herring and Paul Webbink.

31. Raymond V. Bowers, "The Military Establishment," in *The Uses of Sociology,* ed. Paul F. Lazarsfeld et al. (New York: Basic Books, 1967), 251. See also Fox, "Civil–Military Relations Research," 279.

32. Samuel Stouffer et al, *The American Soldier,* vol. 1: *Adjustment during Army Life* (Princeton, N.J.: Princeton University Press, 1949), 5. See also Bowers, "The Military Establishment," 251.

33. See Series III, Subseries A, Box 327, Folder "Social Science Research Council—Analysis of Wartime Studies of Soldier Attitudes 1947-54," CCNY; Lagemann, *The Politics of Knowledge,* 176-77.

34. Nathan Glazer, "'The American Soldier' as Science: Can Sociology Fulfill Its Ambitions?" *Commentary* 8 (1949): 489, 496.

35. Arthur Schlesinger Jr., "The Statistical Soldier," *Partisan Review* 16, no. 8 (1949): 852, 854. On Schlesinger's frequent recourse to such language, see Cuordileone, "'Politics in an Age of Anxiety.'"

36. Quoted in Lagemann, *The Politics of Knowledge,* 176.

37. Mike F. Keen, *Stalking the Sociological Imagination: J. Edgar Hoover's FBI Surveillance of American Sociology* (Westport, Conn.: Greenwood Press, 1999), 157. Keen is quoting from Robert K. Merton and Paul F. Lazarsfeld, eds., *Continuities in Social Research: Studies in the Scope and Method of "The American Soldier"* (Glencoe, Ill.: Free Press, 1950). See also Lagemann, *The Politics of Knowledge,* 177; Robin, *The Making of the Cold War Enemy,* 19-23.

38. Lyle H. Lanier, "The Psychological and Social Sciences in the National Military Establishment," *American Psychologist* 4, no. 5 (1949): 127-47. Lanier was executive director of the Committee on Human Resources in 1947-48. Committee members or consultants included Charles Dollard, John Gardner, Clyde Kluckhohn, Samuel Stouffer, Harold Lasswell, Hans Speier, and Kingsley Davis. Pendleton Herring of the SSRC attended meetings regularly, as did CIA and State Department representatives. Raymond Bowers directed the committee's research staff. See Bowers, "The Military Establishment"; Charles T. O'Connell, "Social Structure and Science: Soviet Studies at Harvard" (PhD thesis, University of California–Los Angeles, 1990), 122, 342-43.

39. I. I. Rabi, "The Organization of Scientific Research for Defense," *Proceedings of the Academy of Political Science* 24, no. 3 (1951): 70-76.

40. William Marvel, "The Carnegie Corporation and Foreign Area Studies Programs in the United States," October 9, 1952, Series III, Subseries A, Box 42, Folder "Area Studies (to) 1961 (I)," CCNY, 1.

41. Michel Foucault, "The Confession of the Flesh," in *Power/Knowledge,* 197.

42. John Gardner, "Area Studies," January 23, 1958, in Series III, Subseries A, Box 42, Folder "Area Studies (to) 1961 (I)," CCNY, 2; Gardner to Donald MacKinnon, June 6, 1947, Folder "Area Studies—Russian 1947-59," in ibid.

43. John Gardner to John W. Macmillan, February 5, 1952, Series III, Subseries A, Box 174, Folder "Human Relations Area Files," CCNY (original emphasis);

"History of the Beginnings of Human Relations Area Files, Inc." October 1949, in ibid, 1.

44. Ford, *Human Relations Area Files: 1949–1969, A Twenty-Year Report* (New Haven, Conn.: Human Relations Area Files, 1970), 12.

45. John W. Gardner, "Are We Doing Our Homework in Foreign Affairs?" *The Yale Review* 37, no. 3 (1948): 400–8.

46. Carnegie Corporation of New York, *Annual Report, 1946,* CCNY, 22, 26.

47. Carnegie Corporation of New York, *Annual Report, 1947,* CCNY, 28.

48. Edward H. Berman, *The Ideology of Philanthropy: The Influence of the Carnegie, Ford, and Rockefeller Foundations on American Foreign Policy* (Albany: State University of New York Press, 1983), 99–125; Inderjeet Parmar, "The Carnegie Corporation and the Mobilisation of Opinion in the United States' Rise to Globalism, 1939–1945," *Minerva* 37 (1999): 355–78; Parmar, "'To Relate Knowledge and Action': The Impact of the Rockefeller Foundation on Foreign Policy Thinking during America's Rise to Globalism, 1939–1945," *Minerva* 40, no. 3 (2002): 235–63.

49. Both the area fellowships program (replaced by a Ford Foundation initiative) and the Committee on World Area Research were terminated in 1953.

50. Norman S. Buchanan, "Notes on Rockefeller Foundation Program in the Social Sciences," August 1955, RG 3.1, Series 910, Box 3, Folder 19, RF, Part II, 2. Buchanan was associate director of the foundation's Social Sciences Division.

51. Julian H. Steward, *Area Research: Theory and Practice* (New York: Social Science Research Council, 1950), xiii. See also Geiger, "American Foundations," 318.

52. Robert B. Hall, *Area Studies: With Special Reference to Their Implications for Research in the Social Sciences* (New York: Social Science Research Council, 1947), 2.

53. Ibid., 10, 25, 18. See also Lagemann, *The Politics of Knowledge,* 174.

54. "Social Science Considerations in the Planning of Regional Specialization in Higher Education and Research," March 10, 1944, RG 3.2, Series 900, Box 31, Folder 165, RF, 5. See also C. B. Fahs, "A Reexamination of Rockefeller Foundation Program in Area Studies," October 24, 1954, in ibid.

55. Hall, *Area Studies,* 24, 49, 42.

56. Rafael, "The Cultures of Area Studies in the United States," 96; Mary Hancock, "Unmaking the 'Great Tradition': Ethnography, National Culture and Area Studies," *Identities* 4, no. 3–4 (1998): 343–88.

57. CBF [Charles B. Fahs], "Area Studies," June 19, 1949, RG 3.2, Series 900, Box 31, Folder 165, RF, 1.

58. Rafael, "The Cultures of Area Studies in the United States," 91, 95–97; Robin, *The Making of the Cold War Enemy,* 20.

59. Charles Wagley, *Area Research and Training: A Conference Report on the Study of World Areas* (New York: Social Science Research Council, June 1948), 5.

60. William Buxton, *Talcott Parsons and the Capitalist Nation-State: Political Sociology as a Strategic Vocation* (Toronto: University of Toronto Press, 1985), 116.

61. Wagley, *Area Research and Training*, 7. See also Steward, *Area Research*, 96. Chaired by Philip Mosely, the conference's Soviet panel included Clyde Kluckhohn, Owen Lattimore, Geroid Robinson, and John Gardner. Panel on the Soviet Union, "Report of the Panel on Its Discussions, November 29, 1947," in Series III, Subseries A, Box 329, Folder "Social Science Research Council—National Conference on Study of World Areas, 1947–1948," CCNY.

62. Lily E. Kay, "Rethinking Institutions: Philanthropy as an Historiographic Problem of Knowledge and Power," *Minerva* 35 (1997): 288. See also Robin, *The Making of the Cold War Enemy*, 6; Herman, *The Romance of American Psychology*, 128.

63. Michel Foucault, *The History of Sexuality*, vol. 1, trans. Robert Hurley (New York: Vintage, 1980).

64. Herring, "The Social Sciences in Modern Society," 2.

65. Wagley, *Area Research and Training*, 5; Robert B. Hall, "Preface," in ibid., iii.

66. Pendleton Herring, "Objectives of Area Study in Colleges," Series III, Subseries A, Box 329, Folder "Social Science Research Council—Committee on World Area Research," CCNY, 3.

67. Problems and Policy Minutes, October 27, 1945, in *Committee on World Regions, 1943*, 61.

68. Herring, "Objectives of Area Study in Colleges," 6, 7. See also Herring, "The Conduct of Cold War," Series III, Subseries A, Box 107, Folder "Cold War: Possible Field of Interest, 1947," CCNY.

69. Fred I. Greenstein and Austin Ranney, "Pendleton Herring, 1903–2004," *Items and Issues* 5, no. 3 (2005), http://www.ssrc.org; Kenton W. Worcester, *Social Science Research Council, 1923–1998* (New York: Social Science Research Council, 2001).

70. Mark Solovey, "The Politics of Intellectual Identity and American Social Science, 1945–1970" (PhD thesis, University of Wisconsin–Madison, 1996), 141.

71. Rafael, "The Cultures of Area Studies in the United States," 94. On the Conference, see George E. Taylor, "Notes on the Second Conference on the Study of World Areas," *Items* 4, no. 3 (1950): 29–32.

72. Wendell C. Bennett, *Area Studies in American Universities* (New York: Social Science Research Council, June 1951).

73. Richard H. Heindel, *The Present Position of Foreign Area Studies in the United States: A Post-Conference Report* (New York: Committee on World Area Research, Social Science Research Council, 1950), 1, 35.

74. A. T. Poffenberger, "The Work of the Social Science Research Council," *Items* 1, no. 1 (1947): 1.

75. Steward, *Area Research*, 2, 7–9, 21–22.

76. Ibid., 55, 72.

77. Minutes, First Meeting of the Joint Exploratory Committee on World Area Research, Washington, D.C., February 23, 1946, in *Committee on World Regions, 1943*, 76.

78. Carnegie Corporation of New York, *Annual Report, 1948*, CCNY, 17, 20.

79. Both quotes are in L. Gray Cowan and Geroid T. Robinson, *History of the Russian Institute, Columbia University, 1946–1953* (New York: Columbia University Press, 1954), 44, 45.

80. The School of International Affairs was founded in 1946, with Russian, East Asian, and European Regional Institutes as building blocks. But it also drew substantially from the Program of Training in International Administration (see chapter 2).

81. L. Gray Cowan, *A History of the School of International Affairs and Associated Area Institutes, Columbia University* (New York: Columbia University Press, 1954), 48. See also Cowan and Robinson, *History of the Russian Institute*, 44–45. Cowan also noted a number of Soviet responses to the Institute, including 1951 and 1952 *Pravda* articles describing a "hotbed of American slanderers, spies and diversionaries" and "arch reactionaries." *A History of the School*, 47.

82. "Report of the Committee on a School of International Affairs and Regional Studies," February 14, 1945, Accession 1.1, Series 200(S), Box 321, Folder 3820, RF.

83. On the SSRC Committee, see "Slavic Studies," *Items* 2, no. 4 (1948): 9.

84. Schuyler C. Wallace to Oliver C. Carmichael, December 2, 1946, Series III, Subseries A, Box 3, Folder "Columbia University 1946–1951," CCNY.

85. Cowan and Robinson, *History of the Russian Institute*, 48. Philip Mosely estimated that 40 percent of Institute graduates had gone into government service; "Interview," January 13, 1954, Accession 1.1, Series 200, Box 322, Folder 3827, RF.

86. "The Russian Institute in New York," RF, 15; Cowan, *A History of the School*, 45–47; Cumings, *Parallax Visions*, 183–84.

87. John N. Hazard, *Recollections of a Pioneering Sovietologist* (New York: Oceana Publications, 1987), 113.

88. Joseph Willets, "Conversation with Geroid Robinson," January 30, 1947, Accession 1.1, Series 200(S), Box 321, Folder 3822 [mislabeled 3823], RF.

89. Panel on the Soviet Union, "Report of the Panel on Its Discussions," 5.

90. Engerman, "The Ironies of the Iron Curtain," 471.

91. "Russian Studies," July 11, 1947, Series III, Subseries A, Box 42, Folder "Area Studies — Russian 1947-59," CCNY. The author of this document is not named, but it is almost certainly Gardner.

92. "Record of Interview, JG [John Gardner] and Philip Mosely," June 2, 1947, Series III, Subseries A, Box 113, Folder "Columbia University — Russian Institute 1946-1957," CCNY.

93. John Gardner, "Review of Area Studies Grants," June 3, 1948, Series III, Subseries A, Box 42, Folder "Area Studies — 1961 (I)," CCNY, 2. See also Diamond, *Compromised Campus*, 66-68. In 1945, Harvard established an interdisciplinary masters' program in Soviet regional studies, where the model was "as much the ASTP as the OSS." Engerman, "The Ironies of the Iron Curtain," 472.

94. "Record of Interview, JG and Philip Mosely," CCNY.

95. Diamond, *Compromised Campus*, 50.

96. A December 12, 1947, memo from a Boston FBI agent confirmed that the "findings and conclusions" of the new Harvard center "will be made available to the State Department and other federal agencies through informal channels long before publication by the university." Quoted in ibid., 55-56. See also "Russian Research Center, Harvard University: Report, 1948-49," Series III, Subseries A, Box 166, Folder "Harvard University — Russian Research Center, 1948-1949," CCNY.

97. Clyde Kluckhohn, "Russian Research at Harvard," *World Politics* 1, no. 2 (1949): 266, 267. On January 7, 1949, Mead presented her research at the RRC, where her "psycho-cultural approach" received a number of testy responses. Minutes of Seminar Meetings (UAV 759.8), Box 1, Records of the Russian Research Center, Harvard University Archives (hereafter RRC). Berlin arrived in the same month — to the alarm of the FBI — and wrote to friends of the "positivist pedantry of American social science." Michael Ignatieff, *Isaiah Berlin: A Life* (Toronto: Viking, 1998), 190. See also Diamond, *Compromised Campus*, 57.

98. Alex Inkeles, "Understanding a Foreign Society: A Sociologist's View," *World Politics* 3, no. 2 (1951): 269 (original emphasis).

99. *Ten-Year Report and Current Projects, 1948-1958* (Cambridge, Mass.: Russian Research Center, Harvard University, January 1958), 2. This description helps to account for "one remarkable fact about the RRC's leadership at Harvard: however expert they were in behavioral sciences, none of the four members of the founding Executive Committee had studied Russian affairs or knew the Russian language." Engerman, "The Ironies of the Iron Curtain," 473.

100. Clyde Kluckhohn, "The Harvard Project on the Soviet Social System: Survey of Research Objectives," October 1, 1951, Folder UAV 759.451 [no Box], RRC, 1. According to the best historian of Sovietology, in its "conclusion that the

Soviet Union was a modern industrial society, in fundamental ways very much like the United States," the Social System initiative made it more difficult to differentiate the two countries and implicitly endorsed restraint, not rollback, as a wiser form of foreign policy. Engerman, "Bernath Lecture," 606.

101. *Ten Year Report and Current Projects,* 7, 58–59. See also O'Connell, *Social Structure and Science,* 455. Of the many publications based on the Soviet Social System project, two stand out: Raymond A. Bauer, Alex Inkeles, and Clyde Kluckhohn, *How the Soviet System Works: Cultural, Psychological, and Social Themes* (Cambridge, Mass.: Harvard University Press, 1956), and Inkeles and Bauer, *The Soviet Citizen: Daily Life in a Totalitarian Society* (Cambridge, Mass.: Harvard University Press, 1959). The former book was the Harvard Project's final report (the quote is from page viii).

102. O'Connell, "Social Structure and Science," 446.

103. Diamond, *Compromised Campus,* 62–63; O'Connell, "Social Structure and Science," 457; Simpson, "U.S. Mass Communication Research," 340.

104. "The Human Resources Research Institute: A Briefing Prepared for the Seventh Meeting of the Advisory Research Council, 25–26 March 1953," MC 188 (Max Millikan Papers), Box 6, Folder 182 ("Human Resources Research Institute, 1953–54"), MIT, 1.

105. O'Connell, "Social Structure and Science," 353.

106. *Report,* Strategic Intelligence Directorate, Human Resources Research Institute, Air University, Maxwell Air Force Base, Alabama, November 30, 1951, in Box 24, Bureau of Applied Social Research Papers, Columbia Rare Book and Manuscript Library, New York (hereafter BASR), 1.

107. Kent, *Strategic Intelligence for American World Policy,* 3–4.

108. *Report,* Strategic Intelligence Directorate, Human Resources Research Institute, BASR, 2.

109. Bowers, "The Military Establishment," 239.

110. A copy of the *Guide* is in Box 25, Folder "Air Interrogation Guide," BASR.

111. O'Connell, "Social Structure and Science," 375; see also p. 332; Diamond, *Compromised Campus,* 83–84.

112. Talcott Parsons, *The Social System* (Glencoe, Ill.: Free Press, 1951).

113. Quoted in Agnew and Corbridge, *Mastering Space,* 51.

114. See Capshew, *Psychologists on the March,* 189; Minutes of Seminar Meetings (UAV 759.8), Box 1, RRC. Kluckhohn's "Comparative Study of Values in Five Cultures Project" (1949–55) at the Laboratory of Social Relations, a search for scientific, universal "moral norms," symbolized anthropology's turn away from the vestiges of relativism. Novick, *That Noble Dream,* 285.

115. Engerman, "Bernath Lecture," 607.

116. Diamond, *Compromised Campus*, 95. See also O'Connell, "Social Structure and Science," 169, 197; Cumings, *Parallax Visions*, 188–89.

117. Buxton, *Talcott Parsons*, 151. According to his student Jesse Pitts, in the early 1950s Parsons's "defense of a young colleague accused of communism and his championing of [Robert] Oppenheimer got him into trouble with the U.S. government." "Talcott Parsons: The Sociologist as the Last Puritan," *American Sociologist* 15, no. 2 (1980): 63.

118. Buxton, *Talcott Parsons*, 119. See also Berman, *The Ideology of Philanthropy*, 106–7; O'Connell, "Social Structure and Science," 226, 238, 483.

119. Talcott Parsons and Edward A. Shils, eds., *Toward a General Theory of Action: Theoretical Foundations for the Social Sciences* (New Brunswick, N.J.: Transaction, 2001), 4. See also Lagemann, *The Politics of Knowledge*, 169.

120. C. Wright Mills, *The Sociological Imagination* (Oxford: Oxford University Press, 1959), 27, 31. *Toward a General Theory of Action* was "controversial within the Department" as a result of its ambitions. Lagemann, *The Politics of Knowledge*, 170.

121. Geroid Robinson, "The Russian Institute," November 27, 1944, Accession 1.1, Series 200(S), Box 321, Folder 3819, RF.

122. Robin, *The Making of the Cold War Enemy*, 4, 5.

123. Stephen F. Cohen, "Scholarly Missions: Sovietology as a Vocation," in *Rethinking the Soviet Experience: Politics and History Since 1917* (New York: Oxford University Press, 1985), 6, 7; see also pp. 10, 23, 25, 27–28; Novick, *That Noble Dream*, 281–82.

124. Engerman, "The Ironies of the Iron Curtain," 478.

125. On Hazard, see the material in Accession 1.1, Series 200, Box 322, Folder 3827, RF.

126. Cohen, "Scholarly Missions," 6; see also pp. 10–11, 17–18.

127. Quoted in Diamond, *Compromised Campus*, 59. Kluckhohn's cooperation may have been assured because the FBI claimed to have information on him that, "if leaked, could have subjected [him] to humiliation" (ibid.); see also David H. Price, "Cold War Anthropology: Collaborators and Victims of the National Security State," *Identities* 4, no. 3–4 (1998): 402–7.

128. Trumpbour, "Harvard, the Cold War, and the National Security State," 51–128. Trumpbour's description of Langer's CIA home as the "research division" (70) is not quite correct. The Office of National Estimates was responsible for the drafting of National Intelligence Estimates. It was one of three units that replaced the inefficient Office of Reports and Estimates in 1950. The other two were the Office of Research and Reports and the Office of Current Intelligence. Woodrow Kuhn, "The Beginning of Intelligence Analysis in the CIA," *Studies in Intelligence* 51, no. 2 (2007), http://www.cia.gov.

129. Ford Foundation, *Report of the Study for the Ford Foundation on Policy and Program* (Detroit: Ford Foundation, November 1949); Berman, *The Ideology of Philanthropy*, 108; Geiger, "American Foundations," 326–28; Lyons, *The Uneasy Partnership*, 278–79. It is also relevant that the Behavioral Science division was closed just as the National Science Foundation began to institutionalize social science funding.

130. Berman, *The Ideology of Philanthropy*, 101.

131. Schorske, "The New Rigorism," 317, 323; Novick, *That Noble Dream*, 300, 304, 311–13.

132. Gabriel A. Almond, *Ventures in Political Science: Narratives and Reflections* (Boulder, Colo.: Lynne Rienner, 2002), 95. Almond consulted for the State Department, the ONR, RAND, the Air Force, and the White House's Psychological Strategy Board in the 1940s and 1950s. From 1958 to 1961, he was a member of a classified Working Committee on Attitudes toward Unconventional Warfare, which also included Max Millikan, Ithiel Pool, and Air Force General Curtis LeMay. The social scientists were "expected to find ways of 'minimizing' unfortunate reactions by target peoples to the use of such [unconventional] weapons." Cumings, *Parallax Visions*, 257 n. 44. By 1968 Almond had severed his ties to government, his reputation secure. Ido Oren, *Our Enemies and US: America's Rivalries and the Making of Political Science* (Ithaca, N.Y.: Cornell University Press, 2003), 13, 152–53.

133. Gabriel A. Almond, "Research in Comparative Politics: Plans of a New Council Committee," *Items* 8, no. 1 (1954): 2.

134. The first book in the series, produced after a June 1959 conference in Dobbs Ferry, New York, was Gabriel A. Almond and James S. Coleman, eds., *The Politics of the Developing Areas* (Princeton, N.J.: Princeton University Press, 1960).

135. Lucian W. Pye, "Political Modernization and Research on the Process of Political Socialization," *Items* 13, no. 3 (1959): 25, 26 (original emphasis). See also Almond, *Ventures in Political Science*, 99–100.

136. Quoted in Nils Gilman, *Mandarins of the Future: Modernization Theory in Cold War America* (Baltimore: Johns Hopkins University Press, 2003), 5 (Lerner's emphasis).

137. Robin, *The Making of the Cold War Enemy*, 42.

138. Michael Adas, "Modernization Theory and the American Revival of the Scientific and Technological Standards of Social Achievement and Human Worth," in *Staging Growth: Modernization, Development, and the Global Cold War*, ed. David C. Engerman et al. (Amherst: University of Massachusetts Press, 2003), 37. See also Gilman, *Mandarins of the Future*, 5.

139. Rostow, *The Stages of Economic Growth*, 134, 142.

140. George Rosen, *Western Economists and Eastern Societies: Agents of Change*

in South Asia, 1950–1970 (Baltimore: Johns Hopkins University Press, 1985), 28; Westad, *The Global Cold War,* 33.

141. Gilman, *Mandarins of the Future,* 5. See also Mirowski, *Machine Dreams,* 256.

142. Quoted in James T. Fisher, "'A World Made Safe for Diversity': The Vietnam Lobby and the Politics of Pluralism, 1945–1963," in Appy, *Cold War Constructions,* 236. In 1966, Ithiel Pool was still referring to Vietnam as "the greatest social-science laboratory we have ever had!" Quoted in Mark Mazower, "Mandarins, Guns and Money," *The Nation,* October 6, 2008, 36.

143. Quoted in Berman, *The Ideology of Philanthropy,* 119. See also Gilman, *Mandarins of the Future,* 18.

144. "A Program of Research in Economic and Political Development," Center for International Studies, MIT, February 1953, AC 236 (Center for International Studies), Series I (Director's Files), Box 2, Folder 1, MIT, 2.

145. Engerman, "Bernath Lecture," 610; George M. Beckmann, "The Role of the Foundations in Non-Western Studies," in *U.S. Philanthropic Foundations: Their History, Structure, Management, and Record,* ed. Warren Weaver (New York: Harper & Row, 1967), 395–409.

146. The published version of Rostow's CIA work was *The Dynamics of Soviet Society* (New York: Norton, 1953).

147. *The Center for International Studies: A Description* (Cambridge: Massachusetts Institute of Technology, July 1955), 7.

148. Cline, *Secrets, Spies and Scholars,* 146. See also Cumings, *Parallax Visions,* 183, 190, 252 n. 1. It was in a Yale seminar run by Richard Bissell that the precocious seventeen-year-old Rostow decided he would "do an answer one day to Marx's theory of history." Quoted in David Milne, *America's Rasputin: Walt Rostow and the Vietnam War* (New York: Hill and Wang, 2008), 25.

149. Allan A. Needell, "'Truth Is Our Weapon': Project TROY, Political Warfare, and Government–Academic Relations in the National Security State," *Diplomatic History* 17, no. 3 (1993): 400. Needell is quoting Justin Miller, president of the National Association of Broadcasters. See also Record of Interview, WM [William Marvel] and Max Millikan, April 13, 1954, Series III, Subseries A, Box 213, Folder "Massachusetts Institute of Technology, 1946–1955," CCNY, 1.

150. Christopher Simpson, *Science of Coercion: Communication Research and Psychological Warfare, 1945–1960* (New York: Oxford University Press, 1994), 8.

151. Needell, "'Truth Is Our Weapon,'" 409, 411. Troy led directly to the creation of the Psychological Strategy Board, which reported to the National Security Council and included representatives from the CIA and the Departments of State and Defense. Robin, *The Making of the Cold War Enemy,* 42–43.

152. Record of Interview, Charles Dollard and Julius A. Stratton, November 30, 1951, Series III, Subseries A, Box 213, Folder "Massachusetts Institute of Technology, 1946–1955," CCNY.

153. Needell, "'Truth Is Our Weapon,'" 417. See also "The Nature and Objectives of the Center for International Studies," August 1953, in AC 4, Box 48, Folder 16, MIT. James Killian "came to regret" the CIA funding; *The Education of a College President: A Memoir* (Cambridge, Mass.: MIT Press, 1985), 67. Under pressure from Students for a Democratic Society, the CIS dissolved its CIA ties in 1965.

154. Needell, "'Truth Is Our Weapon,'" 418, 419. Morison was writing in an annex to the Project Troy Report.

155. Bowers, "The Military Establishment," 243–44; Robin, *The Making of the Cold War Enemy*, 8–10; William E. Daugherty, with Morris Janowitz, *A Psychological Warfare Casebook* (Baltimore: Johns Hopkins University Press, 1958), 425–30. "Operational code" was drawn from Nathan C. Leites, *The Operational Code of the Politburo* (New York: McGraw-Hill, 1951), a study prepared for the RAND Corporation.

156. Bowers, "The Military Establishment," 246.

157. Rose, *Governing the Soul*, vii (my emphasis).

158. Cindi Katz, "Banal Terrorism: Spatial Fetishism and Everyday Insecurity," in *Violent Geographies: Fear, Terror, and Political Violence*, ed. Derek Gregory and Allan Pred (New York: Routledge, 2007), 349–61.

159. Daugherty and Janowitz, *A Psychological Warfare Casebook*, 2–3 (original emphasis). See also Simpson, "U.S. Mass Communication Research," 315; Herman, *The Romance of American Psychology*, 143.

160. John Sharpless, "Population Science, Private Foundations, and Development Aid: The Transformation of Demographic Knowledge in the United States, 1945–1965," in *International Development and the Social Sciences*, 176–200.

161. Memo, Ithiel Pool to J. A. Stratton, "Reasons for the Development of the Behavioral Sciences at M.I.T.," March 1, 1956, AC 132 (Records of the Office of the Chancellor [Stratton] 1949–1957), Box 4, Folder "Center for International Studies," MIT, 4.

162. "A Program of Research in Economic and Political Development," February 1953, MIT, 7.

163. Converse, *Survey Research in the United States*, 1.

164. Schorske, "The New Rigorism," 312.

165. William Warntz, "Trajectories and Co-Ordinates," in *Recollections of a Revolution: Geography as Spatial Science*, ed. Mark Billinge, Derek Gregory, and R. L. Martin (London: MacMillan, 1984), 140. The title of this chapter is chillingly suggestive.

166. Stone, "Geography's Wartime Service," 89–96.

167. Quoted in Miller, "The American Society for Professional Geographers: Goals and Objectives," in *The American Society for Professional Geographers*, 24, 27.

168. George Kish, "American Geography in the 1950s," in *The Practice of Geography*, ed. Anne Buttimer (New York: Longman, 1983), 202.

169. Richard Hartshorne, *The Nature of Geography; A Critical Survey of Current Thought in the Light of the Past* (Lancaster, Pa.: Association of American Geographers, 1939); Barnes and Farish, "Between Regions," 812–13. For an elegant summary of these disputes, see David N. Livingstone, "Statistics Don't Bleed: Quantification and Its Detractors," in *The Geographical Tradition*, 304–46.

170. Eugene Van Cleef, "Areal Differentiation and the 'Science' of Geography," *Science*, June 13, 1952, 654.

171. James, *All Possible Worlds*, 500.

172. Trevor J. Barnes, "Lives Lived and Lives Told: Biographies of Geography's Quantitative Revolution," *Environment and Planning D: Society and Space* 19 (2001): 416.

173. Ackerman, "Geographic Training," 121–43; Katz, *Foreign Intelligence*, 16.

174. These duties, including Hartshorne's position on the Social Studies Panel for the Service Academies Board (Office of the Secretary of National Defense), are noted in "Activities of American Geographers," *Professional Geographer* 1, no. 3 (1949): 38.

175. Quoted in W. L. Baxter, "The Place and Problems of Geography Instruction at the Air Force Academy," *Journal of Geography* 59, no. 9 (December 1960): 412. Hartshorne is listed as a member of the Air Force Planning Board.

176. David Mercer, "Unmasking Technocratic Geography," in Billinge, Gregory, and Martin, *Recollections of a Revolution*, 155–99.

177. R. L. Morrill, "Recollections of the 'Quantitative Revolution's' Early Years: The University of Washington 1955–65," in Billinge, Gregory, and Martin, *Recollections of a Revolution*, 67.

178. R. J. Johnston, *Geography and Geographers: Anglo-American Human Geography since 1945*, 5th ed. (London: Arnold, 1997), 20 (original emphasis).

179. Stone, "Geography's Wartime Service," 93.

180. James and Jones, *American Geography*.

181. Rainger, "Science at the Crossroads," 370. See also Maxwell E. Britton, "Louis Otto Quam (1906–2001)," *Arctic* 56, no. 4 (2003): 425–26; H. Jesse Walker, "Evelyn Lord Pruitt, 1918–2000," *Annals of the Association of American Geographers* 96, no. 2 (2006): 432–39.

182. Edwards, *The Closed World*, 59.

183. Evelyn L. Pruitt, "ONR's Geographic Research Program," *Naval Research*

Reviews April 1960, 1. Pruitt succeeded Quam as Head of the Geography Branch in 1959. See also Walker, "Evelyn Lord Pruitt," 434.

184. Evelyn L. Pruitt, "The Office of Naval Research and Geography," *Annals of the Association of American Geographers* 69, no. 1 (1979): 105–6; Walker, "Evelyn Lord Pruitt," 436–37. Credit for the Foreign Field Research Program was given to Carl Sauer—hardly a quantitative geographer—who was a member of the Geography Branch's Advisory Committee. The ONR and its partner, the National Research Council, "took a hands-off policy," but Pruitt nonetheless considered the program "an important contribution to geography." Pruitt, "The Office of Naval Research and Geography," 105, 106.

185. John Q. Stewart, "Empirical Mathematical Rules concerning Distribution and Equilibrium of Population," *Geographical Review* 37 (1947): 485.

186. John Q. Stewart, "The Development of Social Physics," *American Journal of Physics* 18, no. 5 (1950): 239–53.

187. John Q. Stewart, "Natural Law Factors in United States Foreign Policy," *Social Science* 29, no. 3 (1954): 128.

188. William Warntz, "Contributions toward a Macroeconomic Geography: A Review," *Geographical Review* 47 (1957): 420.

189. William Warntz, "Geography at Mid-Twentieth Century," *World Politics* 11, no. 3 (1959): 447; Warntz, "Progress in Economic Geography," in *New Viewpoints in Geography,* ed. Preston E. James (Washington, D.C.: National Council for the Social Studies, 1959), 58.

190. Allan Sekula, quoted in Shapiro, *Violent Cartographies,* 89.

191. Warntz, "Contributions toward a Macroeconomic Geography," 424.

192. Donald G. Janelle, "William Warntz, 1922–1988," *Annals of the Association of American Geographers* 87, no. 4 (1997): 725. See also Barnes and Farish, "Between Regions," 818–19.

193. George H. T. Kimble, "The Inadequacy of the Regional Concept," in *London Essays in Geography: Rodway Jones Memorial Volume,* ed. L. Dudley Stamp and S. W. Woolridge (Cambridge, Mass.: Harvard University Press, 1951), 159. With Dorothy Good, Kimble later wrote a book on a region that did matter, at least to the Cold War: *Geography of the Northlands* (New York: American Geographical Society and J. Wiley, 1955).

194. Richard U. Light, "Foreword," in "The American Geographical Society: A Statement" (1951), Series III, Subseries A, Box 28, Folder "American Geographical Society (1934–52)," CCNY.

195. Office of the President, Record of Interview, JG [John Gardner] and Richard Harrison, March 30, 1949; Record of Interview, JG and Richard E. Harrison, August 10, 1949; in ibid.

196. Quoted in Neil Smith, "'Academic War over the Field of Geography': The Elimination of Geography at Harvard, 1947–1951," *Annals of the Association of American Geographers* 77, no. 2 (1987): 159. See also Barnes and Farish, "Between Regions," 817.

197. See Allen K. Philbrick, "Principles of Areal Functional Organization in Regional Human Geography," *Economic Geography* 33, no. 4 (1957): 299–336.

198. M. C. Shelesnyak to Vilhjalmur Stefansson, August 7, 1948, in Box 6, Folder "Stefansson (Vilhjalmur and Evelyn) 1946–53," M. C. Shelesnyak Papers, National Library of Medicine, Bethesda, Md.

199. Quoted in Jeff Byles, "Maps and Chaps: The New Geography Reaches Critical Mass," *Village Voice*, July 31, 2001, http://www.villagevoice.com.

200. John K. Rose, "Geography in Practice in the Federal Government, Washington," in *Geography in the Twentieth Century: A Study of Growth, Fields, Techniques, Aims and Trends*, 3rd ed., ed. Griffith Taylor (New York: Philosophical Society, 1957), 586 (original emphasis).

201. Mercer, "Unmasking Technocratic Geography," 186 (original emphasis).

202. N. Katherine Hayles, *How We Became Posthuman: Virtual Bodies in Cybernetics, Literature, and Informatics* (Chicago: University of Chicago Press, 1999).

203. Edward A. Ackerman, "Where Is a Research Frontier?" *Annals of the Association of American Geographers* 53, no. 4 (1963): 430, 435, 436, 437, 440.

4. The Cybernetic Continent

1. Norbert Wiener, *Cybernetics: Or Control and Communication in the Animal and the Machine* (Cambridge, Mass.: Technology Press, 1948), 19.

2. Turing's legendary proof is "Computing Machinery and Intelligence," *Mind* 59 (1950): 433–60.

3. Galison, "The Ontology of the Enemy," 256. See also Geof Bowker, "How to Be Universal: Some Cybernetic Strategies, 1943–70," *Social Studies of Science* 23 (1993): 110–11.

4. Galison, "The Ontology of the Enemy," 261. See also Edwards, *The Closed World*, 20, 146.

5. Quoted in John Clute and Peter Nicholls, eds., *The Encyclopedia of Science Fiction* (London: Orbit, 1993), 287.

6. Steve J. Heims, *Constructing a Social Science for Postwar America: The Cybernetics Group, 1946–1953* (Cambridge, Mass.: MIT Press, 1991), 106.

7. Wiener, *Cybernetics*, 8–9. See also Dana Polan, *Power and Paranoia: History, Narrative, and the American Cinema, 1940–1950* (New York: Columbia University Press, 1986), 159–64; Bowker, "How to Be Universal," 113.

8. Jennifer Light, *From Warfare to Welfare: Defense Intellectuals and Urban Problems in Cold War America* (Baltimore: Johns Hopkins University Press, 2003), 37, 45–46.

9. See, for instance, "The Thinking Machine," *Time,* January 23, 1950, 55–60.

10. Galison, "The Ontology of the Enemy," 256–57; Donna J. Haraway, *Simians, Cyborgs, and Women: The Reinvention of Nature* (New York: Routledge, 1991), 110–11.

11. Russell, "Military Geography," 494–95.

12. Edwards, *The Closed World,* 72; Shapiro, *Violent Cartographies,* 82.

13. Peter Galison, "Computer Simulations and the Trading Zone," in *The Disunity of Science: Boundaries, Contexts, and Power,* ed. Galison and David J. Stump (Stanford, Calif.: Stanford University Press, 1996), 121.

14. See Wiebe E. Bijker, Thomas Park Hughes, and Trevor J. Pinch, eds., *The Social Construction of Technological Systems: New Directions in the Sociology and History of Technology* (Cambridge, Mass.: MIT Press, 1987).

15. E. Laurier and C. Philo, "X-Morphising: Review Essay of Bruno Latour's *Aramis, or the Love of Technology,*" *Environment and Planning A* 31 (1999): 1064.

16. The OSS Assessment Staff, *Assessment of Men: Selection of Personnel for the Office of Strategic Services* (New York: Rinehart, 1948), 3, 124, 31, 467. See also James H. Capshew, *Psychologists on the March,* 111–14.

17. Capshew, *Psychologists on the March,* 5.

18. Herman, *The Romance of American Psychology,* 9, 44–46.

19. See Robert M. Yerkes, "Man-Power and Military Effectiveness: The Case for Human Engineering," *Journal of Consulting Psychology* 5, no. 5 (1941): 205–9.

20. Capshew, *Psychologists on the March,* 50, 145. See also Bowker, "How to Be Universal," 112.

21. The most famous example of such research led to *The American Soldier* (1949), discussed in chapter 3.

22. National Research Council, with the Collaboration of Science Service, *Psychology for the Fighting Man* (Washington, D.C.: Penguin and The Infantry Journal, 1943), 10.

23. Capshew, *Psychologists on the March,* 99. See also Herman, *The Romance of American Psychology,* 22.

24. *Psychology for the Fighting Man,* 24.

25. Walter S. Hunter, "Psychology in the War," *American Psychologist* 1 (1946): 479. See also Capshew, *Psychologists on the March,* 145; Herman, *The Romance of American Psychology,* 128.

26. Quoted in "National Safety and the Universities," *Technology Review* 56, no. 7 (1954): 357. See also J. R. Killian Jr., "Adapting M.I.T.'s Policies and Pro-

gram to the Current National Defense Situation," in F. Leroy Foster, ed., *Sponsored Research at M.I.T., 1900–1968, Volume 3: 1947–1956; Part 1: 1947–1953,* MIT, n.p.

27. Memo, Carl F. J. Overhage to General McCormack, "Notes on Lincoln Laboratory and M.I.T.," February 27, 1959, in MC 365 (Albert G. Hill Papers), Box 14, Folder 9, MIT, 2. See also Memo, Ithiel Pool to J.A. Stratton, "Reasons for the Development of the Behavioral Sciences at M.I.T.," 1.

28. Robert Buderi, *The Invention That Changed the World: How a Small Group of Radar Pioneers Won the Second World War and Launched a Technological Revolution* (New York: Touchstone, 1996), 359.

29. "The Emerging Shield," *Air University Quarterly Review* 8, no. 2 (1956): 56.

30. Report of Proceedings of SAGE Press Conference, January 16, 1956, Box DO.5.3.3/5, Folder "SAGE Press Conference 1956," Lincoln Laboratory Archives, Lexington, Mass. (hereafter LLAB), 21.

31. "SAGE Aircraft Defense," *Technology Review* 58, no. 5 (1956): 230. See also Press Release, January 18, 1956, AC 132, Box 11, Folder "Lincoln Laboratory," MIT, 1.

32. "SAGE Aircraft Defense," 264.

33. Edwards, *The Closed World,* 104. A useful description of a SAGE direction center, originally published in 1957, is Robert R. Everett et al., "SAGE—A Data-Processing System for Air Defense," *Annals of the History of Computing* 5, no. 4 (October 1983): 330–39.

34. Bruno Latour, "Give Me a Laboratory and I Will Raise the World," in *Science Observed: Perspectives on the Social Study of Science,* ed. Karin D. Knorr-Cetina and Michael Mulkay (London: Sage, 1983), 166 (original emphasis).

35. Rouse, *Engaging Science,* 30.

36. Joseph Rouse, *Knowledge and Power: Toward a Political Philosophy of Science* (Ithaca, N.Y.: Cornell University Press, 1987), 22.

37. Fuller, *Thomas Kuhn,* 223.

38. Kent C. Redmond and Thomas M. Smith, *From Whirlwind to MITRE: The R&D Story of the SAGE Air Defense Computer* (Cambridge, Mass.: MIT Press, 2000), 98.

39. *History of the Electronic Systems Division, January–June 1964,* vol. 1, *SAGE—Background and Origins,* LLAB, chap. 3.

40. These included Project Troy, mentioned in chapter 3, and Project East River, discussed in chapter 5.

41. James R. Killian, "The Role of Science in National Security: Three Lectures for Delivery at Harvard University Summer School," August 8–10, 1955, MC 423 (James R. Killian Papers), Box 39, Folder "Harvard Lectures—The Role of Science in National Security—1955 (1/3)," MIT, Lecture II, 12.

42. J. R. Marvin and F. J. Weyl, "The Summer Study," *Naval Research Reviews* 19, no. 8 (1966): 1–7, 24–28.

43. *Problems of Air Defense: Final Report of Project Charles*, vol. 1 (Cambridge: Massachusetts Institute of Technology, August 1, 1951), MIT, Preface. See also Redmond and Smith, *From Whirlwind to MITRE*, 1, 100–2; Gregg Herken, *Counsels of War* (New York: Knopf, 1985), 61–73.

44. *Problems of Air Defense*, vol. 1, MIT, xvii, 203, 3. A month before the launch of Project Charles, MIT chancellor and provost (and future president) Julius Stratton commented to Killian that "Air Defense . . . involves economic and sociological factors quite as important as the purely technological ones and . . . no analysis to date has taken these properly into account." Stratton to Killian, January 2, 1951, AC 132, Box 14, Folder "Project Charles," MIT, 3–4.

45. Summary, Meeting No. 1, Science Advisory Committee, Office of Defense Mobilization, Executive Office of the President, May 12, 1951, AC 4, Box 194, Folder 10 ("Science Advisory Committee, January–August 1951"), MIT, 3.

46. F. B. Llewellyn, "Systems Engineering with Reference to Military Applications," November 1, 1951, AC 4, Box 194, Folder 11 ("Science Advisory Committee, September–December 1951"), MIT, 1.

47. Leslie, *The Cold War and American Science*.

48. J. A. to F. W. Loomis, February 11, 1952, AC 4 (Records, Office of the President, 1930–1958), Box 153, Folder 4 ("Lincoln Laboratory, January–May 1952"), MIT; Valley to Killian, April 10, 1952, in ibid., 7.

49. Memo, Killian to "All Members of the Staff," December 2, 1954, in F. Leroy Foster, *Sponsored Research at M.I.T., 1900–1968, Volume II: 1947–1956, Part II: 1954–1956* (1984), MIT, n.p.

50. Memo, Overhage to President Stratton, Subject: "M.I.T. and Lincoln Laboratory: A New Approach," January 5, 1959, MC 365, Box 14, Folder 9, MIT, 3.

51. *Final Report, 1952 Summer Study Group*, February 10, 1953 (2 Vols.), LLAB; David F. Winkler, *Searching the Skies: The Legacy of the United States Cold War Defense Radar Program* (Langley, Va.: United States Air Force Combat Command, 1997), 26.

52. Lloyd Berkner, "Continental Defense," *Current History* 26 (1954): 262.

53. James R. Killian Jr, Draft of Remarks to American Society of Editorial Writers, October 16, 1953, AC 4, Box 135, Folder 6 ("Lincoln Laboratory, January–October 1953"), MIT, 4.

54. James R. Killian Jr. and A. G. Hill, "For a Continental Defense," *The Atlantic Monthly*, November 1953, 39.

55. Bush to Killian, October 28, 1953, AC 4, Box 4, Folder 6, MIT. The article was Charles J. V. Murphy, "The U.S. as a Bombing Target," *Fortune*, November 1953,

118–21, 219–28. In a letter to Murphy, George Valley complained about being misquoted; Valley to Murphy, October 30, 1953, AC 4, Box 135, Folder 6, MIT.

56. See "Resume of Meeting in Room 6–303," May 27, 1953, AC 4, Box 136, Folder 4 (Lincoln Laboratory — Historical Papers, 1952–1957), MIT, 5. Zacharias may have been referring to another, earlier article by Charles Murphy: "The Hidden Struggle for the H-bomb," *Fortune*, May 1953, 109–10, 230. This piece overstated Robert Oppenheimer's role on the 1952 Summer Study, claiming that he had organized the group to campaign against hydrogen bomb development. Summer Study participants were portrayed as believers not only in the feasibility of a "near-perfect air defense" but also in the moral superiority of a "fortress concept" relative to Strategic Air Command doctrine. Jerrold Zacharias and Myles Gordon, "Military Technology: One of the Lives of Jerrold Zacharias," draft manuscript, October 1, 1986, in MC 31, Box 33, MIT, 167, 170.

57. Quoted in a 1965 Festschrift produced on the occasion of Zacharias's sixtieth birthday, in MC 31 (Jerrold Zacharias Papers), Box 7, MIT, 1.

58. K. H. Stone et al., Appendix G, "Geographic Studies," *Final Report, Summer Study Group*, vol. 2, LLAB, G-1.

59. Kirk H. Stone to A. G. Hill, January 29, 1953, and W. G. Metcalf to M. M. Hubbard, March 4, 1953, "Personnel with Arctic Experience," both in Box DO.5.1.2/4 (Director's Office — Projects and Programs — Project Lincoln Records), Folder "Dew Line — Skull Cliff, Alaska," LLAB.

60. Jerrold Zacharias, "Scientist as Advisor," in MC 31, Box 42, Folder "Scientist as Advisor, 3/29/1961," MIT, 5.

61. "History of Efforts to Establish an Air Defense System Laboratory," July 9, 1951, MC 365, Box 14, Folder 7, MIT, 7. Hill was referring to the Center for International Studies.

62. Massachusetts Institute of Technology, *Defense of North America: Final Report of Project Lamp Light*, vol. 1, March 15, 1955, in MC 420 (Jerome Wiesner Papers), Box 126, MIT, 17. Boxes in the Wiesner Papers held no folders at the time of consultation.

63. Ibid., vol. 3, 11–4; vol. 4, 15–2.

64. Memo to James R. Killian, "Meeting on Air Defense," n.d., AC 4, Box 136, Folder 3 ("Lincoln Laboratory — Historical Papers 1949–1951"), MIT.

65. Louis D. Smullin, "Cooks for the Motor Pool," in "Festschrift," 1965, MC 31, Box 7, MIT, 1.

66. E. J. Barlow, *Active Air Defense of the United States, 1954–1960*, RAND R-250, December 1, 1953, RAND Library, Santa Monica, Calif. (hereafter RL), 2.

67. Herken, *Counsels of War*, 75.

68. Barlow, *Active Air Defense of the United States*, RL, 46.

69. Herken, *Counsels of War,* 75.

70. Press Release, January 18, 1956, MIT.

71. F. N. Marzocco, *The Story of SDD,* RAND SD(L)-1094, October 1, 1956, in MC 075 (Philip Morse Papers), Box 11, Folder 10/14 ("RAND Corp"), MIT, 1.

72. *Operational Plan: Semiautomatic Ground Environment System for Air Defense,* March 7, 1955, LLAB, 19.

73. Lincoln Laboratory Technological Memorandum 20, "A Proposal for Air Defense System Evolution: The Transition Phase," Second Draft, January 2, 1953, LLAB, 5.

74. Mirowski, *Machine Dreams,* 284.

75. Philip E. Mosely, "International Affairs," in *U.S. Philanthropic Foundations: Their History, Structure, Management, and Record,* ed. Warren Weaver (New York: Harper & Row, 1967), 387.

76. J. R. Goldstein, *RAND: The History, Operations, and Goals of a Nonprofit Corporation,* RAND P-2236-1, April 1961, RL, 2.

77. Herken, *Counsels of War,* 26.

78. F. R. Collbohm and Warren Weaver, "Opening Plenary," in *Conference of Social Scientists, September 14–19, 1947 — New York,* RAND R-106, June 9, 1948, RL, 3.

79. David R. Jardini, "Out of the Blue Yonder: The RAND Corporation's Diversification into Social Welfare Research, 1946–1968" (PhD thesis, Carnegie Mellon University, 1996), 42.

80. Der Derian, "The (S)pace of International Relations," 295–310.

81. *Conference of Social Scientists,* RL, vii.

82. Warren Weaver, "Science and Complexity," *American Scientist* 36 (1948): 536.

83. Mirowski, *Machine Dreams,* 170. See also Bowker, "How to Be Universal," 109.

84. Unnamed RAND employee, quoted in Herken, *Counsels of War,* 75.

85. Ibid. Initially, Brodie admired the "scientific strategists," although he grew disillusioned with the enterprise by the early 1960s. For a cautious endorsement, see his "Strategy as a Science," *World Politics* 1, no. 4 (July 1949): 467–88.

86. M. G. Wiener, *War Gaming Methodology,* RAND RM-2413, July 10, 1959, RL, iii.

87. Sharon Ghamari-Tabrizi, "Simulating the Unthinkable: Gaming Future War in the 1950s and 1960s," *Social Studies of Science* 30, no. 2 (April 2000): 163, 164, 169.

88. Kaplan, *The Wizards of Armageddon,* 201.

89. Lincoln P. Bloomfield, "Political Gaming," *U.S. Naval Institute Proceedings* 86, no. 9 (September 1960): 57, 58.

90. Herbert Goldhamer and Hans Speier, "Some Observations on Political Gaming," *World Politics* 12, no. 1 (1959): 73, 74.

91. John T. Rowell and Eugene R. Streich, "The Sage System Training Program for the Air Defense Command," *Human Factors* 6, no. 5 (1964): 537, 538, 539.

92. John L. Kennedy, "The Uses and Limitations of Mathematical Models, Game Theory, and Systems Analysis in Planning and Problem Solution," in Flanagan et al., *Current Trends*, 97, 98, 103. RAND's Social Science Division did not fully move from Washington to Santa Monica until 1956.

93. Robert L. Chapman et al., "The Systems Research Laboratory's Air Defense Experiments," *Management Science* 5, no. 3 (1959): 250. Both operations research and systems analysis enabled scientists to philosophically justify the Air Force demand for value-driven *"preferred instrumentalities and techniques."* Goldstein, *RAND*, RL, 8, original emphasis. See also Kaplan, *Wizards of Armageddon*, 86–87; Malcolm W. Hoag, *An Introduction to Systems Analysis*, RAND RM-1678, April 18, 1956, RL.

94. Kennedy, "The Uses and Limitations of Mathematical Models," 114, 115. See also Robert L. Chapman, *Data for Testing a Model of Organizational Behavior*, RAND RM-1916, March 1, 1960, RL; John Prados, *Pentagon Games: Wargames and the American Military* (New York: Harper and Row, 1987), 20.

95. Ghamari-Tabrizi, "Simulating the Unthinkable," 182. See also Mirowski, *Machine Dreams*, 350.

96. Goldstein, *RAND*, RL, 14.

97. Claude Baum, *The System Builders: The Story of SDC* (Santa Monica, Calif.: System Development Corporation, 1981), 16.

98. Marzocco, *The Story of SDD*, MIT, 1.

99. W. Richard Goodwin, "The System Development Corporation and System Training," *American Psychologist* 12, no. 8 (1957): 526.

100. Rowell and Streich, "The Sage System Training Program," 540. See also Chapman et al., "The Systems Research Laboratory's Air Defense Experiments," 253, 257; William C. Biel, *Description of the Air-Defense Experiments: I. The Physical and Cultural Environments*, RAND P-661, October 17, 1955, RL, 3.

101. John L. Kennedy, "A 'Transition-Model' Laboratory for Research on Cultural Change," *Human Organization* 14, no. 3 (1955): 18. See also Bowker, "How to Be Universal," 123; Ghamari-Tabrizi, "Simulating the Unthinkable," 165; Martin J. Collins, "Planning for Modern War: RAND and the Air Force, 1945–1950" (PhD thesis, University of Maryland, 1998), 19.

102. Isaiah Bowman, "Geographical Objectives in the Polar Regions," *Photogrammetric Engineering* 15 (1949): 9.

103. *Index to the National Geographic Society's Map of the Top of the World* (Washington, D.C.: National Geographic Society, 1949), 7. See also "Top of the

World: The National Geographic Society's New Map of Northlands," *National Geographic*, October 1949, 524–28.

104. "America and Polar Geopolitics," *Army Talk* 173 (April 26, 1947), 7; Office of Armed Forces Information and Education, Department of Defense, *The Arctic: A Hot Spot of Free World Defense* (Washington, D.C.: U.S. Government Printing Office, 1958), 3. Although "polar maps had been included in prewar atlases, they were almost always used to illustrate routes of exploration, highlighting the physical, rather than the political, nature of the Pole." Schulten, *The Geographical Imagination in America*, 231.

105. Stephen B. Jones, *The Arctic: Problems and Possibilities* (New Haven, Conn.: Yale Institute of International Studies, 1948), 1.

106. Quoted in Vilhjalmur Stefansson, "The Arctic," *Air Affairs* 3, no. 2 (1950), 391.

107. M. C. Shelesnyak, "The Arctic as a Strategic Scientific Area," in Box 137, Folder "AINA: General Papers — Conferences and Seminars — Joint AINA-Isaiah Bowman School of Geography, The Johns Hopkins University Seminar: Problems of the Arctic," Henry B. Collins Papers, NAA, 4. Moses Chiam Shelesnyak was head of the Office of Naval Research's Human Ecology branch after the Second World War, where he was involved in defining Arctic research priorities. He left the ONR to become director of the Baltimore–Washington Branch of the AINA and moved to Israel's Weizmann Institute of Science in 1950. Meeting with Carnegie Corporation president Charles Dollard in June of that year, the AINA's executive director Lincoln Washburn disclosed that "hopes of permanent government support for the Arctic Institute have been knocked into a cocked hat by the discovery that the director of the Johns Hopkins end of their show had once registered as a member of the Communist party." It might be necessary, Washburn added, to "separate the man from the organization." This was no doubt a key rationale for the move of the Institute's office from Baltimore to Washington and the installation of Washburn as head of the Washington office (after he resigned as executive director) in 1951. "Record of Interview; Subject: Arctic Institute of North America," June 6, 1950, Series III, Subseries A, Box 42, Folder "Arctic Institute of North America," CCNY.

108. "Proposal for an Arctic Institute of North America," Series III, Subseries A, Box 42, Folder "Arctic Institute of North America," CCNY, 1.

109. Leslie Roberts, "The Great Assault on the Arctic," *Harper's*, August 1955, 37–42.

110. M. C. Shelesnyak, *Across the Top of the World: A Discussion of the Arctic* (Washington, D.C.: Navy Department, August 1947), 47.

111. Ronald E. Doel, "Constituting the Postwar Earth Sciences: The Military's Influence on the Environmental Sciences in the USA after 1945," *Social Studies of Science* 33, no. 5 (2003): 656.

112. R. St. J. MacDonald, ed., *The Arctic Frontier* (Toronto: University of Toronto Press, 1966).

113. Edward Jones-Imhotep, "Disciplining Technology: Electronic Reliability, Cold-War Military Culture and the Topside Ionogram," *History and Technology* 17, no. 2 (2000): 125–75.

114. Shelagh D. Grant, *Sovereignty or Security: Government Policy in the Canadian North, 1936–1950* (Vancouver: University of British Columbia Press, 1988); Sherrill Grace, *Canada and the Idea of North* (Montréal: McGill-Queen's University Press, 2001); Renée Hulan, *Northern Experience and the Myths of Canadian Culture* (Montréal: McGill-Queens University Press, 2002).

115. Major General Guy V. Henry to President Harry Truman, February 26, 1951, RG 59 (General Records of the Department of State), Entry 1176, Box 1, Folder "51/1 Recommendation," NARA.

116. Foucault, *"Society Must Be Defended."*

117. L. B. Pearson, "Canada's Northern Horizon," *Foreign Affairs* 31, no. 4 (1953): 581–91.

118. "Science and Military Power," October 17, 1953, Box 9, Folder "Editorial Writers — October, 1953," Lloyd V. Berkner Papers, LOC, 13.

119. P. H. H. Bryan, "War in the Arctic," *Army Quarterly* 56 (1948): 102.

120. P. D. Baird, "The Arctic Institute of North America," *Polar Record* 8 (January 1956), 22–23. For AINA's 1947 Board of Governors, see "Progress Report, Arctic Institute of North America, June 20, 1947," in RG59, Entry 5245, Box 2, Folder "Arctic Institute of North America," NARA.

121. *A Brief to the Canadian Government for Joint Canadian and United States Support of the Arctic Institute of North America* (Montréal: Arctic Institute of North America, 1949), 3. See also "The Arctic Institute of North America," n.d., enclosed in "Report of Progress to the Carnegie Corporation of New York," August 6, 1952, CCNY.

122. G. Dudley Smith, "The Arctic Institute Roster Project," *Arctic* 2, no. 1 (1949), 43–44.

123. *A Brief to the Canadian Government*, 4.

124. *Pressing Scientific Problems of the North* (Montréal: Arctic Institute of North America, 1958), 1.

125. *Report of the Arctic Institute of North America — Office of Naval Research Arctic Research Advisory Committee* (New York: Arctic Institute of North America, 1957), 1.

126. G. Etzel Pearcy, "Arctic Staff Visit, August 6–14, 1959, Sponsored by Chief of Research and Development, Department of the Army," August 20, 1959, RG59, Entry 5245, Box 1, Folder "Polar Regions — American Travel and Exploration," NARA.

127. *Report of the Arctic Institute of North America*, 8. See also Sanjay Chatur-vedi, *The Polar Regions: A Political Geography* (Chichester, U.K.: John Wiley and Sons, 1996), 83.

128. Department of the Navy, *Naval Arctic Operations Handbook* (Wash-ington, D.C.: Arctic and Cold Weather Coordinating Committee of the Office of the Chief of Naval Operations, 1949), 11, 14, 59 (original emphasis). This handbook was written by the geographer Evelyn Pruitt, working under the direc-tion of Shelesnyak. She received advice from "such Arctic personalities as Vil-hjalmur Stefansson, Sir Hubert Wilkins, and Paul Siple." Walker, "Evelyn Lord Pruitt," 433.

129. Paul H. Nesbitt, "A Brief History of the Arctic, Desert, and Tropic Infor-mation Center and Its Arctic Research Activities," in *United States Polar Explora-tion*, ed. Herman R. Friis and Shelby G. Bale Jr. (Athens: Ohio University Press, 1970), 134–45.

130. On tropicality, see David Arnold, *The Problem of Nature: Environment, Culture and European Expansion* (Oxford: Blackwell, 1996).

131. See, for instance, Martin McGuire, *Survival Geography of South America* (Maxwell AFB, Ala.: Arctic, Desert, Tropic Information Center, Aerospace Studies Institute, 1961).

132. Alex J. La Rocque, "The Role of Geography in Military Planning," *Cana-dian Geographer* 3 (1953): 72.

133. M. C. Shelesnyak, "Comments on United States Naval Interests in Geo-graphical Exploration," n.d., Box 12, Folder "Navy Arctic Program — Propos-als — 1946–48," M. C. Shelesnyak Papers, National Library of Medicine, Bethesda, Md.

134. See, for example, A. Winters et al., *Battling the Elements: Weather and Terrain in the Conduct of War* (Baltimore: Johns Hopkins University Press, 1998).

135. J. D. Sartor, *Evaluation of Environmental Effects on Military Operations*, RAND RM-2080, December 21, 1957, RL, 3. See also Goldhamer and Speier, "Some Observations," 73.

136. Zacharias and Gordon, "Military Technology," MIT, 148.

137. Quoted in Thomas Ray, *A History of the DEW Line*, Air Defense Com-mand Historical Study 31 (Maxwell AFB, Ala.: Air Force Historical Research Agency, n.d.), 8.

138. "Dew Line," MC 365, Box 38, Folder 7, MIT, 18. Berkner is also credited in Zacharias and Gordon, "Military Technology," MIT, 147.

139. *Lamp Light*, vol. 3, MIT, 12–1, 12–7, 13–1, 13–3. As with Project Charles, Kirk Stone wrote an appendix (13-A) of the Lamp Light report, on "Characteris-tics of Proposed Aleutian Radar Sites."

140. Zacharias and Gordon, "Military Technology," MIT, 148–49. Air defense research also exposed differing opinions and animosity among MIT scientists, notably Zacharias and Valley. In a letter to Project Charles head F. Wheeler Loomis, Zacharias complained that he had been subjected to a "four-hour malevolent attack" from Valley and the MIT engineer Gordon Brown. Zacharias to Loomis, January 21, 1952, AC 132, Box 11, Folder "Lincoln Lab." In subsequent years "Zacharias commended Valley for having brought an unenthusiastic Air Force to the point where it would take ground-controlled air defense, not only air offence, seriously." Thomas P. Hughes, "Managing Complexity: Interdisciplinary Advisory Committees," in *Technological Change: Methods and Themes in the History of Technology,* ed. Robert Fox (Amsterdam: Harwood, 1996), 241.

141. Zacharias and Gordon, "Military Technology," MIT, 156.

142. "Projects 572 and 540: Dew Line Engineering Report," October 1, 1957, Radar Collection, Box 172 06 02, Folder 2, AT&T Archives, Warren, N.J., 58.

143. See "A Report to the National Security Council by the Committee on Continental Defense," NSC-159, July 22, 1953, in MC 420, Box 130, Folder "Continental Defense/Strategic Warning," MIT. This document was known as the Bull Report, after the committee's head, retired Army Lieutenant General Harold R. Bull. It followed hard on another effort, led by Mervin J. Kelly of Bell Laboratories, to review the recommendations of the 1952 Summer Study, but the Kelly Committee's May 1953 report "seemed to vindicate both sides" of the offense/defense debate. "In the wake of the Soviet explosion of a hydrogen bomb in August 1953, the National Security Council approved an amended version of the Bull report." Winkler, *Searching the Skies,* 28, 29.

144. Winkler, *Searching the Skies,* 29.

145. Desmond Morton, "Providing and Consuming Security in Canada's Century," *Canadian Historical Review* 81, no. 1 (2000): 19.

146. Berkner, "Continental Defense"; "First Report of the Location Study Group, Distant Early Warning (DEW) Group," November 12, 1954, RG 24 (Department of National Defence fonds), Series D-1-c, Volume 8159, File 1660–67, Part 1 ("Distant Early Warning and Resupply—Arctic Operations"), National Archives of Canada, Ottawa (hereafter NAC). The DEW Line began operating, as scheduled, on July 31, 1957.

147. Pang, "Dome Days," 168, 179, 181.

148. Quoted in "Fuller Future," *Time,* October 20, 1958, http://www.time.com.

149. Francis Bello, "The Information Theory," *Fortune,* December 1953, 136.

150. Ralph Allen, "Will Dewline Cost Canada Its Northland?" *Maclean's,* May 26, 1956, 17.

151. Richard Morenus, *DEW Line: Distant Early Warning: The Miracle of America's First Line of Defense* (New York: Rand McNally, 1957), 31.

152. Scott Kirsch, *Proving Grounds: Project Plowshare and the Unrealized Dream of Nuclear Earthmoving* (New Brunswick, N.J.: Rutgers University Press, 2005).

153. James C. Scott, *Seeing like a State: How Certain Schemes to Improve the Human Condition Have Failed* (New Haven, Conn.: Yale University Press, 1998), 7.

154. "Agreement between Canada and the United States of America to Govern the Establishment of a Distant Early Warning System in Canadian Territory," May 5, 1955, RG 24, Series D-1-c, Volume 8159, File 1660–67, Part 3, NAC. The agreement was reprinted in *Arctic Circular* 9, no. 2 (1956), 23.

155. Trevor Lloyd, "Frontier of Destiny—The Canadian Arctic," *Behind the Headlines* 6, no. 7 (1946): 8.

156. A 1958 document lists "Eskimo" as a job category: "Status of Canadian Personnel on the Dewline, Effective: 31 March 1958," RG 24, Acc. 1983–84/167, Box 6592, File 2–70–99–1, Vol. 1 ("Distant Early Warning Line Co-Ordinating Committee—Agenda and Minutes"), NAC.

157. J. D. Ferguson, *A Study of the Effects of the D.E.W. Line upon the Eskimo of the Western Arctic of Canada* (Ottawa: Northern Research Co-ordination Centre, Department of Northern Affairs and National Resources, April 1957), 3.

158. J. Fried, "Settlement Types and Community Organization in Northern Canada," *Arctic* 16, no. 2 (1963): 94.

159. Interview with J. D. Ferguson, Windsor, Ontario, August 1, 2007.

160. Stone et al., "Geographic Studies," LLAB, G-3.

161. Bowers, "The Military Establishment," 261.

162. Ernest L. McCollum, "The Psychological Aspects of Arctic and Sub-Arctic Living," in *Science in Alaska: Selected Papers of the Alaska Science Conference of the National Academy of Sciences–National Research Council, Washington, November 9–11, 1950*, ed. Henry B. Collins (Washington, D.C.: Arctic Institute of North America, June 1952), 255.

163. Michel Foucault, "Technologies of the Self," in Rabinow, *Essential Works of Foucault,* 1:225.

164. Western Electric Company, *The DEW Line Story* (Paramus, N.J.: Bell Telephone Laboratories, Booklet Rack Service, 1956), 30. The absence of women in primary sources on Arctic militarization accords with the preponderance of masculinity in "northern narrative." Hulan, *Northern Experience,* 12.

165. Office of Armed Forces Information and Education, Department of Defense, *The Arctic,* 8.

166. Laura McEnaney, *Civil Defense Begins at Home: Militarization Meets Everyday Life in the Fifties* (Princeton, N.J.: Princeton University Press, 2000).

167. Winkler, *Searching the Skies*, 21.

168. On Project Charles and the GOC, see Redmond and Smith, *From Whirlwind to MITRE*, 105–6.

169. William D. Hobbs, "The History of the Ground Observer Corps," *Aerospace Historian* 27, no. 3 (1980): 178–79, 186.

170. Bruce D. Callander, "The Ground Observer Corps," *Air Force Magazine*, February 2006, 83. The GOC was reduced to "ready-reserve status" in 1958 and inactivated in 1959, concurrent with the activation of SAGE.

171. Callander, "The Ground Observer Corps," 81; Redmond and Smith, *From Whirlwind to MITRE*, 10; Winkler, *Searching the Skies*, 21.

172. For an example of a GOC manual, see U.S. Department of the Air Force, *Aircraft Recognition for the Ground Observer* (Washington, D.C.: Department of the Air Force, 1955).

173. U.S. Department of Defense, *The Aircraft Warning Service of the U.S. Air Force* (Washington, D.C.: Office of the Secretary of Defense, Civil Defense Liaison, 1950), 2.

174. Peter D. Baird, "Moscow Memories," *American Heritage Magazine*, November 1991, http://www.americanheritage.com. Baird grew up in what was, "according to adolescent folklore . . . unquestionably the number-one Soviet bombing target": Moscow, Idaho.

175. "Flashes," *The Aircraft Flash* 2, no. 1 (1953): 6.

176. Quoted in Callander, "The Ground Observer Corps," 83.

177. Sharon Ghamari-Tabrizi, *The Worlds of Herman Kahn: The Intuitive Science of Thermonuclear War* (Cambridge, Mass.: Harvard University Press, 2005), 100, 98.

178. For RAND researchers, the GOC was, by 1951, "completely incapable of furnishing accurate data rapidly enough for close control of interceptors." *Staff Report for Air Defense Study*, RAND R-225, September 1, 1951, Box B99, Curtis E. LeMay Papers, LOC, 15–16. Members of the Lincoln Laboratory had similar doubts; in 1952 they staged a number of tests that attempted to link GOC spotters with the radar system (known as Cape Cod) under development at the Lab. Results were "dismal." But while the "state of the art had passed the GOC by," technology was not the only element of the cybernetic continent. Redmond and Smith, *From Whirlwind to MITRE*, 256.

179. Interestingly, the official GOC manual was also called "the blue book." A cover image—in this case of the Ground Observers' Guide—can be found in Callander, "The Ground Observer Corps," 80. See also Winkler, *Searching the Skies*, 21.

180. Phil Patton, *Dreamland: Travels Inside the Secret World of Roswell and Area 51* (New York: Villard, 1998).

5. Anxious Urbanism

1. Homi K. Bhabha, *The Location of Culture* (London: Routledge, 1994), 149.

2. Ross, *No Respect,* 46, 45; George Kennan, "The Long Telegram," in Etzold and Gaddis, *Containment,* 63.

3. X [George Kennan], "The Sources of Soviet Conduct," *Foreign Affairs* 25, no. 4 (1947): 582.

4. Ibid.

5. Kennan, "The Long Telegram," 58.

6. Bernard Brodie, "Military Policy and the Atomic Bomb," *Infantry Journal* 59 (1946): 33.

7. "Hiroshima, U.S.A.: Something *CAN* Be Done about It," *Collier's,* August 5, 1950, 16; Beck, *Risk Society.*

8. "Address of Millard Caldwell, Administrator, Federal Civil Defense Administration, before the Philadelphia Bulletin Forum," March 14, 1951, RG 304, Entry 17, Box 37, Folder "Psychological Strategy," NARA, 3, 2. See also Andrew D. Grossman, *Neither Dead nor Red: Civilian Defense and American Political Development during the Early Cold War* (New York: Routledge, 2001), 4–5.

9. Joseph Masco, "'Survival Is Your Business': Engineering Ruins and Affect in Nuclear America," *Cultural Anthropology* 23, no. 2 (2008): 361.

10. Quoted in Peter Conrad, *Modern Times, Modern Places: Life & Art in the 20th Century* (London: Thames and Hudson, 1998), 517.

11. Robert A. Beauregard, *Voices of Decline: The Postwar Fate of U.S. Cities* (Oxford: Blackwell, 1993), 58.

12. Thomas Osborne and Nikolas Rose, "Governing Cities: Notes on the Spatialization of Virtue," *Environment and Planning D: Society and Space* 17 (1999): 740.

13. Light, *From Warfare to Welfare.*

14. Much of the work of the Office of Civilian Defense (OCD), established in 1941, resonates in the records of Cold War civil defense. Still, the OCD was shuttered in June 1945, weeks before the first test of a nuclear weapon in the New Mexico desert—a weapon that dramatically heightened and reconfigured the condition of anxious urbanism.

15. Quoted in Allan M. Winkler, *Life under a Cloud: American Anxiety about the Atom* (New York: Oxford University Press, 1993), 28.

16. Peter B. Hales, *Atomic Spaces: Living on the Manhattan Project* (Urbana: University of Illinois Press, 1997), 362.

17. Quoted in Richard Rhodes, *Dark Sun: The Making of the Hydrogen Bomb* (New York: Simon and Schuster, 1995), 231.

18. Hales, *Atomic Spaces,* 364.

19. Hanson W. Baldwin, "Los Alamos—Capital of the Atomic Age," *New York Times Magazine,* April 24, 1949, 12. See also "Atom City," *Architectural Forum* 83, no. 4 (1945): 103–16; "Model City," *Time,* December 12, 1949, 21. On Levittown's resemblance to the "utopian plan of Oak Ridge, Tennessee," see Peter Bacon Hales, "Levittown: Documents of an Ideal American Suburb," http://tigger.uic .edu/~pbhales/Levittown.html. Oak Ridge had its own precedent in the nearby community of Norris, built by the Tennessee Valley Authority in 1933. The director of planning for Norris was Tracy Augur, who is discussed later in this chapter. Robert H. Kargon and Arthur P. Molella, *Invented Edens: Techno-Cities of the Twentieth Century* (Cambridge, Mass.: MIT Press, 2008), 34–37.

20. Robert W. Seidel, "The National Laboratories of the Atomic Energy Commission in the Early Cold War," *Historical Studies in the Physical and Biological Sciences* 32, no. 1 (2001): 145–62. See also Alvin M. Weinberg, "Impact of Large-Scale Science on the United States, *Science,* July 21, 1961, 161–64; Peter Galison and Bruce Hevly, eds., *Big Science: The Growth of Large-Scale Research* (Stanford, Calif.: Stanford University Press, 1992).

21. Hales, *Atomic Spaces,* 2.

22. Ibid., 27. Hales is quoting Daniel Boorstin.

23. Quoted in Winkler, *Life under a Cloud,* 16.

24. Winfield Riefler, quoted in this chapter's epigraph, was the committee's chairman, and *The Problem of Reducing Vulnerability to Atomic Bombs* (Princeton, N.J.: Princeton University Press, 1947), written by Ansley J. Coale, was prepared for the committee.

25. The reference is to Noel Castree and Bruce Braun, eds., *Social Nature: Theory, Practice, and Politics* (Malden, Mass.: Blackwell, 2001).

26. Herman, *The Romance of American Psychology,* 137, 124.

27. Capshew, *Psychologists on the March,* 181.

28. Margot Norris, *Writing War in the Twentieth Century* (Charlottesville: University Press of Virginia, 2000), 13.

29. Almond, *Ventures in Political Science,* 9, 11, 12. See also Capshew, *Psychologists on the March,* 123.

30. Alexander H. Leighton, "Training Social Scientists for Post-war Conditions," *Applied Anthropology* 1, no. 4 (July–September 1942): 25. See also Leighton, *The Governing of Men: General Principles and Recommendations Based on Experience at a Japanese Relocation Camp* (Princeton, N.J.: Princeton University Press, 1946); Leighton, *Human Relations in a Changing World.*

31. John M. Findlay, "Atomic Frontier Days: Richland, Washington, and the Modern American West," *Journal of the West* 34, no. 3 (1995): 34.

32. Gladwin Hill, "Atomic Boom Town in the Desert," *New York Times Magazine,* February 11, 1951, 14.

33. Mike Davis, "Berlin's Skeleton in Utah's Closet," *Grand Street* 69 (1999): 94.

34. Leo P. Brophy, Wyndham D. Miles, and Rexmond C. Cochrane, *The Chemical Warfare Service: From Laboratory to Field* (Washington, D.C.: Office of the Chief of Military History, Department of the Army, 1959), 185.

35. "At Elm and Main," *Time,* March 30, 1953, 50. After the test, a "dust cloud with its waning radioactivity drifted harmlessly eastward."

36. Federal Civil Defense Administration (FCDA), *Operation Doorstep* (Washington, D.C.: U.S. Government Printing Office, 1953), 2. For a critique of this booklet, see "Operation Doorway," *Time,* July 6, 1953, http://www.time.com.

37. FCDA, *Operation Cue* (Washington, D.C.: U.S. Government Printing Office, 1955); FCDA, *Cue for Survival* (Washington, D.C.: U.S. Government Printing Office, 1956).

38. McEnaney, *Civil Defense Begins at Home,* 54. See also Masco, "'Survival Is Your Business,'" 373. For a superb discussion of Doorstep and Cue, see Tom Vanderbilt, *Survival City: Adventures among the Ruins of Atomic America* (Princeton, N.J.: Princeton Architectural Press, 2002), 69–95.

39. *Operation Cue,* 67; Masco, "'Survival Is Your Business,'" 372, 375.

40. *Operation Doorstep,* 3. See also Rebecca Solnit, *Savage Dreams: A Journey into the Landscape Wars of the American West* (Berkeley: University of California Press, 1999), 5.

41. U.S. Atomic Energy Commission, *Atomic Tests in Nevada* (Washington, D.C.: U.S. Government Printing Office, March 1957), 1, 2, 5. The consequences of these tests are documented in Philip Fradkin, *Fallout: An American Nuclear Tragedy* (Tucson: University of Arizona Press, 1989).

42. Masco, "'Survival Is Your Business,'" 362, 370.

43. George F. Kennan, *Sketches from a Life* (New York: Pantheon, 1989), 130–32. See also Guy Oakes, *The Imaginary War: Civil Defense and American Cold War Culture* (New York: Oxford University Press, 1994), 27.

44. W. R. Burnett, *The Asphalt Jungle* (London: Zomba Books, 1984 [1949]), 177, 250.

45. Christopher Tunnard and Henry Hope Reed, *American Skyline: The Growth and Form of Our Cities and Towns* (Boston: Houghton Mifflin, 1955), 252.

46. Edward Dimendberg, "From Berlin to Bunker Hill: Urban Space, Late Modernity, and Film Noir in Fritz Lang's and Joseph Losey's *M,*" *Wide Angle* 19, no. 4 (1997): 69.

47. Cynthia Enloe, *The Morning After: Sexual Politics at the End of the Cold War* (Berkeley: University of California Press, 1993), 16.

48. Quoted in Marling, *As Seen on TV,* 249. Visits to both Levittown and (as I noted in the Introduction) Disneyland were ultimately dismissed for security reasons.

49. Beauregard, *Voices of Decline,* 3.

50. Quoted in Boyer, *By the Bomb's Early Light,* 239.

51. "The City Under the Bomb," *Time,* October 12, 1950, 12; William Whyte Jr., "Introduction," in *The Exploding Metropolis,* ed. William Whyte Jr. (Berkeley: University of California Press, 1993 [1958]), 8, 9.

52. See Ann Douglas, "Periodizing the American Century: Modernism, Postmodernism, and Postcolonialism in the Cold War Context," *Modernism/ Modernity* 5, no. 3 (1998): 71–98.

53. Norman M. Klein, "Staging Murders: The Social Imaginary, Film, and the City," *Wide Angle* 20, no. 3 (1998): 89.

54. David Reid and Jayne L. Walker, "Strange Pursuit: Cornell Woolrich and the Abandoned City of the Forties," in *Shades of Noir,* ed. Joan Copjec (London: Verso, 1993), 68.

55. Beauregard, *Voices of Decline,* 129, 110. See also Dimendberg, "From Berlin to Bunker Hill," 70.

56. Vivian Sobchack, "Lounge Time: Postwar Crises and the Chronotope of Film Noir," in *Refiguring American Film Genres: Theory and History,* ed. Nick Browne (Berkeley: University of California Press, 1998), 129–70.

57. Beauvoir, *America Day by Day,* 139. De Beauvoir was in the San Francisco area; *The Killers* (1946), based on an Ernest Hemingway story, was directed by Robert Siodmak.

58. Alan Trachtenberg, "The Modernist City of Film Noir: The Case of *Murder, My Sweet,*" in *American Modernism across the Arts,* ed. Jay Bochner and Justin D. Edwards (New York: Peter Lang, 1999), 298; Ralph Willett, *The Naked City: Urban Crime Fiction in the U.S.A.* (Manchester, U.K.: Manchester University Press, 1996), 24.

59. Douglas, "Periodizing the American Century," 83; Elizabeth A. Wheeler, *Uncontained: Urban Fiction in Postwar America* (New Brunswick, N.J.: Rutgers University Press, 2001), 2.

60. Quoted in Rosalyn Baxandall and Elizabeth Ewen, *Picture Windows: How the Suburbs Happened* (New York: Basic Books, 2000), 91. See also Nicholas Christopher, *Somewhere in the Night: Film Noir and the American City* (New York: Free Press, 1997), 50.

61. Baldwin, *The Price of Power,* 256, 257.

62. William F. Ogburn, "Sociology and the Atom," *American Journal of Sociology* 51, no. 4 (1946): 267–75.

63. Peter Conrad, *The Art of the City: Views and Versions of New York* (New York, Oxford University Press, 1984), 297.

64. JoAnne Brown, "'A Is for Atom, B Is for Bomb': Civil Defense in American Public Education, 1948–1963," *Journal of American History* 75, no. 1 (1988): 68–90.

65. Federal Civil Defense Administration, *Home Protection Exercises* (Washington, D.C.: U.S. Government Printing Office, 1953).

66. Kristina Zarlengo, "Civilian Threat, the Suburban Citadel, and Atomic Age American Women," *Signs: Journal of Women in Culture and Society* 24, no. 4 (1999): 931. On the civic garrison, see Grossman, *Neither Dead nor Red.*

67. Stephen J. Collier and Andrew Lakoff, "Distributed Preparedness: The Spatial Logic of Domestic Security in the United States," *Environment and Planning D: Society and Space* 26 (2008): 7–28.

68. McEnaney, *Civil Defense Begins at Home,* 7.

69. U.S. Department of Health, Education, and Welfare, Office of Education, *Education for National Survival: A Handbook on Civil Defense for Schools* (Washington, D.C.: U.S. Government Printing Office, 1956), 1, 42.

70. Quoted in "The City under the Bomb," 12.

71. Quoted in Elaine Tyler May, "Explosive Issues: Sex, Women, and the Bomb," in May, *Recasting America.*

72. Frank Krutnik, "Something More than Night: Tales of the *Noir* City," in *The Cinematic City,* ed. David B. Clarke (London: Routledge, 1997), 83.

73. Both quotes are from Douglas, "Periodizing the American Century," 83.

74. Boyer, *By the Bomb's Early Light,* 14.

75. Baldwin, *The Price of Power,* 252; J. Marshak, E. Teller, and L.R. Klein, "Dispersal of Cities and Industries," *Bulletin of the Atomic Scientists* 1, no. 9 (1946): 13. The economist Jacob Marschak's name was incorrectly spelled in the *Bulletin.*

76. Ogburn, "Sociology and the Atom," 271.

77. *Operation Cue,* 1.

78. "Waiting for September," *Time,* August 14, 1950, http://www.time.com.

79. Osborne and Rose, "Governing Cities," 753.

80. Anthony Vidler, *The Architectural Uncanny: Essays in the Modern Unhomely* (Cambridge, Mass.: MIT Press, 1992).

81. "Naked City," *Time,* November 28, 1949, 66. See also "The City of Washington and an Atomic Attack," *Bulletin of the Atomic Scientists* 6, no. 1 (1950): 29–30.

82. Kevin McNamara, *Urban Verbs: Arts and Discourses of American Cities* (Stanford, Calif.: Stanford University Press, 1996), 176. See also Polan, *Power and Paranoia,* 164–65.

83. P. Morrison, "If the Bomb Gets Out of Hand," in Masters and Way, *One*

World or None, 3. Masters also wrote the famous Cold War novel *The Accident* (1955), based on the experience of Manhattan Project scientist Louis Slotin, who died in 1946 from radiation exposure at Los Alamos.

84. Baldwin, *The Price of Power*, 5; John Hersey, *Hiroshima* (New York: Knopf, 1946).

85. E. B. White, *Here Is New York* (New York: Harper, 1949), 54. As many commentators noted, White's vision became an extraordinarily uncanny one on September 11, 2001.

86. "The 36-hour War," *Life*, November 19, 1945, 27–35.

87. Weart, *Nuclear Fear*, 236.

88. "The Story of This Story," *Collier's*, August 5, 1950, 11. The FCDA also used the Census Bureau to estimate day and night population distribution of city populations and the various types of urban zones where both habitation and "critical industry occur in highest concentration." FCDA, *Annual Report for 1951* (Washington, D.C.: U.S. Government Printing Office, 1952), 47–48; *Population Estimates for Survival Planning* (Washington, D.C.: Department of Commerce, Bureau of the Census, July 1956).

89. "Operation Eggnog," *Collier's*, October 27, 1951, 6. For an immediate critique, see D. F. Fleming, "*Collier's* Wins World War III," *The Nation*, November 10, 1951, 392–95.

90. The standard reference here is Susan Sontag, "The Imagination of Disaster" (1965), in *Against Interpretation and Other Essays* (New York: Farrar, Straus and Giroux, 1967), 209–25. On the imagination of disaster offered by civil defense spectacles and scenarios, see Masco, "'Survival Is Your Business,'" 363.

91. Wheeler, *Uncontained*, 1.

92. Richard Gerstell, *How to Survive an Atomic Bomb* (New York: Bantam, 1950), 91, 127.

93. The NSRB was yet another product of the 1947 National Security Act. It had a checkered and short history. By early 1951, in the midst of the Korean War, its mobilization responsibilities had been transferred to the new and more powerful Office of Defense Mobilization (ODM), while its civil defense mandate had been handed to the FCDA, established as an independent agency in January of that year. The ODM and the FCDA were merged into the Office of Civil and Defense Mobilization on July 1, 1958.

94. Quoted in Oakes, *The Imaginary War*, 39.

95. "Hiroshima, U.S.A.," 66, 67.

96. McEnaney, *Civil Defense Begins at Home*, 149–50.

97. Val Peterson, "Panic: The Ultimate Weapon?" *Collier's*, August 21, 1953, 102. See also Zarlengo, "Civilian Threat," 930–31.

98. Quoted in B. Wayne Blanchard, "American Civil Defense, 1945–1975: The Evolution of Programs and Policies" (PhD thesis, University of Virginia, 1980), 117.

99. "The Attack: Every Man for Himself," *Newsweek*, April 5, 1954, 33.

100. Irving L. Janis, *Air War and Emotional Stress: Psychological Studies of Bombing and Civilian Defense* (New York: McGraw-Hill, 1951), 189.

101. Wadsworth's speech is excerpted in "Memo for State Civil Defense Directors," December 1, 1950, RG 396, Entry 1022, Box 1, Binder "Advisory Bulletins 1–49," NARA, 2, 3, 4.

102. Masco, "'Survival Is Your Business,'" 366–67.

103. FCDA, *Annual Report for 1951*, 15.

104. Tracy C. Davis, *Stages of Emergency: Cold War Nuclear Civil Defense* (Durham, N.C.: Duke University Press, 2007), 24.

105. Robert G. Nixon, "'Enemy' Strikes U.S. Towns; Phantom Bombs Hurled Here," *Atlanta Daily World*, June 15, 1954, 1.

106. "Best Defense? Prayer," *Time*, June 27, 1955, 17; "When Ike 'Fled' Washington," *U.S. News and World Report*, June 24, 1955, 66–69.

107. Grossman, *Neither Dead nor Red*, ix. The gathering resistance to Operation Alert activities, and the crucial role of "mothers against the bomb" in this dissent, is documented in Dee Garrison, *Bracing for Armageddon: Why Civil Defense Never Worked* (Oxford: Oxford University Press, 2006), 93–101.

108. "City CD Primed for Nuclear Hit," *Chicago Defender*, July 11, 1957, 7; "54,500,000 'Die' in CD Test Attack," *Chicago Defender*, July 15, 1957, 7.

109. Masco, "'Survival Is Your Business,'" 370, 378.

110. National Security Resources Board, Civil Defense Office, *Survival under Atomic Attack* (Washington, D.C.: U.S. Government Printing Office, 1950), 17.

111. Irving L. Janis, "Problems of Theory in the Analysis of Stress Behavior," *Journal of Social Issues* 10, no. 3 (1954): 12. See also Lewis M. Killian, "Some Accomplishments and Some Needs in Disaster Study," n.d., Folder "Anthropology and Psychology: Committee on Disaster Studies—Conference on Theories of Human Behavior, February 1955," National Academy of Sciences–National Research Council Archives, Washington, D.C. (hereafter NN), 1. No box numbers are used at this repository.

112. Lloyd V. Berkner, "The Common Aspects of Disasters," Box 9, Folder "Speeches and Papers—LVB: AAAS Symposium on Disaster Recovery: St Louis—12/27/52," Lloyd V. Berkner Papers, LOC, 7.

113. Grossman, *Neither Dead nor Red*, 58.

114. *The Adequacy of Government Research Programs in Non-Military Defense: A Report by the Advisory Committee on Civil Defense* (Washington, D.C.: National Academy of Sciences–National Research Council, 1958), 21, 26.

115. "Proposal for Disaster Studies Program," attached to letter, John R. Wood to M. C. Winternitz, May 29, 1951, Folder "Anthropology and Psychology: Committee on Disaster Studies—beginning of Program, 1951," NN, 1.

116. Glen Finch to Bernard Berelson, December 31, 1956, Folder "Anthropology and Psychology: Committee on Disaster Studies—Sponsors: Ford Foundation," NN, 2.

117. On the "Rescue Street" or "Rescue City" in Olney, Maryland, see "If an Atom Bomb Hits—What Happens to a U.S. City," *U.S. News and World Report,* July 4, 1952, 26–27; David Monteyne, "Shelter from the Elements: Architecture and Civil Defense in the Early Cold War," *Philosophical Forum* 35, no. 2 (2004): 179–99.

118. "The Problem of Panic," June 1, 1954, Folder "Anthropology and Psychology: Committee on Disaster Studies—Subcommittee on Panic: Problem of Panic, 1954," NN (original emphasis). See also Committee on Disaster Studies, National Research Council, "Report to the Surgeons General, Departments of the Army, Navy and Air Force," March 31, 1955, Folder "Anthropology and Psychology: Committee on Disaster Studies—Reports to Sponsors: Department of Defense, 1955," NN.

119. Irving L. Janis, *Victims of Groupthink: A Psychological Study of Foreign-policy Decisions and Fiascoes* (Boston: Houghton, Mifflin, 1972).

120. Stephen B. Withey, "Reaction to Uncertain Threat: A Monograph Oriented towards a Theoretical Organization of Some General Research Findings," October 1957, Folder "Anthropology and Psychology: Disaster Research Group—Studies: Reaction to Uncertain Threat Final Report, 1957," NN.

121. Ralph E. Lapp, "Atomic Bomb Explosions—Effects on an American City," *Bulletin of the Atomic Scientists* 4, no. 2 (1948): 49.

122. Federal Civil Defense Administration, *Battleground U.S.A.: An Operations Plan for the Civil Defense of a Metropolitan Target Area* (Washington, D.C.: U.S. Government Printing Office, 1957).

123. Federal Civil Defense Administration, Technical Manual 8-1, *Civil Defense Urban Analysis* (Washington, D.C.: U.S. Government Printing Office, July 1953), 5.

124. Ibid., 9, 10, 11, 12, 50.

125. "Fifth Meeting, Committee on Disaster Studies, Division of Anthropology and Psychology, National Research Council, in Cooperation with Federal Civil Defense Administration," January 9–10, 1953, Folder "Anthropology and Psychology: Committee on Disaster Studies—Meetings 1952–1957," NN, 11.

126. Fred C. Iklé, *The Social Impact of Bomb Destruction* (Norman: University of Oklahoma Press, 1958), vii, 7, 8.

127. Kingsley Davis, "Urban Analysis Project: A Progress Report, with Future Organization and Research Plans," April 18, 1951, Box 24, BASR.

128. See the sections titled "Sociological and Psychological Components of Intra-Urban Target Analysis" and "Inter-Urban Patterns of Target Analysis" in *Report*, Strategic Intelligence Research Directorate, Human Resources Research Institute, Air University, November 30, 1951, Box 24, BASR.

129. "Inter-Urban Patterns of Target Complexes," in *Report*, Strategic Intelligence Research Directorate, BASR, 4.

130. Lynn Eden, *Whole World on Fire: Organizations, Knowledge, and Nuclear Weapons Devastation* (Ithaca, N.Y.: Cornell University Press, 2004), 97–99.

131. Quoted in Diamond, *Compromised Campus*, 87 (my emphasis). Similar language can be found in "Inter-Urban Patterns of Target Complexes," BASR, 4. This report, likely written by Davis, is dated November 30, 1951, a few months earlier than the document cited by Diamond.

132. "Planning for Wartime Evacuation from American Cities, Progress Report No. 2, September 1 to November 30, 1953," Box 25, Folder "Planning for Wartime Evacuation," BASR.

133. "Committee on Disaster Studies, Division of Anthropology and Psychology, Annual Report, 1952–1953," Folder "Anthropology and Psychology: Committee on Disaster Studies—Reports: Annual, 1952–57," NN.

134. Barnes and Farish, "Between Regions."

135. Memo, William Garrison et al. to Harry B. Williams, n.d., Folder "Anthropology and Psychology: Disaster Research Group—Studies: Civil Defense Evacuation Test—Bremerton: Final Report, 1955," NN, 1. See also *Washington State Survival Plan Studies*, vol. 2 (Seattle: University of Washington, 1956).

136. See "Air Reconnaissance as a Collection Method for Urban Sociological and Psychological Data; A Preliminary Survey," in *Report*, Strategic Intelligence Research Directorate, BASR.

137. Memo, John Balloch et al. to National Research Council, Committee on Disaster Studies, Subject: Spokane Civil Defense Exercise "Operation Walk-Out, April 26, 1954," n.d., Folder "Anthropology and Psychology: Disaster Research Group—Studies: Civil Defense Evacuation Test—Bremerton: Final Report, 1955," NN, 3.

138. "A Report on 'Operation Scat': A 'drive out' Evacuation of a part of Mobile, Alabama," n.d., ibid., 1, 2, 4; "Report on Observation Scat—Participant Observer, 'S,'" n.d., ibid., 6. The NRC-NAS Committee on Disaster Studies was abolished in July 1957, but the work of the Disaster Research Group continued.

139. "Second Press Release," in *Report of the Project East River, Part I* (New York: Associated Universities, 1952), n.p.

140. *Reduction of Urban Vulnerability: Part V of the Project East River* (New York: Associated Universities, July 1952), 1. See also Grossman, *Neither Dead nor Red*, 59–60.

141. *Reduction of Urban Vulnerability,* Appendix V-A, 1a, 6a, 8a.

142. William L. Borden, *There Will Be No Time: The Revolution in Strategy* (New York: MacMillan, 1946), 65.

143. "New Facts about the Atom Bomb," *Reader's Digest,* July 1949, 16.

144. "Civilians: Vital Link in Defense" (1951) was prepared by *Newsweek's* Club and Educational Bureau for its publication *Platform.* A copy is in RG 304, Entry 17, Box 31, Folder "Mrs. Martha Sharp—Civil Defense," NARA, 2.

145. Zarlengo, "Civilian Threat," 936. See also Donald Monson and Astrid Monson, "How Can We Disperse Our Large Cities?" *American City* 65, no. 12 (1950): 90–92, and 66, no. 1 (1951): 107–11; Alexander Hammond, "Rescripting the Nuclear Threat in 1953: *The Beast from 20,000 Fathoms,*" *Northwest Review* 22, no. 1/2 (1984): 181–94.

146. Ogburn, "Sociology and the Atom," 272; Mel Scott, *American City Planning since 1890* (Berkeley: University of California Press, 1969), 369. The argument for dispersal seemingly ran counter to the FCDA's appeal for urban "mutual aid pacts," but some cities were seen as more salvageable than others. Federal Civil Defense Administration, *This Is Civil Defense* (Washington, D.C.: U.S. Government Printing Office, May 1951), 13.

147. William A. Gill, speech to the American Institute of Planners, Washington, D.C., May 14, 1950, Vertical Files Collection, Loeb Library, Harvard University. Gill was the assistant director of the NSRB's Civilian Mobilization Office.

148. Lapp, "Atomic Bomb Explosions," 54.

149. Leslie R. Groves, "Dispersal of Industry in the Atomic Age," *Mechanical Engineering* 77, no. 6 (1955): 487.

150. "The Atomic Bomb and the Future City," *American City* 61, no. 8 (1946): 5. Edward Dimendberg argues that this brief piece "established the equation... between urban concentration and military vulnerability," but he does not mention the USSBS. "City of Fear: Defensive Dispersal and the End of Film Noir," *ANY: Architecture New York* 18 (1997): 15. See also Kargon and Molella, *Invented Edens,* 85–86.

151. Brodie, "Military Policy and the Atomic Bomb," 33. See also Zarlengo, "Civilian Threat," 932. Cluster and linear cities are described in Marshak, Teller, and Klein, "Dispersal of Cities and Industries." The evidence that any industry decentralized *primarily* for protective reasons is ambiguous. See "Spreading Out," *Architectural Forum* 89, no. 4 (1948): 18, 20, 22; Hanson W. Baldwin, "Strategy for Two Atomic Worlds," *Foreign Affairs* 28, no. 3 (1950): 386–97. Of course, the Manhattan Project had already set a new standard. But the call to relocate strengthened with the announcement of a national industrial dispersal policy in August 1951. See U.S. Department of Commerce, Area Development Division, *Industrial Dispersion Guidebook for Communities* (Washington, D.C.: U.S. Government Printing Office, 1952).

152. Louis Wirth, "Does the Atomic Bomb Doom the Modern City?" *New Jersey Municipalities* (April 1946): 25–29.

153. "Planning Cities for the Atomic Age," *American City* 61, no. 8 (1946), 75–76, 123; Tracy B. Augur, "Decentralization Can't Wait," *Appraisal Journal* 17, no. 1 (1949): 107–13. For more on Augur and the Garden City movement, see Kargon and Molella, *Invented Edens*, 34–35, 86–87.

154. Don G. Mitchell, "Social Aspects of Decentralization," *Mechanical Engineering* 70 (1948): 534. See also "Defense Considerations in City Planning: Statement by the American Institute of Planners," *Bulletin of the Atomic Scientists* 9, no. 7 (1953): 268; Scott, *American City Planning,* 449.

155. Martha A. Bartter, "Nuclear Holocaust as Urban Renewal," *Science-Fiction Studies* 13, no. 2 (1986): 153. One prominent example was Walter M. Miller Jr.'s short story/novel *A Canticle for Leibowitz* (1955/1959).

156. Dean MacCannell, "Baltimore in the Morning . . . After: On the Forms of Post-Nuclear Leadership," *Diacritics* 14, no. 2 (1984): 40. See also M. Curry, "In the Wake of Nuclear War—Possible Worlds in an Age of Scientific Expertise," *Environment and Planning D: Society and Space* 3 (1985): 319.

157. "Planning Cities for the Atomic Age," 75; Augur, "Decentralization Can't Wait," 110. See also Tracy B. Augur, "The Dispersal of Cities as a Defense Measure," *Journal of the American Institute of Planners* 14, no. 3 (1948): 29–35; Dimendberg, "City of Fear."

158. Kargon and Molella, *Invented Edens,* 87.

159. National Security Resources Board, *National Security Factors in Industrial Location* (Washington, D.C.: U.S. Government Printing Office, September 1948), 9, 15n.

160. "The City of Washington and an Atomic Bomb Attack," 30.

161. Kermit C. Parsons, "Shaping the Regional City, 1950–1990: The Plans of Tracy Augur and Clarence Stein for Dispersing Federal Workers from Washington, D.C.," Society for American City and Regional Planning History Working Paper (October 1989), 28.

162. Kargon and Molella, *Invented Edens,* 89. The focus on dispersing new or growing industries had a particularly significant impact on research-intensive businesses, "the sectors of the defense complex from which the late twentieth-century high-tech economy grew." Margaret Pugh O'Mara, "Uncovering the City in the Suburb: Cold War Politics, Scientific Elites, and High-Tech Spaces," in *The New Suburban History,* ed. Kevin M. Kruse and Thomas J. Sugrue (Chicago: University of Chicago Press, 2006), 64–65.

163. *Problems of Air Defense: Final Report of Project Charles* (Cambridge: Massachusetts Institute of Technology, June 15, 1951), MIT, VII-I-16. See also Zacharias and Gordon, "Military Technology," MIT, 142–44.

164. "How U.S. Cities Can Prepare for Atomic War," *Life*, December 18, 1950, 77–86.

165. Monson and Monson, "How Can We Disperse Our Large Cities?" 92.

166. "How U.S. Cities Can Prepare for Atomic War," 85, 79. See also Zarlengo, "Civilian Threat," 934–35.

167. Osborne and Rose, "Governing Cities," 749, 750. See also Peter Hall, *Cities of Tomorrow: An Intellectual History of Urban Planning and Design in the Twentieth Century,* updated ed. (Cambridge, Mass.: Blackwell, 1996), 327.

168. Federal Civil Defense Administration, *4 Wheels to Survival* (Washington, D.C.: U.S. Government Printing Office, 1955), n.p.

169. Steven Goddard, *Getting There: The Epic Struggle between Road and Rail in the Twentieth Century* (New York: Basic Books, 1994), 197.

170. Richard Davis, *The Age of Asphalt: The Automobile, the Freeway, and the Condition of Metropolitan America* (Philadelphia: J. B. Lippincott, 1975), 4.

171. Owen Gutfreund, *Twentieth-Century Sprawl: Highways and the Reshaping of the American Landscape* (Oxford: Oxford University Press, 2004), 42.

172. Quoted in Richard F. Weingroff, "Federal-Aid Highway Act of 1956: Creating the Interstate System," *Public Roads* (1996), http://www.tfhrc.gov.

173. Thomas Lewis, *Divided Highways: Building the Interstate Highways, Transforming American Life* (New York: Viking, 1997), 107, 108.

174. Helen Leavitt, *Superhighway-Superhoax* (Garden City, N.Y.: Doubleday, 1970), 187–88; Goddard, *Getting There,* 194.

175. Lewis Mumford, "The Highway and the City" (1958), in *The Highway and the City* (New York: Harcourt and Brace, 1963), 237, 238.

176. Kargon and Molella, *Invented Edens,* 89, 113.

177. Zacharias and Gordon, "Military Technology," MIT, 140–41.

178. Oakes, *The Imaginary War,* 109.

179. Joseph Masco, "Lie Detectors: On Secrets and Hypersecurity in Los Alamos," *Public Culture* 14, no. 3 (2002): 462. See also Valerie Kuletz, *The Tainted Desert: Environmental and Social Ruin in the American West* (New York: Routledge, 1998).

180. *New York Times,* quoted in Oakes, *The Imaginary War,* 60. See also Kargon and Molella, *Invented Edens,* 90.

181. Dimendberg, "City of Fear," 17.

182. Beauregard, *Voices of Decline,* 6.

183. Monson and Monson, "How Can We Disperse Our Large Cities?" 92. See also Donald Monson and Astrid Monson, "A Program for Urban Dispersal," *Bulletin of the Atomic Scientists* 7, no. 9 (1951): 244–50.

184. Matthew Farish, "Panic, Civility, and the Homeland," in *War, Citizenship, Territory,* ed. Deborah Cowen and Emily Gilbert (New York: Routledge, 2007), 97–118.

185. Quoted in Eden, *Whole World on Fire*, 166.

186. Elizabeth W. Mechling and Jay Mechling, "The Campaign for Civil Defense and the Struggle to Naturalize the Bomb," in *Critical Questions: Invention, Creativity, and the Criticism of Discourse and Media*, ed. William L. Nothstine, Carole Blair, and Gary Copeland (New York: St Martin's Press, 1994), 151.

187. Alan Wolfe, "Buying Alone," *New Republic*, March 17, 2003, http://www.tnr.com.

188. *General Report: Part One of the Project East River* (New York: Associated Universities, October 1952), 16.

189. "They Forgot the Southside," *Chicago Defender*, April 20, 1959, 11.

190. MacCannell, "Baltimore in the Morning," 40, 45 (original emphasis).

Conclusion

1. Michael Rogin, "Kiss Me Deadly: Communism, Motherhood, and Cold War Movies," *Representations* 6 (1984): 1–36.

2. Annette Kuhn, "Repressions," in *Alien Zone: Cultural Theory and Contemporary Science Fiction Cinema*, ed. Kuhn (London: Verso, 1991), 92.

3. Bill Nichols, "The Work of Culture in the Age of Cybernetic Systems," in *Electronic Culture: Technology and Visual Representation*, ed. Timothy Druckrey (New York: Aperture, 1996), 127.

4. This short plot summary is my own, but I have been aided by Bill Warren, *Keep Watching the Skies! American Science Fiction Movies of the Fifties: Volume One, 1950–1957* (London: McFarland, 1982), 261–74.

5. Steven Greenblatt, *Marvelous Possessions: The Wonder of the New World* (Chicago: University of Chicago Press, 1991); J. P. Telotte, "Science Fiction in Double Focus: *Forbidden Planet*," *Film Criticism* 13, no. 3 (1989): 25–36.

6. John Trushell, "Return of *Forbidden Planet*?" *Extrapolation* 64 (1995): 85.

7. Manfred E. Clynes and Nathan S. Kline, "Cyborgs and Space" (1960), in *The Cyborg Handbook*, ed. Chris Hables Gray (New York: Routledge, 1995), 29–33.

8. Chris Hables Gray, "An Interview with Manfred Clynes," in ibid., 47, 48. See also Mirowski, *Machine Dreams*, 285.

9. G. Malcolm Brown, "Man in the North," in *Canadian Population and Northern Colonization*, ed. V. W. Bladen (Toronto: University of Toronto Press, 1962), 146.

10. John Johnston, *Information Multiplicity: American Fiction in the Age of Media Saturation* (Baltimore: Johns Hopkins University Press, 1998), 2.

11. Hayles, *How We Became Posthuman*, 105, 110.

12. Donna Haraway, *Modest_Witness@Second_Millenium. FemaleMan©_Meets_ OncoMouse™: Feminism and Technoscience* (New York: Routledge, 1997), 23–24.

13. Quoted in Nigel Thrift, "Inhuman Geographies: Landscapes of Speed, Light, and Power," in *Spatial Formations* (London: Sage, 1996), 296. See also Michel de Certeau, *Heterologies: Discourse on the Other*, trans. Brian Massumi (Minneapolis: University of Minnesota Press, 1986), 200–21.

14. Haraway, *Modest_Witness*, 2.

15. Dale Carter, *The Final Frontier: The Rise and Fall of the American Rocket State* (London: Verso, 1988), 94.

16. Seth Lerer, "Forbidden Planet and the Terrors of Philology," *Raritan* 19, no. 3 (Winter 2000): 73.

17. *Preliminary Design of an Experimental World-Circling Spaceship*, RAND SM-11827, May 2, 1946, RL.

18. Robert W. Buchheim and the Staff of the RAND Corporation, *Space Handbook: Astronautics and Its Applications* (New York: Random House, 1959); Irwin Cooper, *Triangle: Man, Machine, Space*, RAND P-1680, April 29, 1959, RL.

19. Heinz Haber, "Space Satellites: Tools of Earth Research," *National Geographic*, April 1956, 488. On Bonestell, see Vincent Di Fate, *Infinite Worlds: The Fantastic Visions of Science Fiction Art* (New York: Penguin Studio, 1997).

20. James R. Killian Jr., *Sputnik, Scientists, and Eisenhower: A Memoir of the First Special Assistant to the President for Science and Technology* (Cambridge, Mass.: MIT Press, 1977), 3.

21. Joseph Goldsen, "Some Political Implications of the Space Age," February 24, 1958, Accession 1.2, Series 200S, Box 551, Folder 4712, RF, 3.

22. Quoted in Sherry, *In the Shadow of War*, 214 (my emphasis).

23. Foucault, "The Eye of Power," 146–65; Richard K. Ashley, "The Eye of Power: The Politics of World Modeling," *International Organization* 37, no. 3 (1983): 495–535.

24. Michael J. Shapiro, "That Obscure Object of Violence: Logistics and Desire in the Gulf War," in *The Political Subject of Violence*, ed. David Campbell and Michael Dillon (Manchester, U.K.: Manchester University Press, 1993), 118.

25. Wily Ley, *The Conquest of Space* (New York: Viking, 1949).

26. John Cloud, "Crossing the Olentangy River: The Figure of the Earth and the Military-Industrial-Academic Complex, 1947–1972," *Studies in the History and Philosophy of Modern Physics* 31, no. 3 (2000): 371–404; Cloud, "Imagining the World in a Barrel: CORONA and the Clandestine Convergence of the Earth Sciences," *Social Studies of Science* 31, no. 2 (2001): 231–51; Deborah Jean Warner, "From Tallahassee to Timbuktu: Cold War Efforts to Measure Intercontinental Distances," *Historical Studies in the Physical and Biological Sciences* 30, no. 2 (2000): 393–415. CORONA's first successful mission was not until August 18, 1960.

27. Manzione, "'Amusing and Amazing and Practical and Military,'" 50.

28. Allan A. Needell, *Science, Cold War, and the American State: Lloyd V. Berkner and the Balance of Professional Ideals* (Amsterdam: Harwood, 2000), 147, 297.

29. "Science in Space," in Box 18, Folder "LVB Speeches and Papers: RI [Rockefeller Institute] Sigma Xi Lecture, 3 March 1959," Lloyd V. Berkner Papers, LOC, 2.

30. Berkner, "Man Attempts to Understand His Environment," 53.

31. Lloyd V. Berkner, "Geography and Space," *Geographical Review* 59, no. 3 (1959): 305–14.

32. The statement is attributed to Donald Menzel, who in 1953 was the acting director of Harvard's College Observatory. He was also a prolific writer on space in the 1950s.

33. Robert A. Heinlein, "Baedeker of the Solar System," *Saturday Review of Literature*, December 24, 1949, 9–10.

34. Killian, *Sputnik*, 124, 289. The full "Introduction to Outer Space" is reprinted in ibid., appendix 4.

PUBLICATION HISTORY

Parts of the Introduction, chapter 1, chapter 3, chapter 4, and the Conclusion are present in Matthew Farish, "Targeting the Inner Landscape," in *Violent Geographies: Fear, Terror, and Political Violence*, ed. Derek Gregory and Allan Pred (New York: Routledge, 2007), 255–71. Reprinted with the permission of Taylor and Francis.

Part of chapter 2 is based on Matthew Farish, "Archiving Areas: The Ethnogeographic Board and the Second World War," *Annals of the Association of American Geographers* 95, no. 3 (2005): 663–79. Reprinted with the permission of Taylor and Francis.

Passages from chapters 2 and 3 appear in Trevor J. Barnes and Matthew Farish, "Between Regions: Science, Militarism, and American Geography from World War to Cold War," *Annals of the Association of American Geographers* 96, no. 4 (2006): 807–26. Reprinted with the permission of Taylor and Francis.

Parts of chapter 4 appear in Matthew Farish, "Frontier Engineering: From the Globe to the Body in the Cold War Arctic," *Canadian Geographer* 50, no. 2 (2006): 177–96; and in P. Whitney Lackenbauer and Matthew Farish, "The Cold War on Canadian Soil: Militarizing a Northern Environment," *Environmental History* 12, no. 4 (2007): 921–50. Reprinted with the permission of Wiley-Blackwell and the Forest History Society respectively.

Earlier versions of chapter 5 are published in Matthew Farish, "Another Anxious Urbanism: Defence and Disaster in Cold War America," in *Cities, War, and Terrorism: Towards an Urban Geopolitics*, ed. S. Graham (Oxford: Blackwell, 2004), 93–109; and Matthew Farish, "Disaster and Decentralization: American Cities and the Cold War," *Cultural Geographies*, no. 3

(2003): 125–48. Reprinted with the permission of Wiley-Blackwell and Sage Publications respectively.

Passages from chapter 5 appear in Matthew Farish, "Panic, Civility, and the Homeland," in *War, Citizenship, Territory,* ed. D. Cowen and E. Gilbert (New York: Routledge, 2007), 97–118. Reprinted with the permission of Taylor and Francis.

INDEX

Program of Training in International Administration, 69, 292n80; Russian Institute, xxvi, 121–23, 129; School of International Affairs, 121, 292n80; School of Military Government, 69, 72; Teacher's College, 264n47. *See also* Bureau of Applied Social Research, Columbia University

Commager, Henry Steele, 42–43

Committee for National Morale, 79–80

Committee on the Present Danger, 39. *See also* NSC-68

communism, xiv, xxv–xxvi, 43–44, 93, 121, 122, 142, 147, 242; area studies and, 102, 129, 131–32, 137, 142; cities and, 204, 206; containment and, 30–32, 193–94; science fiction and, 239. *See also* McCarthyism

comparative politics, 53, 69, 80, 111, 118, 122, 124, 131–32, 137, 145, 200, 294n114. *See also* modernization theory; Social Science Research Council

Compass of the World (Weigert and Stefansson), 6–9. *See also* Harrison, Richard Edes

Compton, Karl, xxi, 35, 38–39, 58. *See also* Massachusetts Institute of Technology

computers, 4–5, 46, 146, 147–48, 150, 157, 163, 192, 239, 245; in *Forbidden Planet*, 241; at RAND, 164, 168–70; SAGE, 154–58, 160, 164, 169–71, 189, 313n170; Whirlwind, 158, 161

Conant, James, 35, 38–39, 271n146; Committee on the Present Danger and, 39; geography and, 145. *See also* Harvard University

Congress for Cultural Freedom, 49

containment, xiv, xvi, xxi, 2–3, 28, 48, 78, 212; as biology, 193–94, 206, 239; George Kennan and, 23, 30–33, 193–94, 268n105; science fiction and, 239; urban, xxii, 194, 206, 212, 228

continental defense, xxvi, 148, 150–51, 154–57, 161, 176, 178, 181, 184, 188–89, 192, 193, 196, 235, 236, 244; in Arctic, 162, 175–77, 181–86, 188–89; cybernetics and, 150, 154, 188–89, 313n178; Eisenhower administration and, 183; Ground Observer Corps and, 191, 313n178; Project Charles and, 157–59, 161–62, 170, 189, 230, 304n44, 311n140; RAND and, 163–64, 170–73, 313n178; SAGE and, 154–58, 160, 164, 169–71, 189, 313n170; Truman administration and, 177, 183. *See also* Distant Early Warning (DEW) Line

Coon, Carleton, 92

Cooper, Irwin, 245

Coordinated Investigation of Micronesian Anthropology (CIMA), 67, 69

CORONA satellite, 247, 327n26

Cosgrove, Denis, 1–3

Cross-Cultural Survey (CCS), 61, 64–67, 71, 86–87, 277n52; classification and, 64; expansion into Human Relations Area Files, 66; United States Navy and, 65–67, 69, 85

culture and personality studies. *See* national character

Cumings, Bruce, 103

Matthew Farish is assistant professor of geography at the University of Toronto.

CPSIA information can be obtained
at www.ICGtesting.com
Printed in the USA
LVHW031809151118
597259LV00002B/206/P

9 780816 648436